Peter Fratzl, Michael Friedman, Karin Krauthausen, Wolfgang Schäffner (Eds.)

Active Materials

AF696012

Also of Interest

Biomimetics
A Molecular Perspective
Jelinek, 2021
ISBN 978-3-11-070944-5, e-ISBN 9783110709490

Intelligent Materials and Structures
Abramovich, 2021
ISBN 978-3-11-072669-5, e-ISBN 978-3-11-072670-1

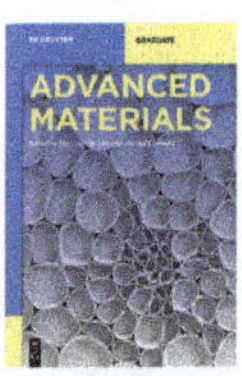

Advanced Materials
van de Ven, Soldera (Eds.), 2020
ISBN 978-3-11-053765-9, e-ISBN 978-3-11-053773-4

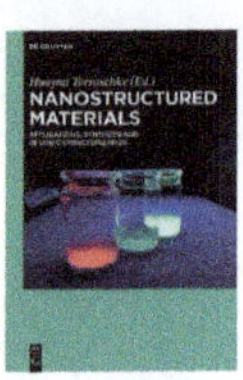

Nanostructured Materials
Applications, Synthesis and In-Situ Characterization
Terraschke (Ed.), 2022
ISBN 978-3-11-045829-9, e-ISBN 978-3-11-045909-8

Active Materials

Edited by
Peter Fratzl, Michael Friedman, Karin Krauthausen,
and Wolfgang Schäffner

DE GRUYTER

Editors
Prof. Dr. h.c. Peter Fratzl
Max Planck Institut of Colloids and Interfaces
Potsdam Science Park
Am Mühlenberg 1 OT Golm
14476 Potsdam
Germany
Peter.Fratzl@mpikg.mpg.de

Dr. Michael Friedman
Humboldt Universität zu Berlin
Exzellenzcluster
Matters of Activity. Image Space
Material
Unter den Linden 6
10099 Berlin
Germany
michael.friedman@hu-berlin.de
friedmanm@tauex.tau.ac.il

Dr. Karin Krauthausen
Humboldt Universität zu Berlin
Exzellenzcluster
Matters of Activity. Image Space Material
Unter den Linden 6
10099 Berlin
Germany
karin.krauthausen@hu-berlin.de

Prof. Dr. Wolfgang Schäffner
Humboldt-Universität Berlin
Philosophische Fakultät III
Institut für Kulturwissenschaft
Georgenstr. 47
10117 Berlin
Germany
schaeffner@culture.hu-berlin.de

The editors would like to acknowledge the support of the Cluster of Excellence 'Matters of Activity. Image Space Material' funded by the Deutsche Forschungsgemeinschaft (DFG, German Research Foundation) under Germany's Excellence Strategy – EXE 2025 –390648296.

ISBN 978-3-11-056181-4
e-ISBN (PDF) 978-3-11-056206-4
e-ISBN (EPUB) 978-3-11-056220-0
DOI https://doi.org/10.1515/9783110562064

BY-NC-ND

This work is licensed under the Creative Commons Attribution-NonCommercial-NoDerivatives 4.0 International License. For details go to http://creativecommons.org/licenses/by-nc-nd/4.0/.

Library of Congress Control Number: 2021943130

Bibliographic information published by the Deutsche Nationalbibliothek
The Deutsche Nationalbibliothek lists this publication in the Deutsche Nationalbibliografie; detailed bibliographic data are available on the Internet at http://dnb.dnb.de.

© 2021 Walter de Gruyter GmbH, Berlin/Boston
The book is published open access at www.degruyter.com.

Cover image: Norway spruce wood taken from the lower side of a branch (compression wood) and cut along the fibre direction. Image was taken with an Environmental Scanning Electron Microscope (ESEM). © Michaela Eder – MPIKG
Typesetting: Integra Software Services Pvt. Ltd.
Printing and binding: CPI books GmbH, Leck

www.degruyter.com

Contents

Introduction

Interviews: Scientists on Active Matter

Essays: Part 1: Cultural, Scientific, and Design Perspectives

Essays: Part 2: Historical and Philosophical Perspectives

Introduction

Michael Friedman, Karin Krauthausen

Materials Matter: Introduction

"materials are a boundary concept"[1]

Why should materials be considered a boundary concept? First, materials are increasingly seen as simultaneously both specific and general; second, they have become historically loaded with a range of metaphysical and empiricist, scientific and artisanal interpretations. Taking this boundary character into consideration, it would not be an understatement to claim that in the second half of the twentieth century our understanding of 'materials' has changed decisively. Starting in the middle of the twentieth century, alongside well-established 'raw' materials such as iron, wood, or ceramics, one starts to see the appearance of "materials by design" [Bensaude-Vincent 2011b, p. 119]. But what are these 'materials by design'? In contrast to raw materials, these new materials are designed for complex and specific tasks, and for that reason stand not at the beginning but at the end of the research and design process. This remarkable shift in our understanding and handling of materials should also be seen against the background of the considerable growth in the importance of materials for industry and engineering. At the latest since the establishment of materials science departments (beginning in the 1960s), 'materials' have become "characteristic ontological creatures of a new style of reasoning" [ibid., p. 108]. And yet materials science is also a symptom of an even more comprehensive transformation: the emergence of a "materials way of thinking" [ibid., pp. 107–108]. The aim of this volume is to explicate and explore this novel way of thinking.

Of course, the extraction and consumption of materials is fundamental in all societies. The relation of humans to their environment (including what is called 'nature') is

1 Epigraph taken from: [Bensaude-Vincent 2011b, p. 117]. "Boundary concept" is probably a paraphrase of the notion of 'boundary objects,' which according to Susan L. Star and James R. Griesemer are "scientific objects which both inhabit several intersecting social worlds [...] *and* satisfy the informational requirements of each of them" [Star and Griesemer 1989, p. 393]. Their boundary character is shown by the fact that they may be simultaneously specific and general, "abstract or concrete." See also: [Star 2010].

Acknowledgments: The editors acknowledge the support of the Cluster of Excellence 'Matters of Activity. Image Space Material' funded by the Deutsche Forschungsgemeinschaft (DFG, German Research Foundation) under Germany's Excellence Strategy – EXC 2025 – 390648296.

We would also like to thank our authors and interview partners for their wonderful contributions and for their patience. We thank also warmly Benjamin Carter for his careful copyediting and translations and Elisabeth Rädler for her extremely helpful support during the last stages of preparing the manuscript, Richard Weinkamer for helpful discussions and Michaela Eder for her research image which is now on the cover of this book.

Open Access. © 2021 Michael Friedman, Karin Krauthausen, published by De Gruyter. This work is licensed under the Creative Commons Attribution-NonCommercial-NoDerivatives 4.0 International License.
https://doi.org/10.1515/9783110562064-001

structured and secured via materials and the 'techniques' of extracting and processing these materials.[2] This is all the more so in highly industrialized and technologized societies. It is hardly surprising, therefore, that materials in these societies have become an important object of research, as well as, consequently, high-performance products. Linked with this development, however, are two notable changes: On the one hand, with the founding of the materials science, material becomes a worthy epistemic object and simultaneously a generic category – despite the material still being understood until late into the twentieth century as a specific, largely individuated substance. Unlike the traditional physical category *matter* (with its claim to universality), metal, ceramics, wood, and other things constitute a "zoology of materials" [Bensaude-Vincent 2011b, p. 110], and thus a multiplicity of materials is studied by different disciplines and linked with different techniques.[3] With this, not only the individual material migrated into the natural sciences but also a practical knowledge: technology became even more tightly linked with epistemology and ontology.

On the other hand, since the 1980s at the latest, the aspect of 'performance' came to the fore in materials science, and thus an 'activity' that is attributed to the material itself – to the extent that this is inherent to the material's structures and properties. In this understanding, 'material' implies an autonomous agency that should be analyzed and, as far as possible, generalized using the methods of natural science (theories, models, and tools from physics, chemistry, and mathematics) – which ideally should then lead to new basic knowledge as well as, of course, to new materials and innovative design. This attentiveness to 'active' materials has grown considerably in importance both in and beyond the natural sciences (in particular, in engineering, architecture, and design). One's focus in the materials science is the analysis of 'biological materials'[4] and 'biomaterials' and thus on the one hand of complex configurations of structure and process with outstandingly functional properties, which can be found not only in plants and animals but also at the microscale in cell tissue; and on the other hand of engineered substances, which are developed for among other things medical-therapeutic or -diagnostic interaction with components in living systems. Since the mid-1990s in physics and in particular in condensed matter physics, however, another field called *active matter* has emerged. In this field, activity is understood as a basic property of matter, but one that requires

2 That is the approach of André Leroi-Gourhan, a historian of pre- and early history, who describes historical societies via techniques of the extraction and processing of material [Leroi-Gourhan 1971 (1943)].

3 Metallurgical research was linked to physics, the science of wood to biology, and the knowledge of ceramics to chemistry – and all three were frequently boundary fields to the extent that they stimulated the transfer to application, for instance, in engineering.

4 For 'biological materials' the term 'nature's materials' is also in circulation in the materials sciences. These terms are actors' terms, that is, technical terms used by the protagonists of this research. For 'biological materials' see, for example: [Meyers et al. 2008]. For 'nature's materials' see, for example: [Fratzl and Weinkamer 2007; Fratzl 2007; Fratzl and Barth 2009].

certain conditions in order to appear. The research in this field aims explicitly at the universal category of matter and, in doing so, works intensively with mathematical modeling and simulations. Here too, however, the research object is frequently related to 'self-propelled' or 'self-driven' entities that are classified as 'living matter,' with examples extending from the nano- and the micro- to the macroscale (e.g., cell tissue, bacteria, bird flocks, and human dance behavior) [Ramaswamy 2010; Menon 2010; Gompper et al. 2020; Popkin 2016]. The scientific description of this living matter is oriented to the theories and equations of statistical physics and hydrodynamics that, from a formal perspective, ignore the opposition living/dead. Thus, bird flocks, for instance, are modeled analogue to the ferromagnetic interaction, when the spins are aligning in the same direction – with the difference, however, that the cause of the movement in the case of active matter is inherent to the (non-equilibrium) system, and the aligning refers to the direction of motion [Vicsek et al. 1995, p. 1226; Keller 2016; Vehlken 2012; Grote 2019, pp. 56–110, 186–193].[5]

Since the 1980s, and increasingly in the 2000s, therefore, the question of the activity of materials, and more fundamentally of matter, has gained enormously in significance. This is due not only to the possibilities of mathematical modeling and the new instruments of investigation; the promise of active materials goes beyond epistemological procedures and concerns. The hope is to create 'better' – since more sustainable and/or efficient – materials. Drawing on the model of biological materials, artificial materials should be developed that are not only controllable but also capable of similarly complex performances. This performance specification can be read from the characterization of the corresponding research objects: 'self-propelled' materials, 'self-assembling' materials, 'self-healing' materials, 'responsive' materials, 'adaptive' materials, 'smart' materials, 'intelligent' materials, and so on.[6] Those characteristics are given to both biological materials and artificial (i.e., man-made) materials. A few examples are: Andreas Walther describes the "humidity-induced actuation in a pine cone" as a "responsive system," the "dynamic camouflaging on cephalopods" as a "dynamic adaptation of color and topography," and the "signal-induced actuation in a venus fly trap" as a "single trajectory adaptation" [Walther 2019, p. 3]; Giovanni Noselli et al. point out that the construction of "[s]mart helical structures" is "inspired by the pellicle of euglenids" [Noselli, Arroyo, and DeSimone 2019, p. 234]; Barbara Mazzolai et al. note that "[p]lants or plant parts, such as roots or leaves, offer countless cues for making innovation in technology" and that one may summarize some "of the main principles studied in the plant roots" as the "capacity of growth and movement," "sensory

5 To the extent that the source and cause of the self-movement lies in the system itself, the need to include an external impetus in the physical consideration is eliminated. In the case of self-propelled particles, the ability to move is among the requirements of the material system.

6 We refer to the interviews in this volume, which exemplify these various characterizations. See also the contribution by Peter Fratzl and Wolfgang Schäffner for the difference between programmable, adaptive, and self-learning materials.

capabilities," and "emergent behavior" [Mazzolai, Beccai, and Mattoli 2014, p. 2]; and last but not least Thomas J. Wallin et al. have developed, using 3D printing, "fluidic elastomer actuators," a "material system [which] permits rapid autonomic self-healing via sunlight" [Wallin et al. 2017, p. 6253].

In the meantime, however, this search for new kinds of materials is no longer merely a technical but in part also an ecological concern. The enormous consumption of resources by highly industrialized and technologized societies is leading to a destabilization of the ecological balance and the rapid depletion of materials such as wood, coal, oil, and ore – and this fatal spiral should now also be countered by these new materials.

Against this background, the urgency of the field of active matter research becomes apparent, whereby three directions can be discerned: (1) 'active materials' is a research field with a strong internal dynamics that has the potential to significantly expand the natural sciences; (2) a promise arises from this research, and from the prospective development of related technologies, of a 'better' treatment by humans of the environment; and (3) the research on active matter changes how we understand 'material' and 'matter' – not only in the natural sciences but also in engineering, architecture, design, and finally within the whole culture. This edited volume responds to this urgency and takes the recent growth in research on active matter as a starting point for disciplinary and interdisciplinary analyses. Thus, the state of natural-scientific research is surveyed through interviews with central protagonists on their representative research objects (see the interviews with Joanna Aizenberg, Nikolaus Correll, John Dunlop, Ramin Golestanian, Jean-François Joanny, Barbara Mazzolai, Rob Shepherd, and Thomas Speck). Parallel to this ongoing research on active materials is also multiply contextualized: the transfer of approaches in the materials science to design practice is carried out and reflected on (see the contribution by Ianis Lallemand); the interdisciplinary negotiation of active materials is opened up on the one hand between physics, materials science, and cultural studies (see the contribution by Mohammad Fardin Gholami, Lorenzo Guiducci, Susanne Jany, and Khashayar Razghandi), and on the other hand between biology and philosophy (see the contribution by Sonja Dheur and Sven J. Saupe); and the topos of the activity of material and matter is complemented by perspectives on significant historical discourses from art history, visual and literary studies, and philosophy (see the contributions by Horst Bredekamp, Stephan Kammer, Sylvie Kleiman-Lafon, and Charles T. Wolfe). In addition, these interviews and contributions are preceded by a programmatic position in which Peter Fratzl and Wolfgang Schäffner jointly consider active materials' characteristic features from two different but related perspectives: materials science and the cultural history of knowledge.

This introduction in turn will address in a first part three significant *historical* material scenes (from Georgius Agricola in the sixteenth century to eighteenth- and nineteenth-century chemistry to the material sciences of the twentieth and twenty-first centuries) in order to highlight the different ways material has been conceived and handled in technology and science. The second part focuses on central aspects of *current* research on active materials, thus primarily the field of active matter and the promise of 'bioinspired' materials. This is outlined by a quick look at the prehistory of active materials research (which is still partly in force today): the suggestion of a 'dematerialization,' which is linked to pliable, high-performance, and essentially passivized materials such as iron, particularly in nineteenth- and twentieth-century architectural theory.

1 Technology and Science

1.1 "Man's Needs"

From a modern Western historical perspective, the meaning and importance of materials for the development of culture seems to be clear, since the broad historical periods of mankind are demarcated according to the materials that were used and available: the Stone Age, the Bronze Age, or the Iron Age. However, these demarcations are very much influenced by a conception of nature as an (infinite) repository of materials (or just one material: stone, bronze, or iron) waiting to be used and shaped. This conception of the material as a resource for technological-cultural developments has parallels with the classical division between matter and form. The ancient Greek term ὕλη (*hyle*) forms the generic root of all material phenomena – a term which was translated into Latin as *materia* or *materies*, but which originally meant 'dead wood,' 'brushwood,' or 'timber.' The Greek term ὕλη (*hyle*) stood in opposition to the concept of form. For Aristotle, matter is abstract possibility, δύναμις (*dynamis*), which has to be shaped and in a sense 'realized' by form. According to this understanding, material too is largely passive or inert and shaped by the active and external imposition of form. Accordingly, hylomorphism postulates an asymmetry between passive matter on which a form is imposed – a view that has had (and still has) a widespread influence. In hylomorphism, form and matter are treated as separate principles, whereby matter is thought of as a receptive substance (and in this sense as a mere 'possibility') and form as a causing and fulfilling 'reality' [Detel et al. 1980].[7] As Tim Ingold notes, following Gilbert Simondon's interpretation of hylomorphism, the problem here is that man (from the early modern period on, as natural philosopher or later, as scientist) is presented as the one who "stands outside the works and sees what goes in and what comes out but nothing of what happens in between, of the actual processes whereby materials of diverse kinds come to take on the forms they do. It is as though, in form and matter, he could grasp only the ends of

7 "Because matter is a precondition of form, the two behave like *possibility* and *reality*." ("Weil die Materie Vorbedingung der Form ist, so verhalten sich beide wie *Möglichkeit* und *Wirklichkeit*" [Baeumker 1890, p. 261 (emphasis spaced out in the original)].)

two half-chains but not what brings them together […]" [Ingold 2013, p. 25]. Instead of the hylomorphic model, Ingold suggests shifting the point of view toward materials, that is, forgetting what we know from the sciences about the chemical composition of materials and following instead the artisan's "desire to see what the *material* can *do*, by contrast to the scientist's desire to know what it *is*" [ibid., p. 28].[8] Ingold is of course not alone when he stresses this research direction. To give only two examples, Pamela Smith and Pamela O. Long also emphasize the artisan's knowledge with respect to materials and the techniques of their processing [Long 2011; Smith 2004; Smith 2018]. A similar attention to techniques of processing has also been suggested by André Leroi-Gourhan (already in 1943). Material is explained by the historian of pre- and early history via "techniques de fabrication" ("techniques of fabrication") [Leroi-Gourhan 1971 (1943), p. 161], which include the elements fire, water, and air, as well as the use of elementary forces (the "Moyens élémentaires d'action sur la matière" ("elementary means of acting on material/matter") [ibid., p. 43]). But in order to appreciate the significance of *la matière* – used in Leroi-Gourhan in the double sense of 'matter' and 'material' – one also has to take production into account. Material/matter becomes intelligible and understandable only in connection with gestures and techniques that put it in relation to man. Thus, while Leroi-Gourhan speaks in connection with material of solids (*solides*), and thus uses a term that is also used in solid-state physics (a subfield of condensed matter physics),[9] when he divides the solids, on the basis of their state, into six categories (stable, fibrous, semi-plastic, plastic, supple, and fluid),[10] the detailed description of the individual materials relates to the way they were extracted and processed, that is, via techniques.[11] For Leroi-Gourhan, therefore, material/matter (*la matière*) stands for physical-concrete components of the world (*solides*), which he understands in a broad sense (as *matériaux* (materials)), but in respect to human needs and practices, that is, in their technical-instrumental and to a certain extent 'nurturing' function. Accordingly the individual materials cannot be determined in a solely formal-abstract way, for instance only via the classification of 'bodies' (solids) according to their physical states. The materials exhibit certain properties, which in turn can be linked

8 See also [ibid., p. 28]: "What do we mean when we speak of materials? […] To understand the meaning of materials for those who work with them – be they artisans, craftsmen, painters or practitioners of other trades – I believe we need, as art historian James Elkins recommends, to take a 'short course in *forgetting* chemistry'" (emphasis M.F./K.K.).

9 In physics, the term solids is used to describe material in a certain (solid) state. The research on these states contributes to a general theory of matter.

10 See the titles of the subchapters in Chapter IV, "Les techniques de fabrication": "Solides stables," "Solides fibreux," "Solides semi-plastiques," "Solides plastiques," "Solides souples," "Fluides" [ibid., pp. 162, 174, 191, 206, 234, 296]. Leroi-Gourhan's division of the solids does not represent a physical classification, but is founded in his epistemic perspective: an anthropology of techniques.

11 Thus metal, for example, belongs to the "solides semi-plastiques," hence, bodies that in part can be 'formed,' whose extraction from the earth and stone through mining, smelting, and processing in smithies is described by Leroi-Gourhan [ibid., pp. 191–206].

with certain functions – each material is characterized by its own particular possibilities and limitations (analogue to Ingold's "what the *material* can *do*" quoted above), which in turn are to be thought in relation to the possibilities and limitations of humans, thus their techniques, their knowledge, and their needs (as it were, what *humans* can *do*).

Simondon's rejection of the traditional matter/form or material/form opposition goes in a similar direction, but puts a greater emphasis on the logic inherent to the technological aspect [Simondon 1995 (1964), p. 53]. 'Nature' 'offers' itself to the engineer (but also to the artisan and the artist) not as an abstract substance that can be mastered and controlled but in its structured materiality. As Henning Schmidgen notes, technicians must always deal with "matter-forms" ("Materie-Formen") to meet human needs – to that extent humans constantly move in a "prefigured region" ("präfigurierten region") [Schmidgen 2012, p. 128] with which they enter into an exchange mediated by tools and techniques.[12] In short, solely the interaction between man and material can provide information about what is to be understood under 'material,' and this approach does not lead to an ontology (the knowledge of what material *is*), but to a technology, that is to a knowledge of the necessary gestures, procedures, and techniques that determine "the material exchange with nature" ("den Stoffwechsel mit der Natur" [Schmidgen 2012, p. 127]).[13]

1.2 The Sixteenth Century: Action on Materials

If one looks at the history of the changing attitudes to a 'material,' then the approach of Leroi-Gourhan (or of Ingold and Simondon) is well founded. If one consults Georgius

12 On the knowledge of materials through tool use, see Simondon: "Knowing how to use a tool is not only to have acquired the practice of the necessary gestures; it is also to know how to recognize, via signals that come to man through the tool, the implicit emerging form of the matter in the precise place that the tool strikes." ("Savoir utiliser un outil, ce n'est pas seulement avoir acquis la pratique des gestes nécessaires; c'est aussi savoir reconnaître, à travers les signaux qui viennent à l'homme par l'outil, la forme implicite de la matière qui s'élabore, à l'endroit précis que l'outil attaque" [Simondon 1995 (1964), p. 51].) Unless stated otherwise, all translations from German and French were made by Benjamin Carter.

13 Presupposed here is Simondon's understanding of technology as an unavoidable medium. See the elegant exposition by Schmidgen: "Simondon is much more cautious when speaking of the technical objects as mediators (*médiateurs*) between nature and man. The consideration of technical objects is thereby transferred to a medial register that [...] is oriented to the problem of production. Since Marx, production is described as a material exchange with nature, which through man's activity is 'mediated, regulated, and controlled.'" ("Deutlich vorsichtiger spricht Simondon von den technischen Objekten als Vermittlern (*médiateurs*) zwischen Natur und Mensch. Die Betrachtung des technischen Objekts wird damit in ein mediales Register überführt, das [...] sich am Problem der Produktion orientiert. Seit Marx ist Produktion als ein Stoffwechsel mit der Natur bestimmt, der durch die Tätigkeit des Menschen 'vermittelt, geregelt und kontrolliert' wird" [Schmidgen 2012, p. 127].) Here Schmidgen is quoting [Marx 1972 (1869), p. 192].

Agricola's *De re metallica libri XII* (1556) – and thus a famous early treatise on the methods and machinery of mining and smelting metal – then it becomes clear that the work on material (the extraction and further processing of material), as well as the resistance that the material offers to this work, essentially belongs to the clarification of what a material at a certain time 'is.' This is particularly true of Agricola's 'metal' to the extent that during his time the term 'metal' is used in the broad sense that the term had in the sixteenth century, namely to designate silver, gold, and iron, but also other useful 'earths.' Agricola's study does not simply recapitulate traditional philosophical and/or alchemical knowledge on metals, but nor does it merely bring together the observed tools and recipes.[14] Rather he systematizes the mining practices and systematizes at the same time the causal relations of smelting procedures (which temperatures and catalyzers for which metals) as well as the geometrical recording of the metal-bearing veins, the passages to be constructed, and the construction of diverse lifting, ventilation, and crushing machines. *De re metallica libri XII* was intended not only as a learned treatise (as evidenced by the choice of Latin and the historical remarks) but also as an empirical study of contemporary machines and techniques, and in sum an encyclopedic work *avant la lettre* (i.e., before the publication of Denis Diderot and Jean Le Rond d'Alembert's famous *Encyclopédie* in the eighteenth century). The novelty of Agricola's work is the combination of theory and practice that aims to optimize the techniques of acquiring resources through the collection and dissemination of explicit and tacit knowledge. Thus, *De re metallica* not only represents the gradual transition from natural philosophy to the natural sciences but can also be understood as a prefiguration of the twentieth century applied sciences. This is seen clearly in the fourth and fifth books of *De re metallica*, which present geometrical considerations and constructions that – at least according to Agricola – are essential for mining.[15] The introduction of practical geometry for the research of artisanal and other practices is one of the characteristics of the early modern period, but as Thomas Morel notes, there is "a gap between the uses of geometry in early modern mines and their presentation in the *De Re Metallica*" [Morel 2020, p. 42].[16]

Notwithstanding the imaginary mathematization of mining techniques, the treatise focuses on among other things the extraction of metals. But while this extraction is extremely useful, and thus valuable, the Latin term *materia* – which generally stands for a substance from which something emerges – is rarely used. Agricola speaks rather of *terra* (earth or soil in the most general sense), which, with its components (soil, stones,

14 On this and the subsequent discussion on Agricola, cf. [Krauthausen 2022]. See [ibid.] also for references to further secondary literature.

15 Hoover and Hoover's English translation, [Agricola 1912, p. 117]: "[...] the miner who is not ignorant of geometry can calculate from the other mines the depth at which the *canales* of a vein bearing rich metal will wind its way through the rock into his mine."

16 See also the entire paper for a thorough survey of the cultural background of Agricola's practical geometry and the mathematical practices introduced by him for mining.

and sediments), hides the metals from sight and makes them "unformed" (Latin: *informe*; in the German translation: *ungestalt*). He emphasizes the active extraction of the metal by smelting in book IX, which begins as follows:

> Since I have written of the varied work of preparing the ores, I will now write of the various methods of smelting them. Although those who burn, roast and calcine the ore, take from it something which is mixed or combined with the metals; and those who crush it with stamps take away much; and those who wash, screen and sort it, take away still more; yet they cannot remove all which conceals the metal from the eye and renders it crude and unformed [*ac efficit informe*] [...]. Wherefore smelting is necessary, for by this means earths, solidified juices, and stones are separated from the metals so that they obtain their proper colour and become pure [*purum*], and may be of great use to mankind in many ways. When the ore is smelted, those things which were mixed with the metal before it was melted are driven forth, because the metal is perfected by fire in this manner [*metallum igni quodammodo perficitur*].[17] [Agricola 1912, p. 353]

For Agricola, material does not mean an accessible, natural resource, but something obscured, indeterminate, without a form, which has to be 'completed' (Latin: *perficitur*) by human intervention in order to be made visible and useable (i.e., present in its utility for further uses). This is to be seen in the glossary of terms in Latin and German that Agricola adds to his treatise. Here, he uses the term *materia* only in combination with various activities: "Materiam metallicam discenere a terris &c." (literally 'to separate/distinguish the metallic matter/material from earth') and translates this Latin phrase with a single German word, a verb, "ertzscheiden"[18] (in the sense of 'to separate the ore'). The 'metallic material' must be actively separated from the

17 Hoover and Hoover's English translation. For the Latin original, see [Agricolae 1556, p. 285 (book 9)]: "Scripsi de diverso venarum praeparandarum opificio, nunc scribam de varia earundem excoquendarum ratione. Quanquam enim qui venas urunt, & torrent & cremant, aliquid detrahunt de his, quae cum metallis mista vel composita esse solent: multum, qui tundunt pilis: plurimum, qui lavant, cribrant, discernunt, omne tamen id quod metallorum speciem ab oculis removet, ac efficit informe quiddam & rude adimere non possunt: quocirca necessario inventa est excoctio, qua terrae, succi concreti, lapides sic separantur a metallis, ut suus cuique color insideat, ut purum fiat, ut multis in rebus homini magno usui sit. Cum autem excoctio sit eorum, quae, anteaquam venae excoquerentur, cum metallis erant permista, secretio, quodque metallum igni quodammodo perficitur." (The Latin has been slightly conventionalized.)

18 Agricola's extensive glossary, which records the nomenclature of mining and metallurgy of his time in Latin and German, is present in the Latin first edition from 1556, but is not retained/adapted in either the German or the English translations. That is regrettable since this glossary was important for the contemporary reader from mining practice to the extent that they spoke little Latin. The glossary, like the figures, mediates between the scholarly audience and the practitioners. In addition, the author inserts the list of terms twice: once following their naming in the books ("Rei metallicae. Nomina latina graeca qve germanice reddita, et ex ordine, quo quodque primo occurit, collocata.") and once in alphabetical order ("Index secvndvs continens eadem rei Metallicæ nomina Latina, Græcaque Germanice reddita, sed in Lectoris gratiam, secundum Alphabeti ordinem digesta."). The terms in the two glossaries do not completely overlap. See [Agricola 1556, index without page numbers]; here all from "Libro Octavo" and correspondingly under 'M' in the second alphabetic index.

earth (*a terris*) before it can be brought to its proper (distinguished) appearance with the help of the fire of the smelter. And even when a little later in the glossary, Agricola lists "Materia metallica" (metallic matter/material) as a substantive, and not as an activity, this is translated to German as 'das werck' [Agricola 1556, index],[19] a term meaning "tätigkeit, wirksamkeit, arbeit" (activity, agency, work), or occasionally "anstrengung, mühe" (endeavor, effort).[20] If one follows Agricola's own lexical explanation, then the Latin *materia* (in the context of mining and smelting) denotes a skillful and purposeful as well as laborious activity in which the useful metal is obtained and brought to visibility. The result, the artfully 'perfected' metallic resource, can take on a variety of appearances (silver, iron, gold, etc.). But Agricola's use of the Latin *metallum*, despite this plurality, does not mean a general category, but denotes specific modes of being – or, more precisely, specific modes of *coming into being*.

The early-modern understanding of metals (and in this sense of material) thus remains linked to the procedures of extraction and processing (see Fig. 1). Due to the laborious extraction of metals, their origin, the earth (*terris*), is as meaningful as the human techniques that detect, expose, and complement the material. Agricola's description of the *materia metallica* suggests an understanding of material that locates this midway between nature and culture – an understanding that, according to Bernadette Bensaude-Vincent, has been forgotten in the current conception of materials in the natural sciences: "material combines *phusis* and *technē*, it is a hybrid notion referring to an alliance between natural beings and man's needs [...]" [Bensaude-Vincent 2011b, p. 109][21]

In Agricola's study on the mining of 'metal,' technique is deployed even before establishing a function (tools and weapons from iron, coins from gold, etc.), namely, during the mining of the metal itself. Nevertheless, these techniques still remain linked to the given world, hence, *phusis* and *technē*. Even when the metals in Agricola's treatise, due to the systematization of the practices and considerations, already grow in abstraction, the understanding of 'material' remains tied to the specifics of the respective process of uncovering, and in this way to the concrete process of becoming material. However, the activity that Agricola attributes to the *materia metallica* (metallic material/matter) in his Latin–German glossary with the help of the old German term *werck* (activity, agency, work), and in this way a compound of nature

19 Shortly afterward in [ibid.] again "Materia, werck."

20 For the meaning of the German 'werck,' cf. the article "Werk" in [Grimm and Grimm 1960, Sp. 347, Z. 13] – here the Latin equivalent is given by the terms "opus," "opificium, operatio," thus not '*materia*.' The semantic shift that Agricola carries out in his register (by translating 'materia' by 'werck,' and thus with a word that is strongly related to activity) is significant for his understanding of metal, and that means from today's perspective of material.

21 On the extraction of wood from trees, see: [ibid.]: "Trees were extracted from nature, separated from their natural environment and became wood in relation to the design of a specific artefact." Material thus designates a transformation process: the tree becomes wood because man decontextualizes and functionalizes it.

Fig. 1: The mining and smelting of metals in the sixteenth and seventeenth centuries.
From: [Löhneysen ca. 1660, plate 6]. Zentralbibliothek Zürich, NG 132 | G, DOI: 10.3931/e-rara-50189. Public Domain Mark.

and human techniques, is, in his view, increasingly located solely on the side of man. Already for Agricola, man's systematic approach, with the help of tools, machines, and knowledge, brings forth the 'pure metal' and in this way the specific material (iron, silver, or gold, etc.). The more, in the following centuries, this epistemic and technical competence is formed into a separate field, the more the activity is attributed solely to the intervention of man – and the material to a merely passive counterpart.

1.3 The Eighteenth and Nineteenth Centuries: A Science of Substances and Its Alphabetization

Agricola posited the miner in the role not only of an experienced practitioner but also of a proto-scientist – that is, as the one who can decipher nature's composition and work according to a plan, thereby uncovering from the earth the various particular materials and processing them to make them pure. As such, in its 'purest' form, processed according to the wishes of the miner, material becomes perfected – and in this way more controllable and docile, but also *passive*. Crucial to this development was on the one hand the improvement of the techniques of extracting material, and on the other the compiling and systematization of analytical knowledge (as opposed to focusing on authorities, as in the scholarly tradition). Added to the extraction and smelting techniques are thus epistemic techniques that along with the emerging natural sciences are developed into a separate field. In his treatise, Agricola dissociated knowledge about which ores could be smelted and broken down (e.g., through the addition of substances, through temperature) from alchemistic explanations. Nevertheless, chemistry had not yet been invented, which is why he still has to introduce the knowledgeable handling of various ores as practical knowledge. Only with the introduction of chemistry does one see the beginning of the systematic scientific understanding of the various substances of nature and their composition – which gradually prompts a new understanding of materials. Thus, in the eighteenth and nineteenth centuries, one sees a growing tendency to treat materials 'alphabetically,' and that means as more or less equivalent elements in a formal system.[22] This is done by deciphering their pure chemical compound, which in turn led, with synthetic chemistry (in the twentieth century), to a specific conception of materials as something that can be duplicated. To explicate, when considering chemistry and the research of materials, during the eighteenth century – and here one should note that chemists at that time tended to use the term 'substances' rather than 'materials' to denote materials – "there was a comparatively small group of substances that chemists identified and classified according to their chemical composition, and a much larger group that they identified and classified according to their natural origin, mode of extraction, perceptible properties, and practical uses" [Klein and Lefèvre 2007, pp. 299–300]. In short, in the eighteenth century, the systems of classification of substances were heterogeneous – there was neither a single taxonomic system nor a "single conceptual umbrella or paradigm" for their classification [ibid., p. 3]. Moreover, the theories of eighteenth-century chemistry

22 In such a formalization, the materials become abstract values – and the eligibility conditions (the resistance of the materials and the techniques for their extraction, which stand to the fore in Agricola) disappear completely from view. For this symbolic negotiation of materials, Ingold's "*forgetting* chemistry" can be asserted, thus that procedure with which he wants to return to the practices in the handling of material and thereby also to the activity of the materials [Ingold 2013, p. 28].

"evolved not primarily around 'atoms' or 'corpuscles' but chemical substances" [ibid., p. 6], which points to the fact that the conception of treating materials 'alphabetically' was neither an obvious nor a dominant one during this period, although it certainly announced itself with the rise of the chemical 'paper tools.'[23] In the nineteenth century, there was a shift regarding the experimental study of structure: "The stabilization of quantitative analysis and the use of chemical formulae by nineteenth-century chemists privileged knowledge of composition and structure over sensible properties, but also extended the chemists' tools box for identifying and classifying chemical substances" [ibid., p. 304]. In this sense, one can speak of an "ontological shift that reconfigured the range of substances accepted as objects of inquiry in organic and inorganic chemistry" occurring in nineteenth-century chemistry with respect to the investigation of materials [ibid., p. 305]. This shift embedded the research of materials in a new scientific milieu "of experiments, precision measurement, and work on paper with chemical formulae" [ibid., p. 304]. As noted above, with the nineteenth-century chemical 'paper tools,' as Ursula Klein notes, there was a shift with respect to the understanding of 'organic' materials in terms of their chemical formula:

> *Organic* no longer meant a particular natural origin of chemical materials from the organs of plants and animals. [...] The new concept of the 'organic' was defined mainly by the elemental composition of the substances [...]. In the experimental culture of carbon chemistry, an 'organic' substance was, first, a pure chemical compound rather than any kind of extracted vegetable and animal material. [Klein 2003, p. 221]

With the insights delivered by crystallography (during the nineteenth century), on the one hand, and stereochemistry (starting in the second half of the nineteenth century) on the other (by, among others, René-Just Haüy and Jacobus Henricus van't Hoff, respectively), it became clear that not only the chemical formula played a role in determining the properties of the materials but also the three-dimensional structure of the crystal or molecule. The twentieth century saw the advance of chemical synthesis, which reinforced, with the production of various synthetic chemicals, a conception of materials as something that "can always be replaced or substituted" [Bensaude-Vincent 2011b, p. 111], which in turn favors a conception of nature as a "huge library of resources gathered through random processes of combination" [Bensaude-Vincent and Newman 2007, p. 18]. These processes, and especially, one may suggest, chemical synthesis, underline a slow process of 'passivization' of materials, which might be understood as a return to the Aristotelian understanding of *hyle* in contrast to form. These passivization processes may therefore be conceived via a 'textualization' or 'alphabetization' of (chemical) materials, presented as a set of letters and three-dimensional structures, which can later be combined and recombined, produced and reproduced. The shift toward alphabetization and 'readability' reached its

23 We refer here to the title of [Klein 2003]: *Experiments, Models, Paper Tools*. On the history of the metaphor of the readability of nature and the world, see: [Blumenberg 1981].

zenith in biology, one may argue, as Lily Kay does, with the discovery of DNA and the acceptance of its associated discourse as the 'code' of life [Kay 2000].

1.4 The Twentieth and Twenty-First Centuries: Materials Science

Bensaude-Vincent has pointed out that, until the middle of the twentieth century, material was always understood as a specific, individuated substance. The different materials – metal, ceramics, wood, and so on – formed a "zoology of materials" (i.e., a variety of material species) researched by different disciplines [Bensaude-Vincen 2011b, p. 110]. Hence, if the natural sciences aim at generality, abstraction, and systematics, as can be seen with the 'alphabetization' and 'code' paradigms and discourses of the twentieth century, there can be no basic science of *singular* materials, Bensaude-Vincent claims. This is already implicit in the conception of chemical synthesis – as if *any* material can be reduplicated, hence ignoring its singularity, which leads to materials being understood as an abstract ontological and epistemic category similar to that of matter in physics. In the mid-twentieth century, however, a fundamental shift took place regarding the difference between materials (as individual and technological) and matter (as universal and ontological). With the emergence of *materials science* as a discipline, materials have become a scientific category and, moreover, a generic concept. The new discipline is the expression and motor of a 'science of materials' (i.e., a systematics of the individuated materials), and thus of a hybrid of the singular and the generic, that is, not a 'science of matter,' as seen in the classical disciplines [ibid., p. 112].

Materials science emerged in the USA in the 1960s,[24] primarily in metallurgy departments (which were increasingly renamed, e.g., 'metallurgy and materials science' departments) before establishing itself as an independent subject a few years later [ibid., p. 114]. What characterizes metallurgy in the twentieth century is the coupling of structure and property, a characterization that is at the root of materials science. With the help of x-ray crystallography during the first decades of the twentieth century, for example, it was possible to analyze the microstructure of crystals and metals.[25] Terms such as 'crystal lattices,' 'dislocation,' or 'defect' have been used to describe the micro- and nanostructure of a material and in this way to help to understand its macroscopic properties. The connection of microstructure and mechanical properties is tested and the physicists' models help to design new materials [Hoddeson et al. 1992]. Indeed during the 1960s, "[t]he focus on structure-sensitive properties in the study of crystals

24 We follow here mainly: [Hentschel 2011; Bensaude-Vincent 2011b].

25 To explicate: "Not only instrumental techniques – from x-ray diffraction to STM and AFM – opened the way to microstructures and nanostructures, but also they helped create a scientific community" [Bensaude-Vincent 2011b, pp. 121–122].

has been identified as the main route leading to materials science" [Bensaude-Vincent 2011b, p. 114].

It is around this time that there was a shift in the conception of materials, which later, in 1989, consisted of four related aspects: properties, structure, processing, and performance (see Fig. 2). This tetrahedron of 'fundamental aspects' was and is considered – already before its presentation in 1989 but mainly after it – as a paradigm for materials science [Hentschel 2011, p. 26]. Bensaude-Vincent notes that the novel "notion of material" – as a plural entity (i.e., materials) – was not only thought from the beginning as an interdisciplinary one. It also "requires that structure and properties be coupled with functions or performance. It is only with respect to functionalities that one can gather under the same umbrella such dissimilar stuffs as wood, concrete, paper, polymers, metals, semiconductors, and ceramics. [...] they are all materials *for* ..." [Bensaude-Vincent 2011b, p. 115].

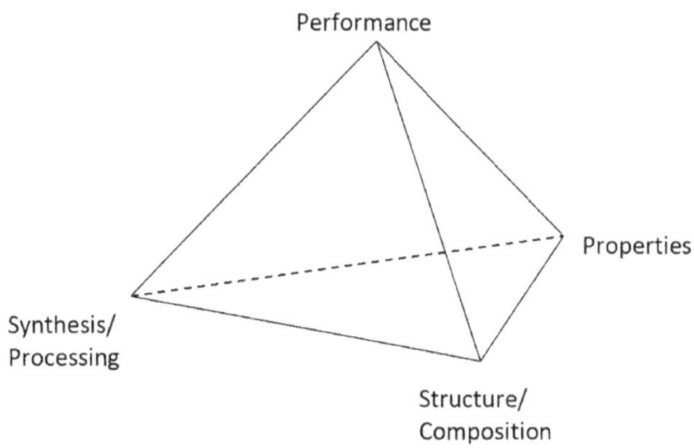

Fig. 2: Tetrahedron of the fundamental aspects of materials science. See [COSMAT 1989, p. 29]. © Graphic redrawn by M.F.

The history of materials science consists, then, starting in the 1960s, of heterogeneous research directions: prompted by research on polymers (a research that began at the end of the nineteenth century and continued to be developed during the first decades of the twentieth century) [Bensaude-Vincent 2013; Bensaude-Vincent 2007, pp. 294–297], there was a shift to the production of "composite materials with desirable properties" [Bensaude-Vincent 2011a, p. 105] – an approach known as 'materials by design' – rather than of "conventional materials with standard specification" [Bensaude-Vincent 2013, p. 21], and since the 1990s one has seen a rise in the importance of the nanosciences, followed (starting in the 2000s) by a focus on bioengineering and bioinspired materials. While the first generation of materials science dealt with relations between structure and properties, the second dealt with the tetrahedral relations presented above. Smart materials – as "systems responding to their environment" [Bensaude-Vincent

2011a, p. 108] – are named explicitly as such during the 1980s, whereas the general conception of these newly developed materials – whether bioinspired, smart, or programmable – considers them as end products: "it testifies to the changing status of materials, which used to be a priori constraints for engineering projects and eventually turned into the *end product* of a design process" [Bensaude-Vincent 2011b, p. 117].

With the establishment of materials science departments, a decisive change in the natural-scientific understanding of material occurs, one that (1) strengthens the individuality of the materials (high-end materials or materials by design) and thereby (2) brings clearly into view the techniques of production, and yet (3) explicates materials as a quasi-generic category, and in this way leads them to converge with the category of matter. Linked to this, however, is a further important aspect: to the extent that the 'performance' of materials is declared – alongside properties, structure/composition, and synthesis/processing – to be a core concern of scientific analysis, the activity of the materials comes explicitly into view. Whereas in the chemistry of the nineteenth century, materials were understood as formalizable and combinable units, and thus respond to the question of what a material *is*, the materials sciences are concerned with what materials can *do* (if in a slightly different sense to Ingold). Techniques and technology again come to the fore, but this time right at the center of the sciences' claim to universality.

2 Active Materials

Due to the new focus on materials of the second generation of materials science, the activity of materials is also taken into consideration. But this newly formed focus on active materials was reinforced as well as intertwined with the almost parallel and by now related rise and development of a much more specialized domain in physics: active matter.[26]

2.1 Matter in Action

In the 2000s, the research on active matter attracted considerable interest in and beyond the natural sciences, with particular attention being given to a new understanding of

26 Here it is essential to emphasize that, while it seems that this domain of research seems to offer a new theory of matter, the theory (or theories) of matter developed within the framework of active matter neither include (or cover) nor aim to include the entirety of the discoveries and the developments of the twentieth and twenty-first centuries with respect to the question of what matter is. To give two prominent examples: neither what may be seen as the derived theory of matter of quantum mechanics (or quantum mechanics itself) is dealt with in the discourse and investigations of *active matter*; nor the fact that, starting in the second half of the twentieth century, a variety of the produced *materials* are in fact 'electronic,' that is, composed of semiconductors and transistors.

bird flocks. Following a paper by Tamás Vicsek et al., bird flocks could be modeled as spontaneous formations, whereby this form is precisely not initiated or guided by a lead bird (see Fig. 3) [Vicsek et al. 1995]. Rather the emergence of a flock depends solely on the density of the cluster of birds, while the bird's flight behavior is determined by the interaction between neighboring birds. This physical description of a biological phenomenon was remarkable in a number of ways: Above all, the flock is not understood via consciousness and intention or communication and hierarchy between individuals. While the totality of the birds and the transition from disorder (random cluster of birds) to order (flock) come into view, this agent-based modeling need not refer to the fact that at issue here is living beings. Rather the behavior of the birds is compared to the motion of mere particles and in this sense to a 'lifeless' material entity – insofar as a form of mathematical modeling is employed on the flock that is known from the description of Brownian motion or ferromagnetic formations.[27] However, Vicsek's model was not the first successful attempt to model such flocking movement – one should also mention the 1987 model developed by Craig W. Reynolds. In order that a structure of a cohesive flock exists, Reynolds' model, in which "the *simulated* flock is an elaboration of a particle system, with the simulated birds being the particles," imposes the restriction that each animal should avoid collision with its neighbors, follow

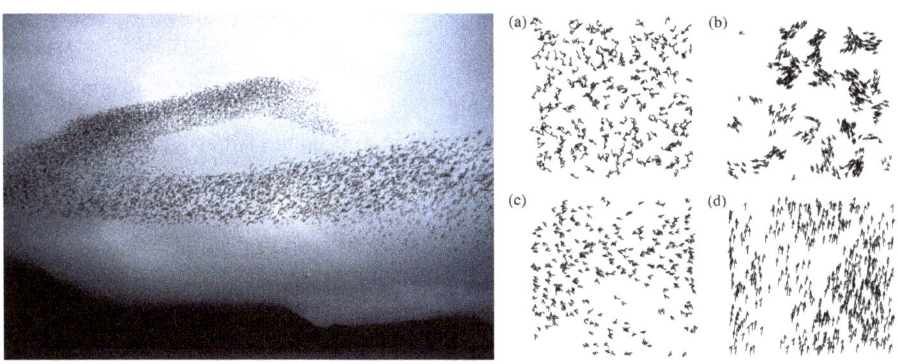

(a) (b)

Fig. 3: (a) Auklet flock, Shumagin Islands 1986. Wikimedia Commons, the free media repository, 2020. Available from https://commons.wikimedia.org/wiki/File:Auklet_flock_Shumagins_1986.jpg (accessed June 28, 2021). (b) Vicsek et al.'s model of flock behavior: "In this figure the velocities of the particles are displayed for varying values of the density and the noise" [Vicsek et al. 1995, p. 1227]. Reprinted with permission from: [ibid.]. © 2021 by the American Physical Society.

27 See [Ramaswamy 2010, p. 327]: "Like the continuous-spin magnets that they resemble, the Vicsek family of models display a well-defined phase transition from a disordered phase to a coherent flock as η is decreased or the number density is increased [...]." Although always with the difference that in active matter it is a question of *self*-movement.

its species, and move in the same direction as the rest of the group, where 'group' here means the animal's six or seven immediate neighbors [Reynolds 1987, p. 25, 28].[28]

Nevertheless, the formalization conceptualized by Vicsek et al. as an 'analogy' of ferromagnetism differs in one essential way from what it was modeled on: "In this sense our model is a [...] nonequilibrium analog of the ferromagnetic type of models, with the important difference that it is inherently dynamic" [Vicsek et al. 1995, p. 1226] – thus expanding the focus of their research to include nonequilibrium phenomena. From a scientific point of view, this approach has been extremely productive, since it has allowed further deductions on the emergence of collective behavior and above all a broadening in scope to include phenomena of movement at different scales and in different contexts. This broad claim of active matter research, however, is at odds with the relatively narrow definition formulated, for instance, by the mathematician Gautam I. Menon in 2010:

> The term *active matter* describes diverse systems [...] [which] are often idealizable in terms of collections of individual units, referred to as active particles or self-propelled particles, which take energy from an internal replenishable energy depot or ambient medium and transduce it into useful work performed on the environment, in addition to dissipating a fraction of this energy into heat. [...] [The] active particles can exhibit remarkable collective behavior as a consequence of these interactions, including non-equilibrium phase transitions between novel dynamical phases, large fluctuations violating expectations from the central limit theorem.
> [Menon 2010, p. 193]

The physics of flocking was extended not only through descriptions such as John Toner and Yuhai Tu's continuum field-theoretic approach [Toner and Tu 1995] but also through the creation of additional subject areas: through apolar flocks, which in contrast to the directed, 'polar ordered' states of bird flocks are understood as "flocks that go nowhere" (e.g., active polar states in granular matter) [Ramaswamy 2010, p. 331]; through the so-called 'swimmers' who designate the movements of living entities in a fluid medium (from bacteria and algae to schools of fish – thus through different scales); through 'active gels' (e.g., the cytoskeleton), to the extent that these follow the same hydrodynamic equations;[29] and finally through the collective movement of living cells during growth and healing (tissue dynamics).

What unites the objects from different scales and milieus (such as flying birds and swimming fish, bacteria and cells, tissues and gels, and real entities and hypothetical model systems) under the umbrella term 'active matter' is the use of theories and tools

---

28 The media history of the swarm during the twentieth century as well as the problem of representation and modeling of it is described extensively in: [Vehlken 2012].

29 The cytoskeleton is understood as a "suspension of filaments endowed with active internal forces" and active matter research aims at the "natural mechanisms that promote the alignment of neighboring filaments, through excluded volume as well as activity" [Ramaswamy 2010, p. 325].

of statistical physics and hydrodynamics or liquid crystal hydrodynamics (thus soft matter physics).[30] These include active nematic theory (i.e., the descriptions of symmetry breaking and topological defects), conservation laws, and agent-based standard models, as well as analytical tools such as simulations, field-theoretical methods, and "dynamic density functional theory" [Gompper et al. 2020, p. 3]. The overarching field to which active matter research is explicitly assigned [Ramaswamy 2010; Menon 2010] is condensed matter physics, which brings together the multiplicity of matter's manifestations (solids, liquids, plasmas, etc.) in their macroscopic and microscopic properties under a collective roof – insofar, for instance, as basic physical principles are applied to the states of matter.

According to Sriram Ramaswamy, the research on active matter takes up the comprehensive approach of condensed matter physics and should therefore develop a "systematic theory" by expanding the principles of statistical physics and hydrodynamics to include living entities and thus the objects of biology [Ramaswamy 2010, p. 323]. "[L]iving, metabolizing, spontaneously moving matter" is a paradigmatic object for activity and should therefore be integrated as 'active matter' into condensed matter research [ibid., p. 326]. That is an ambitious claim that in the final analysis might mean that 'living matter' will be classified using descriptions and explications that closely relate to those of 'nonliving' matter – it remains to be seen, however, whether the claim of the research field active matter can be realized in this way.

Clearly the project of active matter research also recalls nineteenth- and early-twentieth-century biophysics, as it is prominently represented by D'Arcy Wentworth Thompson. The approach of active matter research, however, is not concerned with the forms of living beings (e.g., anatomy and shape) or the evolution of these forms but with a 'matter in action,' understood here as a spontaneously emergent collectivization and form construction,[31] which calls on a different physical regime: nonequilibrium thermodynamics or statistical physics. Living systems are a prime example of this, since they are open systems that operate far from equilibrium and successfully accommodate the related instabilities. Examples are found, however, beyond living systems and at all scales. The aim of developing a comprehensive physics of life based on nonequilibrium thermodynamics has already – that is, before the research

Here it should be borne in mind that these active gels are not considered as a medium but as a "distinct type of material" and thus as a particular manifestation of matter [ibid., p. 339].

30 And not only that, but also as Menon emphasizes the "philosophy" of active matter, one does not begin with distinct units, but with a continuum model: "The philosophy of these approaches is the following: Rather than begin from a microscopic model for a swimmer or individual moving particle and then generalize from the microscopics to realize symmetry-allowed equations of motion for the fluid velocity field and for the local concentration of swimmers, one can start with a coarse-grained continuum model for a physical viscoelastic gel which is driven by internally generated, non-equilibrium sources of energy" [Menon 2010, p. 213].

31 See [Menon 2010, p. 194]: "Such [active matter] systems are generically capable of emergent behavior at large scales."

on active matter – been formulated by Ilya Prigogine. Nevertheless, he introduced these themes much more broadly in physics and chemistry[32] and in addition looked for support in philosophy. Already in 1984, in their book *Order Out of Chaos*, in a chapter titled "Active Matter," Prigogine and the philosopher Isabelle Stengers note programmatically that "[a]t all levels, be it the level of macroscopic physics, the level of fluctuations, or the microscopic level, nonequilibrium is the source of order. Nonequilibrium brings 'order out of chaos.'" [Prigogine and Stengers 1984].

What is particularly difficult about nonequilibrium systems is that for their explication one cannot have recourse to classical physical principles such as temporal reversibility but is forced to have recourse to the challenging theories of nematic physics and liquid crystal hydrodynamics. Thus, active matter researchers see their domain as a radical expansion: "a new area of fundamental physics" [Ramaswamy 2010, p. 341]. In 2020, this statement from 2010 has to be supplemented and corrected: active matter research is explicitly no longer the concern of physics alone but can be considered a "truly interdisciplinary endeavor at the interface of biology, chemistry, ecology, engineering, mathematics, and physics" [Gompper and Winkler 2020, p. 2].

In order to summarize the understanding of activity being developed in the context of the research on active matter, one must emphasize three aspects: (1) activity is linked with movement (motility) but cannot be equated with movement, rather it is distinguished from other driven systems to the effect that here (2) internal energy is converted and consumed; and (3) this movement can give rise to phase transitions and to the emergence of order, as for instance in the remarkable collective behavior of the bird flock or the repair work of cells in tissue. This characterization can be formulated for active matter research. However, this by no means covers the full range of current research on active materials.

2.2 Inherent Activity

One can ask oneself why the research on the activity of materials has grown so significantly since the 1990s. That materials – and in particular organic materials – have a life of their own was not unknown. The carpenter or instrument maker has long been aware that wood expands and shrinks and that it continues to change shape long after it has been worked. Nevertheless, especially in the nineteenth and early twentieth centuries, a conception of an ideal material becomes prevalent that associates this material with passivization and plasticity – or that even wants a material to disappear almost completely. Here architecture becomes an exemplary field, since, with iron construction in the nineteenth century, a changed structural deployment of

32 See his focus on self-organization and his concept of dissipative structures in: [Prigogine and Nicolis 1977].

materials begins. The bulkiness of stone construction is exchanged for lightness and modularity, but this leads above all to the impression of a dematerialization and to a claim of overcoming matter. The research on active materials, on the other hand, radically distances itself from this utilization and understanding of materials by investigating, for instance, their internal architecture in its function for movement. This is linked above all with a new interest in 'nature.' Now the organisms of nature are interrogated for the potential in their material structure for movement and other activities.

2.2.1 'Dreams of Dematerialization'

Agricola describes the *materia metallica* neither as homogeneous nor as passive. While metal is a natural resource, this must be laboriously extracted from the ground through the deployment of people and machines and then be gradually revealed in the smelting works through additional techniques. Only when the desired properties are manifested can the metal be made available for more precise functions and be turned for instance into gold coins, silver jewelry, tools, or weapons. However, the greater the number of new techniques and facilities that were developed in the following centuries (e.g., when the steam engine in the eighteenth century improved the bellows and thus the performance of the furnace), and the more new knowledge was acquired about chemical composition and physical behavior, the more reliably specific materials could be created and accurately processed. As Bensaude-Vincent notes, in the nineteenth century iron became a "single-class" material that could be treated mathematically as "pure deformable continua" and well controlled both formally and in the concrete processing [Bensaude-Vincent 2011b, p. 114]. She therefore describes iron as a "model material" for the scientific (and engineering-based) handling of material in the nineteenth and early twentieth centuries [ibid.]. The material iron can be considered passive (insofar as it "does not work by itself; it is put to work") and also allows mathematical treatment, since it is homogeneous and isotropic [Guillerme 1994, p. 233].[33] Indeed, in the architecture of the nineteenth century, iron became a highly functional material. It allowed a 'lighter' method of construction that is capable of surprisingly tall towers (such as the Eiffel Tower (1889) in Paris), of long bridges (the Brooklyn Bridge (1883) in New York), and of vast buildings (the Crystal Palace (1851) in London). But metal-rod construction also changed the perception of load-bearing structures. The architectural theorist Gottfried Semper speaks of an "almost invisible material" ("gleichsam unsichtbaren Stoffe"), a dematerialization of architecture that he finds regrettable [Semper 1849, p. 521].[34] Nevertheless, already in the

33 Quoted in [Bensaude-Vincent 2011b, p. 113].

34 See the full quotation: "However, this much is clear, that iron, and indeed every hard and tough metal used as slender rods and sometimes as cables as befits its nature, due to the slighter surface which is offered by these forms withdraws all the more from the eye the more perfect the construction

nineteenth century and fully developed in the twentieth century, lightweight construction is the object of a new aesthetic and the motor of a new architecture. The architect and theorist Richard Buckminster Fuller designates the new possibility of "structural lightness" [Fuller 1979, p. 175] in architecture as "ephemeralization" [Fuller 1938, p. 284], which he describes as follows: "Doing the most with the least – segregated compression and tension members, flexible joints, stabilized force triangles – net scientific structure in time annihilating transportation, communication, and power harnessing" [Fuller 1932, p. 36].[35] Therefore ephemeral buildings should not only reduce the amount of material used but also develop and present a new type of load-bearing structure. Fuller wants to modularize, network, and dynamize the structures, as for example in his most famous geodesic dome, the American Pavilion at the 1967 International and Universal Exposition in Montreal, Canada (see also the variant in Fig. 4). The dematerialization leads in Fuller's case to an understanding of architecture as a highly efficient lightweight structure – here too the material seems to recede behind the structure, or to become one with it, but in this particular case it is already a matter of the dynamization of structures.

Fuller's innovative concept of ephemeralization, however, belongs to the mid-twentieth century. In the nineteenth century, dematerialization was associated with the idea of a plastic and passive material. If one takes into account the discussion on the various materials invented during the nineteenth century – rubber, celluloids, the development of plastics – a discussion accompanied, on the one hand, by a demand to bring justice or "truth to materials" ("Materialgerechtigkeit"), one cannot ignore that this discussion culminated in a demand to develop a single material that would be appropriate to the desires and needs of man. In 1902, Henry van de Velde claimed that "our dream of a supple material that follows our intentions as easily as our speech follows our thoughts will come true."[36] No longer imitation or perfection of nature,[37] a

is, and therefore architecture, which accomplishes its effects on the spirit through the organ of vision cannot admit of this as it were invisible material when it is a matter of effects of mass and not merely of light accessory parts." ("Doch so viel steht fest, daß das Eisen, und überhaupt jedes harte und zähe Metall, als konstruktiver Stoff seiner Natur entsprechend in schwachen Stäben und zum Teil in Drähten angewendet, sich wegen der geringeren Oberfläche, welche es in diesen Formen darbietet, dem Auge umso mehr entzieht, je vollkommener die Konstruktion ist, und daher die Baukunst, welche ihre Wirkungen auf das Gemüt durch das Organ des Gesichts bewerkstelligt, mit diesem gleichsam unsichtbaren Stoffe sich nicht einlassen darf, wenn es sich um Massenwirkungen und nicht bloß um leichtes Beiwerk handelt" [Semper 1849, p. 521].) On dematerialization and ephemeralization as aesthetic experience and technological program in the nineteenth and twentieth centuries, cf. [Krausse 2001]. On 'dreams of dematerialization,' cf. [Bensaude-Vincent 2013].

35 Quoted in [Krausse 2001].

36 "Denn unser Traum von einem geschmeidigen Material, das unseren Absichten so leicht folgt, wie die Sprache unseren Gedanken, wird in Erfüllung gehen" [Van de Velde 1902, p. 31].

37 On how this new image of man emerged during the early modern age, who neither imitates nor perfects nature (a conception of the relations between man and nature which is highly influenced by Aristotle), see: [Blumenberg 2000].

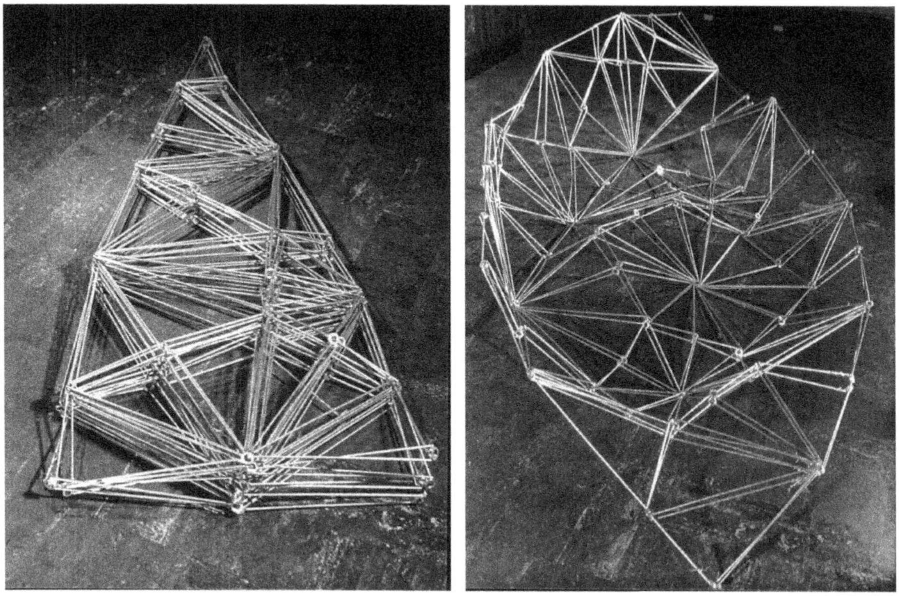

Fig. 4: R. Buckminster Fuller's necklace dome (1950) is one of the first geodesic, dome-shaped structures constructed by Fuller. The folded structure emphasizes the modularity and dynamic character of these buildings. © Courtesy, Estate of R. Buckminster Fuller.

wish emerges of a future invention of an almost formless material that will fulfill all of man's needs.[38] At the beginning of the twentieth century, a similar development occurred regarding wood. As Cecil Elliott notes, "German companies in 1933 began manufacturing synthetic resin glues," and while, as "linear structural members, timbers performed well [...] at connections the more complex forces were limited by the lateral weakness of the material. By gluing crossed layers of veneer, plywood balanced these capabilities of wood, so long as it was used as a sheet" [Elliott 1992, p. 21]. This shift to the mass production of plywood can be seen as an effort to treat it as a "passive material" [Eder et al. 2020, p. 3]. Such a passivization was clearly formulated, for example, in Alfred Gotthold Meyer's reflections on iron and iron constructions. In his posthumous 1907 book *Eisenbauten: Ihre Geschichte und Ästhetik* (*Iron Constructions: Their History and Aesthetics*), Meyer (1864–1904), a professor of the history of decorative arts in Berlin, examined iron as a building material and its effects on the development of style. According to Meyer, iron was celebrated as what does not have any constraint, in contrast to wood or stone. It enables architects and

38 See also: [Wagner 2004, 2015]. In this sense, the call during the twentieth century to a (new) "conversation with materials" can be seen as a continuation of the discussion on *Materialgerechtigkeit* during the nineteenth century. See: [Schon 1992].

engineers to achieve their mutual goal: "they [architects and engineers] [...] have a common goal: the victory over matter [*Materie*]."[39] This victory enables one to free oneself from the constraints of the materials and to see iron constructions as an "embodiment of what is abstractly recognized as necessary."[40] The reduction to the necessary must be understood here with regard to the seamless mathematical treatment of material and form: "And with the unlimited plasticity of iron it is possible to create the statically most 'rational' form [...] artificially."[41] Emphasizing the role of mathematics and geometry for architecture in general, Meyer notes that iron constructions enable one to "transfer the problems of mechanics from the area of arithmetic operations and algebraic forms into the depictions of graphic formations" [Meyer 1907, p. 29–49].[42] In this sense of dematerialization and unlimitedness, iron constructions allow much more freedom than stone or wooden ones, since the former are also more easily mathematically calculable.

2.2.2 Dreams of Architecture

Meyer's "victory over matter" calls for an understanding of matter that grasps this via properties such as mass, strength, rigidity, and indivisibility – as is found prominently in Isaac Newton's mechanical philosophy, but which remained in force into the nineteenth century.[43] This points to an underlying conception of matter as passive and shapeless. But it is worth recalling that, against the mechanical approach

39 "Architekten und Ingenieure haben [...] ein gemeinsames Ziel: den Sieg über die Materie" [Meyer 1907, p. 4].

40 "[...] die Verkörperung des abstrakt als notwendig Erkannten" [ibid., p. 5].

41 "Jeder Teil einer Eisenkonstruktion übertrifft einen gleichgroßen Holzbalken an Festigkeit zehnfach [...]. Und bei der unbeschränkten Bildsamkeit des Eisens wird es möglich, die statisch 'rationellste' Form [...] künstlich zu schaffen" [ibid., p. 13].

42 "Ist doch eine Eisenkonstruktion schon an sich eine besonders sinnfällige Verkörperung der im Bau wirksamen statischen Kräfte; sie steht dadurch gewissermaßen nur am Ende jenes synthetischen Weges, der die Probleme der Mechanik aus dem Bereich arithmetischer Operationen und algebraischer Formeln in die Anschauung graphischer Gebilde überträgt" [ibid., p. 43].

43 One may also recall the earlier approach of René Descartes, of a homogenous matter consisting of extension alone. But as Andrew Pyle notes, to view the history of matter theory in the seventeenth century "in terms of the rejection of Scholastic Aristotelianism and its replacement by one version or other of the mechanical philosophy (Cartesian or atomist) would be oversimplified, to say the least. As Leibniz saw, neither the Cartesians nor the atomists had a satisfactory account of material substance" [Pyle 2017, p. 440]. Nevertheless, mechanical philosophy in the natural sciences was progressing during the seventeenth century, accompanied by a growing mathematization of various phenomena. During the eighteenth century, the advancement of the mechanization of matter theory certainly left its mark on how matter theory was reshaped: it became "reductionist rather than essentialist, [and] it embraced quantification where possible [...] and in reaction to its association with mechanism, it emerged very much as an experimental, as opposed to a speculative, discipline" [Gaukroger 2014, p. 689].

of inert matter, the seventeenth- and eighteenth-century vitalists – though far from being a homogenous group – posited matter as active,[44] objecting to the world being reduced to a heap of dead, passive matter.[45] As the nineteenth century continued, however, vitalism was gradually abandoned and rejected, to be replaced by chemical and physical explanations as theories of matter. Nevertheless, in the second half of the twentieth century, two shifts occurred with respect to the latter conception of matter: first, as noted above, the dream of finding one single material for all purposes and needs, which is encapsulated in Meyer's claim of a victory over matter, is replaced by an extensive research on materials – as can be seen with the formation of the multidisciplinary materials science out of metallurgy departments; second, the research on active matter expanded the physical understanding of matter to include nonequilibrium systems – and thus also living systems (the prime example of nonequilibrium systems). Thus, next to solid matter and soft matter (or the overarching condensed matter) comes living matter, whereby the physical theories of statistical physics and hydrodynamics as well as the accompanying challenging mathematical tools produce points of intersections between the different fields of matter.[46] Here 'active' means not an activity imposed from outside, for instance, a movement induced by an impact – in active matter research, driven systems are always self-driven systems; that is, the activity is attributed to the matter itself.

Within the – in comparison to active matter – broader field of research in the materials science, the activity of materials is explicated less via the theory of nonequilibrium systems than via the internal architecture of living or dead materials. Indeed here too one frequently has recourse to exemplary configurations from living nature. A well-known example is the research on the capacity of bones to adapt to stress, since bones are capable of adapting to concrete environmental and living conditions through growth and remodeling [Weinkamer and Fratzl 2011]. And yet the 'dead' (or, more precisely, nonliving) components of organisms are also considered for the understanding of active materials insofar as these exhibit an ability to move and react based on their internal architecture. A well-known example is the pine cone, which consists largely of cellulose and thus of a dead substance, but still has the ability to open its scales (when dry) or to close them (when wet) in order to release its seeds into the environment only with the appropriate environmental conditions (dry weather;

44 For the various approaches to vitalism and their reactions to the mechanical theories of the eighteenth century, see: [Wolfe 2014].

45 To give only one example, Charles Wolfe notes that John Toland, in his *Letters to Serena* from 1704, "is explicit that 'Matter neither ever was nor ever can be a sluggish, dead and inactive Lump, or in a state of absolute repose' [...] he also denies that '[...] Matter is or ever was an inactive dead Lump in absolute Repose, a lazy and unwieldy thing'" [Wolfe 2014, p. 99]

46 See: [Gompper and Winkler 2020, p. 2]: "Fundamental biological processes, such as morphogenesis and tissue repair, require collective cell motions [...] Tissues are nature's active materials, and are therefore very interesting as blueprints for synthetic active materials."

see Fig. 5) [Reyssat and Mahadevan 2009; Dunlop, Weinkamer, and Fratzl 2011].[47] This reversible movement is not motivated by internal energy balance but is based on the internal anisotropic structure, that is, the specific arrangement of the cellulose fibers – it is therefore the internal architecture (together with the environmental conditions) that enables the activity, and not an external active force that imposes movement on an actually inert object. Similar functional arrangements of cellulose are found in the trunk and branches of trees. Although here it is not a matter of reversible movements, as in the pine cone, but of a combination of rigid and minimally flexible tissue adapted to stresses and environmental conditions so that, in the appropriate places, either the load-bearing capacity or elasticity is promoted. For Michaela Eder et al., living trees are therefore, also in their dead tissue parts, anything but homogeneous and passive: "woody material [...] does not obey this idea of neutral matter" – rather, "wood shows an intrinsic activity" [Eder et al. 2020, p. 3].

Fig. 5: Three-dimensional rendering of x-ray microtomography images of the same pine cone (*Pinus sylvestris*) in the wet (left) and dry (right) states. Note the two different tissue types, light yellow and orange, that can be seen in the base of the pine cone scales. © MPIKG and Michaela Eder.

The activity of wood and other nonliving materials determined by their internal architecture is not an experience of the naked human eye. Rather, it is made visible in the experimental setups of natural scientists. It is first of all the epistemic procedures that enlighten us about the 'properties' of the components of a tree – and via the scientific generalization of these findings, these properties are to a certain extent ontologically established. At the same time, however, something else occurs: the tree (a living organism) becomes in this perspective the carrier of a variety of 'materials,' which it

47 Compare also the structure of the seed capsule of an ice plant, which is discussed in several of the contributions in this volume (Fratzl and Schäffner; Gholami, Guiducci, Jany, and Razghandi; Dunlop and Krauthausen) and earlier biological research [Haupt 1977]. Wolfgang Haupt even terms this movement, induced by the structure alone, as "structural work" ("Strukturarbeit") [ibid., 137].

has produced itself, which it exploits, and whose entirety it consists of. The talk of 'biological materials' or of 'nature's materials' is not self-evident and should be thought in all its consequences: 'nature' is harnessed to a logic of engineering and in this way of technology. It is now indeed the supplier of not only renewable resources but also materials that are complex in themselves. And yet the equation of organic entities with materials understands nature as a 'manufacturer' or 'inventor.' Only from this perspective advocated by (but not only by) the discipline of the materials science can one consider, for example, that the tree is above all a producer of wood and its goal the highest possible efficiency, thus optimal performance.

2.2.3 Dreams of Bioinspiration

For active matter research, living matter is the privileged object to acquire not only new knowledge but also better applications of this knowledge in the invention of new materials and other technical things. For the diverse materials science research on active materials (in all their shades, whether smart, self-propelled, self-assembling, adaptive, or responsive), it is thus biological materials that are investigated.[48] The aim is to develop basic knowledge on structure–property–performance–synthesis relations, to arrive at the construction of artificial 'active' materials that are suited to complex tasks. This orientation of the materials science is described by the protagonists themselves as 'bioinspired' (or 'biomimetic'), and encompasses two different approaches: in the first approach, one has direct recourse to biological materials, but in order to alter them for specific applications, for instance when the researchers "embed functionality into the wood structure and thereby generat[e] wood *materials* with new property profiles and responsiveness" [Eder et al. 2020, p. 10 (emphasis M.F./K.K.)]; the second approach, on the other hand, attempts for example to draw "inspiration from the activity of wood" [ibid., p. 12], that is, not to add properties to natural materials but rather to be inspired by their given structure, whereby 'nature' – in this example the tree – is considered as an object in the untouched realm of reality, which ideally can and should be viewed undisturbed, as if one can indeed only be inspired by it.[49]

The inspiration paradigm is clearly aligned with the "view of nature as an unrivaled engineer, [a view which] underlies the attempts at making artificial materials with characteristics analogous to the variety of properties offered by natural materials [...]" [Bensaude-Vincent and Newman 2007, p. 18]. This view is not new, however. Already in his book *On Growth and Form* published in 1917, Thompson compares the

48 On this see also the interviews with the natural scientists in this volume.

49 See also the statement by Stanislav Gorb, who notes that one of the recent foci of the latest developments in the field of materials sciences is the one given to bioinspired materials, when one "[is] [d]rawing on living Nature as an endless source of inspiration" [Gorb 2020, p. 57].

structure of the bone and its resulting properties to the work of the engineer.[50] Bensaude-Vincent plays down the implicit metaphysical implications and aspirations to be found in the contemporary 'inspiration' discourse: while the "goal is neither to produce a faithful copy, nor to reproduce the appearance of the biological model [...] [c]ontemporary materials scientists are content with picking up local models as solutions to their current technological problems [...]." [Bensaude-Vincent 2007, p. 304]. Nevertheless, there are crucial distinctions to be made between late-nineteenth-century biophysics (to which Thompson relates) and the current bioinspired research. In a recent volume on materials research published in 2020 by the *Deutsche Akademie der Technikwissenschaften*, the focus is explicitly on nature as "a source of and inspiration for inventions and innovations" [Spath and Schraudner 2020, p. 7]. But the difference to Thompson and others lies in the fact that it "is now possible to make use of our increasing understanding of biological structures and growth processes to develop new materials" [Fratzl et al. 2020, p. 14]. Bioinspiration is thus based on an expanded basic knowledge:

> The [c]ore aspects of this application of Nature's principles to modern materials development are the multicomponent structure across hierarchical levels, which [...] makes it possible to encode new and specific functions in the material structure, [and] the selection of 'intelligent' (functional) structures by computer simulation and modelling. [Ibid.]

Here it becomes clear what is meant by the investigation of nature. This is certainly not a romantic return to nature, rather it focuses on very specific (biological) materials[51] with several structural hierarchies with the aim of "encod[ing] specific functions in the material structure," followed by a selection of various structures via "computer simulation and modelling" [Fratzl et al. 2020, p. 14]. If we concentrate again on wood, only by this "encoding"[52] – that is, by the investigation of the different materials by various instruments, modeling, and simulations – can one even say that "wood shows an intrinsic activity" [Eder et al. 2020, p. 3], and only then can one lay claim to

---

50 See: [Thompson 1917, pp. 680–683] concerning the comparison between Hermann Meyer's research of the trabeculae of bone and Karl Culmann's research on engineering of structures and cranes.

51 Besides an overview of the current research directions in bioinspired materials, the domains and examples discussed in this volume are, among others, "bio-inspired elastic cement," "bio-inspired modification of wood," "cellulose-based optical materials," or "organic iontronic devices for neuromorphic computing." See [Bensaude-Vincent 2013, p. 26]: "Despite their admiration for nature's achievement, biomimetic chemists are not inclined to revive natural theology and its celebration of 'the wonders of nature.' Rather, biomimicry proceeds from a technological perspective on nature." While Bensaude-Vincent stresses the practical point of view of the work of biomimetic chemists, the statements cited above do point toward such a celebration, which is accompanied with an unclear distinction between the natural and the artificial.

52 See also [Eder et al. 2020, p. 3]: "[...] in seed awns one can perceive strong active torsion by drying or elongation through humidity as a reversible operation. This plant device acts according to its inner structure as a sort of *intrinsically coded matter*" (emphasis M.F./K.K.).

"harvesting the activity of wood" [ibid., p. 9], that is, to producing, for example, "wood materials with magnetic properties or transparent wood" [ibid., p. 10].

While 'nature' is one umbrella term used occasionally quite vaguely in the discourse of the materials sciences and of bioinspired materials, the term 'inspiration' is another such term. 'Inspiration,' which since antiquity has been used as a religious and artistic term – it is enough to recall the origin of this word in the Latin *inspirare* ('to breathe or blow into'), which has given rise to the word 'spirit' – becomes a blanket term covering a complex range of scientific procedures.[53] In turn, these procedures decontextualize the biological materials – not only by ignoring their genetics, or their evolutionary processes (i.e., how the hierarchical structures found in these biological materials evolved and in which ways), but also by isolating them from their milieu, their surroundings – hence by examining them first as an experimental and 'epistemic object' and then – after the internal structure is understood – turning them into a 'technical object.'[54] Only then is the material scientist able to select and isolate specific structures in order to "convert[] them into inorganic functional and structural materials" [Fratzl et al. 2020, p. 17].[55] No longer conceiving nature as such as a "library of resources" [Bensaude-Vincent and William R. Newman 2007, p. 18], it is the biological materials themselves that are now considered one by one as unique libraries of possible structures and solutions. Hence, the term 'inspiration' refers to an expansion of the universe of libraries, whereby biological materials now 'offer' unique solutions but only after a laborious experimental investigation. The resulting developed materials (i.e., the bioinspired materials) are not natural resources but complex artifacts and belong to the category of 'materials by design' that Bensaude-Vincent declares to be archetypical of materials in the late twentieth and early twenty-first centuries [Bensaude-Vincent 2011b, p. 119]. To repeat what was highlighted at the beginning of this introduction, these materials by design are *not* the materials one starts with (as passive materials waiting to be shaped) but the end product; they are high-performance materials that are often made for very specific purposes, that is, truly singular phenomena to be examined at the end of the research

53 For example, see the following citation from 2018, which presents natural materials as a source of inspiration, that is, as a repository of sources to be inspired from: "Although natural materials have long been a prime source of inspiration for engineered ones, with our modern tools from nanoscience that allow inspection and construction at the scale of molecules, research in this direction is rapidly expanding" [Eder, Amini, and Fratzl 2018, p. 543].

54 Hans-Jörg Rheinberger distinguishes between the 'epistemic things' of the experimental sciences that advocate for an open question and an unknown terrain (what one does not yet know) and 'technical things,' into which the epistemic things frequently transform when they are described and understood. These technical things designate instrumental facilities and applications of the erstwhile obscure, since unknown, research object [Rheinberger 1997].

55 The authors of the introduction to the volume on materials research published in 2020 refer to the process of biotemplating: "Biotemplating is another possible approach, which takes biopolymer structures and converts them into inorganic functional and structural materials" [Fratzl et al. 2020, p. 17].

and production processes. If the bioinspired materials want to achieve the complexity of their models, then this will depend on how the 'encoding' of specific functions and the 'operativity' of these new artificial materials are understood.[56] The question thus arises of what is meant by 'code.' Is it in fact another name for a non-static, dynamic structure that interacts with its environment due to its specific and special geometrical and topological properties? Is not what lurks behind this term and its derivatives a return to a discourse of the readability of materials or their alphabetization? In his recent edited volume *Active Matter*, Skylar Tibbits claims that when we "redefine our relationship with matter," either by embedding "properties of the digital [...] in the physical world like logic, reprogrammability [...]," or by embedding "properties from the natural world [...] in the synthetic world, like growth, repair [...]," then "[t]hese principles are now fundamentally available to read/write within *matter* itself." [Tibbits 2017, p. 16].[57] Here one should ask whether there is in fact another form of reading (and writing) of 'nature.' Following the developments in materials science, it can be suggested that this 'encoding' is no longer an alphabetization of nature or of materials in general, as if one scheme (or one alphabet) would suffice for an understanding of the entire "zoology of materials" [Bensaude-Vincent 2011b, p. 110] or for "read[ing]/writ[ing] within *matter* itself" [Tibbits 2017, p. 16], but rather a much more concentrated effort of making *specific materials*, and hence the activity of certain specific biological materials, 'readable.' This can be seen in several current research projects and directions in materials science, which carefully select "local models as solutions" – although, as we note, not necessarily or not only "to their current technological problems" [Bensaude-Vinvent 2007, p. 304], but rather as a motor to construct new materials. This is the epistemological landscape in which one can situate the research on active materials, in which materials themselves emerge as a unique category, as a boundary concept.

Bibliography

Agricola G. De re metallica libri XII. Basel, Froben, 1556.

Agricola G. De re metallica: Translated from the First Latin Edition of 1556 with Biographical Introduction, Annotations and Appendices. Trans. Hoover HC, Hoover LH. London, Mining Magazine, 1912.

Baeumker C. Das Problem der Materie in der griechischen Philosophie: Eine historisch-kritische Untersuchung. Münster, Aschendorffsche Buchhandlung, 1890.

56 On code and the operativity of materials, cf. the programmatic contribution by Peter Fratzl und Wolfgang Schäffner in this volume.

57 Although what is meant here by "reading" or "writing" is not historically reflected in the above-mentioned volume. Moreover, in the same edited volume, the single contribution offering a philosophical reflection on active matter concentrates mainly on *New Materialism*, Manuel DeLanda and Félix Guattari/Gilles Deleuze [Leach 2017]. Though essential to the philosophical understanding of active matter, we claim in this volume that a broader perspective is needed.

Bensaude-Vincent B. Reconfiguring Nature through Syntheses: From Plastics to Biomimetics. In: Bensaude-Vincent B, Newman WR, eds. The Artificial and the Natural: An Evolving Polarity. Cambridge, London, MIT Press, 2007, 293–312.
Bensaude-Vincent B. Materials as Machines. In: Carrier M, Nordmann A, eds. Science in the Context of Application. Dordrecht, Heidelberg, London, New York, Springer, 2011a, 101–111.
Bensaude-Vincent B. The Concept of Materials in Historical Perspective. NTM Zeitschrift für Geschichte der Wissenschaften, Technik und Medizin 2011b, 19, 107–123, DOI: 10.1007/S00048-010-0043-0.
Bensaude-Vincent B. Plastics, Materials and Dreams of Dematerialization. In: Gabrys J, Hawkins G, Michael M, eds. Accumulation: The Material Politics of Plastic. Routledge, 2013, 17–29.
Bensaude-Vincent B, Newman WR. Introduction: The Artificial and the Natural: State of the Problem. In: Bensaude-Vincent B, Newman WR, eds. The Artificial and the Natural. An Evolving Polarity. Cambridge, London, MIT Press, 2007, 1–18.
Blumenberg H. Die Lesbarkeit der Welt. Frankfurt am Main, Suhrkamp, 1981.
Blumenberg H. 'Imitation of Nature': Toward a Prehistory of the Idea of the Creative Being. Trans. Wertz A. Qui Parle 2000, 12 (1), 17–54.
COSMAT. Materials Science and Engineering for the 1990's. Report by the Committee on Materials Science and Engineering. Washington, D.C, National Research Council, 1989.
Detel W, Schramm M, Breidert W, Borsche T, Piepmeier R, Hucklenbroich P. Materie. In: Ritter J, Gründer K, eds. Historisches Wörterbuch der Philosophie, Vol. 5. Basel, Schwabe Verlag, 1980, 870–924, DOI: 10.24894/HWPh.5260.
Dunlop JWC, Weinkamer R, Fratzl P. Artful Interfaces within Biological Materials. Materials Today 2011, 14 (3), 70–78, DOI: 10.1016/S1369-7021(11)70056-6.
Eder M, Amini S, Fratzl P. Biological Composites – Complex Structures for Functional Diversity. Science 2018, 362 (6414), 543–547, DOI: 10.1126/science.aat8297.
Eder M, Schäffner W, Burgert I, Fratzl P. Wood and the Activity of Dead Tissue. Advanced Materials 2020, 2001412, 1–15, DOI: 10.1002/adma.202001412.
Elliott C. Technics and Architecture. The Development of Materials and Systems for Building. Cambridge, The MIT Press, 1992.
Fratzl P. Biomimetic Materials Research: What Can we Really Learn from Nature's Structural Materials?. Journal of the Royal Society Interface 2007, 4 (15), 637–642, DOI: 10.1098/rsif.2007.0218.
Fratzl P, Barth FG. Biomaterial Systems for Mechanosensing and Actuation. Nature 2009, 462 (7272), 442–448, DOI: 10.1038/nature08603.
Fratzl P, Jacobs K, Möller M, Scheibel T, Sternberg K. Introduction. In: Fratzl P, Jacobs K, Möller M, Scheibel T, Sternberg K, eds. Materials Research: Inspired by Nature Innovation Potential of Biologically Inspired Materials. Munich, acatech, 2020, 14–18.
Fratzl P, Weinkamer R. Nature's Hierarchical Materials. Progress in Materials Science 2007, 52 (8), 1263–1334, DOI: 10.1016/j.pmatsci.2007.06.001.
Fuller RB. Universal Architecture: Essay 3. Shelter 1932, 2 (4).
Fuller RB. Nine Chains to the Moon. Philadelphia, J.B. Lippincott & Co, 1938.
Fuller RB Synergetics. Vol. 2. Explorations in the Geometry of Thinking. New York, London, MacMillon, 1979.
Gaukroger S. Philosophy and the Physical Sciences. In: Garrett A, ed. The Routledge Companion to Eighteenth Century Philosophy. Oxon, New York, Routledge, 2014, 687–705.
Gompper G, Winkler RG. Introduction. In: Gompper G, Winkler RG, Speck T, et al. The 2020 Motile Active Matter Roadmap. Journal of Physics: Condensed Matter 2020, Vol. 32, 193001, DOI: 10.1088/1361-648X/ab6348.
Gompper G, Winkler RG, Speck T, et al. The 2020 Motile Active Matter Roadmap. Journal of Physics: Condensed Matter 2020, 32, 193001, DOI: 10.1088/1361-648X/ab6348.

Gorb S. Hairs with Unlimited Adhesion. In: Fratzl P, Jacobs K, Möller M, Scheibel T, Sternberg K, eds. Materials Research: Inspired by Nature Innovation Potential of Biologically Inspired Materials. Munich, acatech, 2020, 56–57.

Grimm J, Grimm W. Werk. In: Grimm J, Grimm W. Deutsches Wörterbuch, Vol. 14, 1,2. Widrig – Wiking. Leipzig, S. Hirzel, 1960, Digitalized version at https://www.dwds.de/wb/dwb/werk (accessed June 28, 2021).

Grote M. Membranes to Molecular Machines: Active Matter and the Remaking of Life. Chicago, London, The University of Chicago Press, 2019.

Guillerme A. Bâtir la ville. Révolutions industrielle dans les matéraux de construction, France, Grande Bretagne (1760–1840). Paris, Champ-Vallon, 1994.

Haupt W. Bewegungsphysiologie der Pflanzen. Stuttgart, Georg Thieme, 1977.

Hentschel K. Von der Werkstoffforschung zur Materials Science. NTM Zeitschrift für Geschichte der Wissenschaften, Technik und Medizin 2011, 19 (5), 5–40, DOI: 10.1007/s00048-010-0046-x.

Hoddeson L, Braun E, Teichmann J, Weart S, eds. Out of the Crystal Maze: Chapters from the History of Solid State Physics. Oxford, New York, Oxford University Press, 1992.

Ingold T. Making: Anthropology, Archaeology, Art and Architecture. Oxon, Routledge, 2013.

Kay LE. Who Wrote the Book of Life? A History of the Genetic Code. Stanford, Stanford University Press, 2000.

Keller EF. Active Matter, Then and Now. History and Philosophy of the Life Sciences 2016, 38, 11, DOI: 10.1007/s40656-016-0112-3.

Klein U. Experiments, Models, Paper Tools. Stanford, Stanford University Press, 2003.

Klein U, Lefèvre W. Materials in Eighteenth-Century Science: A Historical Ontology. Cambridge MA, The MIT Press, 2007.

Krausse J. Ephemer. In: Barck K, Fontius M, Schlenstedt D, Steinwachs B, Wolfzettel F, eds. Ästhetische Grundbegriffe: Historisches Wörterbuch in Sieben Bänden. Stuttgart, J. B. Metzler, 2001, 240–260.

Krauthausen K. Hütte werden. Zurich, Berlin, Diaphanes, 2022, forthcoming.

Leach N. Matter Matters: A Philosophical Preface. In: Tibbits S, ed. Active Matter. Cambridge MA, MIT Press, 2017, 18–24.

Leroi-Gourhan A L'Homme et la Matière. Évolution et Techniques 1. Paris, Éditions Albin Michel, 1971 [1943].

Löhneysen GE. Bericht vom Bergwerck, wie man dieselben bawen und in guten Wolstande bringen sol, sampt allen dazu gehörigen Arbeiten, Ordnung und rechtlichen Processen [...], [no location], [no publisher], [ca. 1660], DOI: 10.3931/e-rara-50189.

Long PO. Artisan/Practitioners and the Rise of the New Science, 1400–1600. Corvallis, Oregon State University Press, 2011.

Marx K. Das Kapital. Vol. I. In: Marx K, Engels F. Werke. Vol. 23. Berlin, Dietz Verlag, 1972 [1869].

Mazzolai B, Beccai L, Mattoli V. Plants as Model in Biomimetics and Biorobotics: New Perspectives. Frontiers in Bioengineering and Biotechnology 2014, 2, 2, DOI: 10.3389/fbioe.2014.00002.

Menon GI. Active Matter. In: Krishnan JM, Deshpande AP, Kumar PB, eds. Rheology of Complex Fluids. New York, Springer, 2010, 193–218.

Meyer AG. Eisenbauten. Ihre Geschichte und Aesthetik. Ed. and completed Tettau W Freiherr von. Esslingen, Paul Neff, Max Schreiber, 1907.

Meyers MA, Chen P-Y, Lin AY-M, Seki Y. Biological Materials: Structure and Mechanical Properties. Progress in Materials Science 2008, 53 (1), 1–206, DOI: 10.1016/j.pmatsci.2007.05.002.

Morel T. De Re Geometrica: Writing, Drawing, and Preaching Mathematics in Early Modern Mines. Isis 2020, 111 (1), 22–45, DOI: 10.1086/707640.

Noselli G, Arroyo M, DeSimone A. Smart Helical Structures Inspired by the Pellicle of Euglenids. Journal of the Mechanics and Physics of Solids 2019, 123, 234–246, DOI: 10.1016/j.jmps.2018.09.036.

Popkin G. The Physics of Life. Nature 2016, 529, 16–18, DOI: 10.1038/529016a.
Prigogine I, Nicolis G. Self-Organization in Non-Equilibrium Systems: From Dissipative Structures to Order through Fluctuations. New York et al., John Wiley & Sons, 1977.
Prigogine I, Stengers I. Order out of Chaos. Toronto et al., Bantam Books, 1984.
Pyle A. The Theory of Matter. In: Kaufman D, ed. The Routledge Companion to Seventeenth Century Philosophy. Oxfordshire, New York, Routledge, 2017, 410–446.
Ramaswamy S. The Mechanics and Statistics of Active Matter. Annual Review of Condensed Matter Physics 2010, 1, 323–345, DOI: 10.1146/annurev-conmatphys-070909-104101.
Reynolds CW. Flocks, Herds, and Schools: A Distributed Behavioral Model. Computer Graphics 1987, 21 (4), 25–34.
Reyssat E., Mahadevan L. Hygromorphs: From Pine Cones to Biomimetic Bilayers. Journal of the Royal Society Interface 2009, 6 (39), 951–957, DOI: 10.1098/rsif.2009.0184.
Rheinberger H-J. Toward a History of Epistemic Things: Synthesizing Proteins in the Test Tube. Stanford CA, Stanford University Press, 1997.
Schmidgen H. Das Konzept der Maschinen: Simondons politisches Programm. Zeitschrift für Medien- und Kulturforschung 2012, 3 (2), 117–134, DOI: 10.28937/1000106372.
Schon DA. Designing as Reflective Conversation with the Materials of a Design Situation. Research in Engineering Design 1992, 3, 131–147, DOI: 10.1007/BF01580516.
Schon DA. Designing as Reflective Conversation with the Materials of a Design Situation. Research in Engineering Design 1992, 3, 131–147, DOI: 10.1007/BF01580516.
Semper G. Der Wintergarten. Zeitschrift für praktische Baukunst 1849, 9, 516–526.
Simondon G. L'Individu et sa genèse physico-biologique. Grenoble, Jerôme Millon, 1995 [1964].
Smith PH. The Body of the Artisan. Art and Experiment in the Scientific Revolution. Chicago, The University of Chicago Press, 2004.
Smith PH. Epistemology, Artisanal. In: Sgarbi M, ed. Encyclopedia of Renaissance Philosophy. Cham, Springer, 2018, DOI: 10.1007/978-3-319-02848-4_1182-1.
Spath D, Schraudner M. Foreword. In: Fratzl P, Jacobs K, Möller M, Scheibel T, Sternberg K, eds. Materials Research: Inspired by Nature. Innovation Potential of Biologically Inspired Materials. Munich, acatech, 2020.
Star SL. This is Not a Boundary Object: Reflections on the Origin of a Concept. Science, Technology, & Human Values 2010, 35 (5), 601–617, DOI: 10.1177/0162243910377624.
Star SL, Griesemer JR. Institutional Ecology, "Translations" and Boundary Objects: Amateurs and Professionals in Berkeley's Museum of Vertebrate Zoology, 1907–39. Social Studies of Science 1989, 19, 387–420, DOI: 10.1177/030631289019003001.
Thompson DW. On Growth and Form. Cambridge, Cambridge University Press, 1917.
Tibbits S. An Introduction to Active Matter. In: Tibbits S, ed. Active Matter. Cambridge MA, MIT Press, 2017, 11–17.
Toner J, Yuhai. T. Long-Range Order in a Two-Dimensional Dynamical XY Model: How Bird Fly Together. Physical Review Letters 1995, 75 (23), 4326–4329, DOI: 10.1103/PhysRevLett.75.4326.
van de VH. Kunstgewerbliche Laienpredigten. Leipzig, Hermann Seemann Nachfolger, 1902.
Vehlken S. Zootechnologien. Eine Mediengeschichte der Schwarmforschung. Zürich, Diaphanes, 2012.
Vicsek T, Czirók A, Ben-Jacob E, Cohen I, Shochet O. Novel Type of Phase Transition in a System of Self-Driven Particles. Physical Review Letters 1995, 75 (6), 1226–1229, DOI: 10.1103/PhysRevLett.75.1226.
Wagner M. "Materialgerechtigkeit": Debatten um Werkstoffe in der Architektur des 19. und frühen 20. Jahrhunderts. In: Petzet M, Pursche J, eds. ICOMOS-Tagung: Historische Architekturoberflächen. München, 2004, 135–139.
Wagner M. Das Material als Akteur – oder: "Eine Schleimmasse, die einen Willen Hat". In: Feist U, Rath M, eds. Et in Imagine Ego: Facetten von Bildakt und Verkörperung. Berlin, Akademie Verlag, 2015, 481–491, DOI: 10.1524/9783050094175.481.

Wallin TJ, Pikul JH, Bodkhe S, et al. Click Chemistry Stereolithography for Soft Robots that Self-Heal. Journal of Materials Chemistry B 2017, 5 (31), 6249–6255.
Walther A. Viewpoint: From Responsive to Adaptive and Interactive Materials and Materials Systems: A Roadmap. Advanced Materials 2019, 32 (20), 1905111, DOI: 10.1002/adma.201905111.
Weinkamer R, Fratzl P. Mechanical Adaptation of Biological Materials – The Examples of Bone and Wood. Materials Science and Engineering: C 2011, 31 (6), 1164–1173, DOI: 10.1016/j.msec.2010.12.002.
Wolfe C. Materialism. In: Garrett A, ed. The Routledge Companion to Eighteenth Century Philosophy. Oxon, New York, Routledge, 2014, 91–118.

Michael Friedman is a historian of mathematics and a senior lecturer at the Cohn Institute at the Tel Aviv University and a research associate at the Cluster of Excellence 'Matters of Activity. Image Space Material' at the Humboldt-Universität zu Berlin. The focus of his research is on how material and visual practices prompt and influence symbolical-mathematical knowledge, with a particular interest in the early modern period, as well as in the 20th and the 21st century. He is the author of: A History of Folding in Mathematics: Mathematizing the Margins (Birkhäuser, 2018), and the coeditor of volumes on Martin Heidegger, psychoanalysis, and mathematics, including (with Wolfgang Schäffner) On Folding (transcript, 2016). Papers available at: https://hu-berlin.academia.edu/MichaelFriedman.

Karin Krauthausen is a literary scholar and a historian of culture at the Cluster of Excellence 'Matters of Activity. Image Space Material' at the Humboldt-Universität zu Berlin, where she works on 3D-Writing. The focus of her research is on the production of reality in literature, the unsettled constellation of art and sciences, and the media and materialities of artistic design ('Entwurf'). She is the coeditor of volumes on structural realism (together with Stephan Kammer: Make it real. Für einen strukturalen Realismus. Diaphanes, 2021), the media of the German author Hubert Fichte (together with Stephan Kammer: Hubert Fichtes Medien. Diaphanes, 2014), and the techniques of taking notes in arts and sciences (together with Omar W. Nasim: Notieren, Skizzieren. Schreiben und Zeichnen als Verfahren des Entwurfs. Diaphanes, 2010).

Peter Fratzl, Wolfgang Schäffner

On the Activity of Materials

The idea of physical reality usually refers to the rigidity and resistance of materials that are regarded as the fundamental stuff out of which our world is built or the ground on which it rests. Above all, solid materials relate to a more general notion of matter. This classical idea is based on the fundamental dichotomies of matter and form, and matter and activity, whereby form appears to be imposed from the outside on amorphic matter, transforming it into objects, and matter, as a fundamentally passive substance, seems to be externally activated, and in this way to undergo a transformation from rest to process.[1] Following these principles – above all, since industrialization – our technology, instruments, objects, and devices have been designed based on materials with controllable behaviors that can be standardized as properties. Hence, solid and rigid materials such as iron, steel, or concrete were considered suitable as stable and reliable components for machinery, bridges, or buildings, whereas elastic materials such as rubber, fiber tissues, or plastics permitted the design of flexible structures. In all these cases, form, activity, or special functions are externally implemented and define materials as their passive carrier.

But materials are also omnipresent in natural systems, from rocks and minerals to wood and wool or silk. Even we ourselves are built from a variety of materials such as bones, muscles (flesh), or hair. Seen in this way, it is obvious that all these natural objects have a function or exhibit some activity. The processes and activities related to these objects are quite complex occurrences. But are the materials these objects are formed of in the same way passive carriers of activities, or do they show a different mode of activity and form-building? Do biological materials perform by themselves what human activity and design implement from outside in the case of artifacts?

These questions challenge the classical paradigm of separating passive matter – which seems to be the essence of a material – from the form, activity, or function the object might have. This separation also echoes the basic dichotomy of matter and mind,

1 The dichotomy of matter and form in the sense of hylomorphism focuses on matter as a passive carrier of form. This is evident in the usage of materials. In the nineteenth century, iron is seen as the most formless material that can be brought into any form [Meyer 1907]. Clearly, this tradition did not begin in the nineteenth century (see the introduction to this volume). In his *Kritik der Urteilskraft* (Critique of Judgment), the philosopher Immanuel Kant, for example, noted in 1790: "But the possibility of a living matter is quite inconceivable. The very conception of it involves self contradiction, since lifelessness, inertia, constitutes the essential characteristic of matter" [Kant 2007, p. 222]. On the difference between passive and active matter, cf. [Keller 2016].

The authors acknowledge the support of the Cluster of Excellence 'Matters of Activity. Image Space Material' funded by the Deutsche Forschungsgemeinschaft (DFG, German Research Foundation) under Germany's Excellence Strategy – EXC 2025 – 390648296.

Open Access. © 2021 Peter Fratzl, Wolfgang Schäffner, published by De Gruyter. BY-NC-ND This work is licensed under the Creative Commons Attribution-NonCommercial-NoDerivatives 4.0 International License.
https://doi.org/10.1515/9783110562064-002

which can be regarded as one of the founding principles of occidental culture.[2] Reframing the relationship between matter and activity in terms of active materials forces us to look closely at basic assumptions connected with the idea of materials and their properties, with the dynamics of energy, information, and activity in a broader sense.

1 Action, Reaction, and Interaction

A very basic physical principle is that there is no action without reaction. The activity of an object (or a material or body) must necessarily be defined with respect to some reference. The existence of an outside reference implies the existence of a boundary of the object and the object's activity corresponds to an interaction with its outside world.

The word 'interaction,' commonly referring to all the forces that govern matter in the most general physical theory, already encapsulates the fact that action is mutual, from the object to its environment and vice versa. Considering gravitation, for example, a small mass falling toward the Earth experiences the same gravitational force from the Earth as the Earth experiences from the small mass (Fig. 1, left). We are just deceived by the fact that the Earth moves much less under the influence of these forces than the so much lighter mass m_2. When this mass touches the Earth's surface, the impenetrability of the Earth creates a solid reaction force that exactly compensates the gravitational force. Hence, the equilibrium of a seemingly inactive

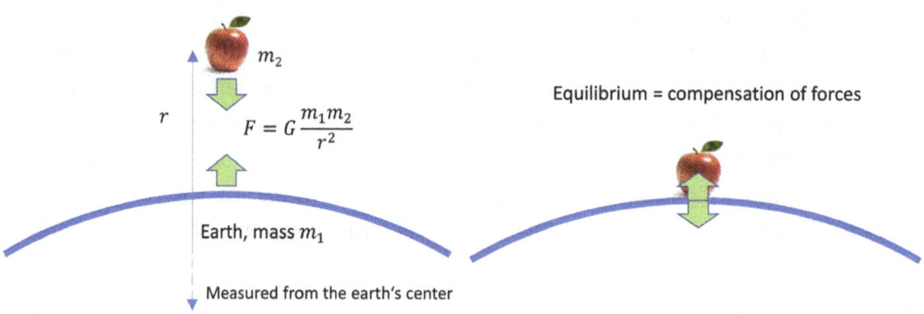

Fig. 1: Gravitational forces between the Earth and a small mass m_2. (Left) The Earth and the small mass exert exactly the same force on each other, just in the opposite direction (with G being the gravitational constant). (Right) When the small mass touches the Earth's surface, the gravitational force is compensated by the solid reaction of the surface, which has exactly the same magnitude. © Graphic by the authors (P.F./W.S.).

2 The most famous version of this dichotomy is Cartesian dualism; see: [Descartes 1641]. One of the most evident modern versions of this dichotomy is classical robotics with its separation of CPU-based information and externally controlled material periphery. Soft robotics challenges this separation by emphasizing the intrinsically coded material, referred to as physical intelligence [Sitti 2021].

object lying on the Earth's surface is not the absence of activity but the exact compensation of interactive forces between the object and its environment (i.e., the Earth's surface in this case). Therefore, the perceived inactivity of any object is just an equilibrium of forces between the object and its environment.

2 Disequilibrium

On Earth, any kind of object is exposed to gravity, and is therefore part of an activity field that affects everything, everywhere, and at any time: a falling body, a suspended body, an erected building, or even the solid ground. There is no difference between their status in motion or at rest insofar as both states result from activities depending on their mass and their environment. When the environment exerts an equivalent compensation force, the body is at rest; in all other cases, a downward-oriented force produces motion, since the Earth itself is the bigger mass compared to all the bodies on the Earth. A building collapses when the static forces lose their equilibrium. If a body is sufficiently small, other forces such as wind can easily lift it in the opposite direction.

The interaction between two bodies or between a body and its environment is due to a disequilibrium; it acts only as a relationship between them. The lever as a classical tool of mechanics puts this disequilibrium into action (see Fig. 2).

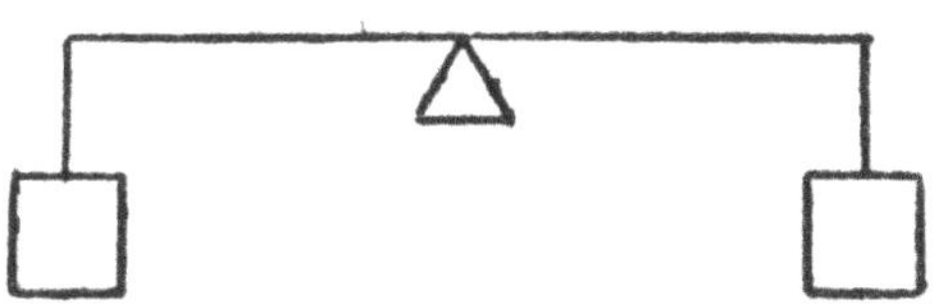

Fig. 2: The lever as a classical tool of mechanics from: [Mach 1897, p. 12, Fig. 2].

The lever beam represents the relational character of activity; it is the connection and the boundary between two objects. It can be considered as a tool where the two bodies are literal 'objects' – directed against each other – so that one is always the environment of the other and vice versa. The downward gravitational force of each body is inverted by the lever into a lifting force upon the other body. When they have the same mass, they are in equilibrium and their mutual activity appears as rest; their different mass, on the other hand, will result in motion. In the latter case, by varying the geometric proportion of the lever beam, equilibrium can be reestablished. The static equilibrium is the result of opposing activities, not the absence of activities.

Architecture also exemplifies this activity of a seemingly static state. Composition and building (Lat.: *struere*) using stones, logs, or steel beams make this permanent activity obvious. To achieve stability, the gravitational forces of the building materials have to be calculated. The classical example is the keystone of an arch that brings all the stones of the arch into a relationship and thus generates their static equilibrium as rest. Architectural statics does not mean the absence of activity but the equilibrium of all activity involved. The building structure describes the diagram of forces in a state of equilibrium. Any fundamental change transforms the inherent activity into an instability that makes the building collapse, thus generating a new equilibrium of the involved elements.

What is important here is that the activity of two objects, or – what is the same – a body and its environment can only be defined in a relational manner. This kind of original difference makes evident that any object and its activity have to be regarded together, as an elementary pair.

3 Material Property: Predictability

Very often, when the environment of an object (a material) is not considered explicitly but described only with respect to the challenge to the material, the symmetry of physical interactions is broken. Taking the example of Fig. 1, it is easy to combine all the effects of the environment into a single challenge to the object that we call 'input' (in this example, the gravitational field strength g), which will result in the gravitational force (see Fig. 3), which we might call 'output' in this case.

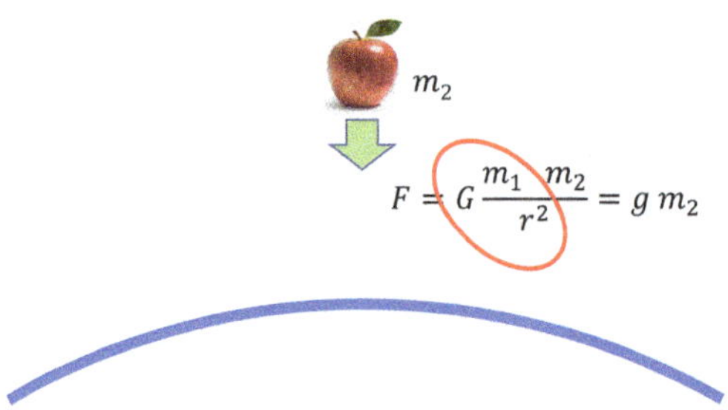

Fig. 3: The influence of the Earth on the small mass (see Fig. 1) is grouped into a single 'input' to the little mass m_2 and the gravitational field strength g. In this formulation, the symmetry of the interaction between the object (the small mass) and its environment (the Earth) is no longer obvious, although it of course still exists. This 'input' multiplied by the mass of the small object yields the 'output,' that is, the force moving this small object. © Graphic of the authors (P.F./W.S.).

In this way, the input from the environment is converted through a property of the object alone (its mass) into an output, the gravitational force. Usually, the intrinsic properties of a material would be expressed per unit of volume so that the relevant material parameter for this interaction is the mass density. For any given input (here, the gravitational field strength at a given position on Earth) would lead to different outputs in terms of gravitational force depending on the mass density of the material (e.g., in grams per centimeter cubed, mass density would be 0.95 for an apple, 1.3 for typical plastic films, 2.8 for granite rock, 8 for steel, and 11.3 for lead).

Material properties are therefore operational quantities that define how an input from the environment is transformed into an output that acts back on the environment. Of course, materials may possess a large number of material properties that transform all kinds of input. To give a few examples: resistivity is a property that transforms electric voltage into current, the elastic Young's modulus transforms uniaxial pressure into deformation, thermal conductivity transforms a temperature difference into a heat flux, and color is the result of the way by which a white light spectrum is reflected. Some material properties also define the limits within which materials operate in the way described. Material strength, for example, describes the limiting force that can be applied to an object before it fractures.

Materials should therefore be considered as operators transforming various inputs from the environment into well-defined outputs. Material property charts have been developed to help designers choose the right material for each type of application [Ashby 1999]. Figure 4, for example, shows areas of the diagram where Young's modulus and density are defined for whole material classes. It is relatively easy to find stiff materials (that resist deformation) with high density (metal alloys and ceramics), but there are very few stiff materials with low density, a well-known challenge for lightweight construction.

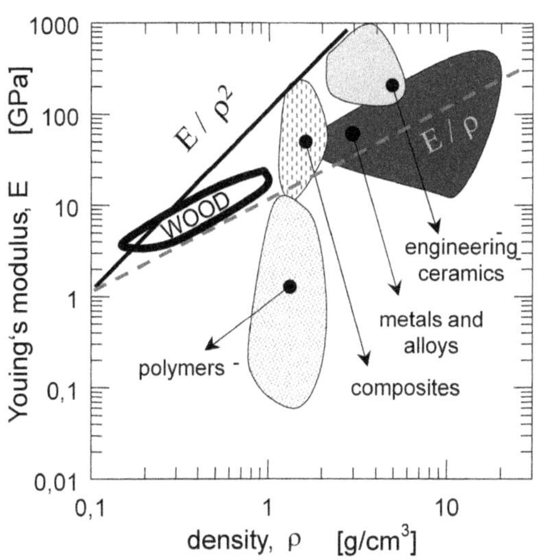

Fig. 4: Ashby map for whole material classes showing Young's modulus (i.e., resistance against deformation) against the mass density. Such charts imply that material properties (and thus the operability of the material) are fully predictable. The two lines in the graph show positions with equal values of either E/ρ or E/ρ^2, parameters essential for mechanical design purposes [Ashby 1999]. Reproduced from: [Fratzl and Weinkamer 2007, p. 1282, Fig. 17]. License: CC BY-NC 4.0 (https://creativecommons.org/licenses/by-nc/4.0/).

In principle, if one knows all the relevant material parameters, one can predict how a material will behave based on various inputs from the environment. The more precisely the material parameters are known, the better the predictability and the better the materials will fulfill the technical requirements.

4 Material, Environment, Machine

The perspective changes fundamentally if one considers materials not as passive elements but as operators. In general, a property is a stable characteristic of a material, a state of being that can be standardized and used as a reliable building block. If one considers a material as a device that acts with respect to its environment, all its properties are transformed into the operational elements of a processing unit. The material processes according to its inherent structure, thus transforming an input into a different output.

This involves a decisive shift in which materials are no longer conceived as bodies exposed to forces but are understood as modes of processing, transmitting, and storing activities. It is a shift from physics to machinery. The classical model of machinery, as it was developed by Jean-Victor Poncelet in the nineteenth century, is composed of three parts: (1) a motor or receiver (*receveur*), (2) a communicator or modificator, and (3) a tool (*outil*), where the special output of work is done [Poncelet 1845, p. 15]. Between receiver and working tool, the machine processes the energy of the motor into a special type of movement (rotation, step movements, etc.). Thus, the special types of material property, such as resistivity, elasticity, stiffness, thermal conductivity, or hydro-reactivity, describe inherent 'gear' mechanisms that transform temperature or humidity into color change or mechanical work in a predictable way.

In this context, the fundamental relationship between the material and its environment can be regarded as an input into a chain of internal transmission and processing, and as its corresponding output. Therefore, both elements – material and environment – are extrinsically as well as intrinsically related to each other as a pair of elements. Black-boxing all these activities dates back to the times of thinking matter as solid and taking materials as passive substances defined by their specific properties. Looking more closely into these black boxes allows us to focus on the material's inner structure and its operational character.

5 Predictability and Entropy

Let us now consider all the possible states in which a material can exist for a given set of conditions imposed by the environment (such as temperature, pressure, electric field, and mechanical load). These states can be shape, color, smell, texture, and so on, including all material properties. If the material is perfectly predictable, then there will be only one state for a given set of conditions. However, if several states are possible, then we define these as degrees of freedom. Traditionally, the number W of degrees of freedom is measured through the Boltzmann entropy S that is directly related to the logarithm of W, through a famous formula engraved on Boltzmann's gravestone in the Vienna Central Cemetery (see Fig. 5). A situation where there is only one state ($W = 1$), therefore, has an entropy of zero.

A robotic arm, or an even more complex device, has many possible states based on the number of articulations. In this case, external control is needed to reduce the number of degrees of freedom (and thus reduce the entropy) in order to generate a predictable behavior of the robotic arm, for example. Many degrees of freedom reduce the predictability or require extreme outside control. This is why stiff elements connected by few articulations are easier to control than a soft body that can deform in many ways. This need for control has led very naturally to the selection of materials that are as predictable as possible, stiff, and without any degrees of freedom. This separation between, on the one hand, passive and fully predictable materials and, on the other, active (usually digital) control systems is a hallmark of our digital age. Mechanical systems that were the state of the art in the nineteenth-century technology use mechanical information transmission, and so do many natural systems (see Fig. 6).

Fig. 5: Gravestone of Ludwig Boltzmann with his formula for entropy. Wiener Zentralfriedhof, Austria. From: https://de.wikipedia.org/wiki/Datei:Zentralfriedhof_Vienna_-_Boltzmann.JPG. License: CC BY-SA 3.0 (https://creativecommons.org/licenses/by-sa/3.0/).

What is striking about the movement of the awn of a plant seed (Fig. 6B) is the fact that the material is by no means rigid. Nevertheless, its internal structure is such that the geometric shape is predictable based on the outside air humidity. In this sense, the internal structure of the cellulose fibers confers to this material a property that relates air humidity (input) to shape (output) in a way that is totally analogous to mass in the case of gravitational forces (Figs. 1 and 3). The gear in Fig. 6A would normally be considered a system and not a material, while the awn is composed of woody cells and is thus very similar to wood, which would normally be considered a material. The complexity of the awn's movement is, however, more reminiscent of a system. What if we coated the gear in Fig. 6A with a skin, not revealing the inside? Could it then be a material in the same right as the awn in Fig. 6B? Does the distinction between material and system even make sense? In fact, both are only defined by their interactions with the environment, and setting the boundary is what defines them. Hence, each material is a system, and each system can be a material for the construction of larger entities. This hierarchical principle is a hallmark of natural materials [Fratzl and Weinkamer 2007] and can be directly generalized to all materials and systems.

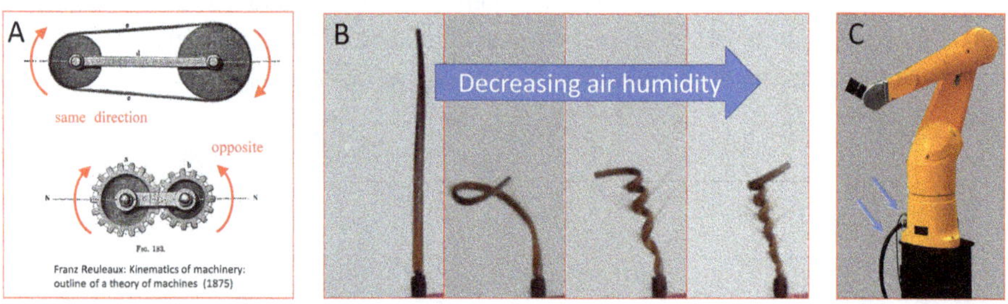

Fig. 6: Degrees of freedom and energy supply in mechanical systems. (A) The right wheel in the system governed by a belt or a gear has no degree of freedom. It has to turn either in the same direction or in the opposite direction to the left wheel. The energy for the movement is transferred directly from the movement of the left wheel. Adapted from: [Reuleaux 1876, p. 262, figs. 182, 183]. (B) The various states of the *Erodium* awn depend on its water content, which in turn depends on air humidity. © Rivka Elbaum (Hebrew University of Jerusalem) and Peter Fratzl (MPI-CI). As with any material property, the geometric appearance (output) is a direct function of air humidity (input). Hence, the geometric state is completely predictable from the input, in the same way as the rotation direction of the right wheel in (A) is completely predicted by the movement of the left wheel. The corresponding material property has been programmed by the plant during the growth of the awn by a complex arrangement of cellulose fibrils [Abraham et al. 2020]. The energy for the shape change is directly taken from the environment (via water absorption or evaporation). (C) In contrast to (A) and (B), the robotic arm in (C) has several degrees of freedom materialized by its articulations (the photo was edited by Peter Fratzl). The control is delegated to an outside processor and the information needs to be imported through a cable (blue arrow), as does the energy for the movement (second blue arrow).

6 Symbolic Dimension: Constraints and Information

The operational character transforms the material from a static solid body into a series of possible states. Thus, the material as a realm of possibilities also acquires a symbolic character. The degrees of freedom define a space of possibilities that can in principle be realized.[3] Under certain conditions, therefore, every state has a certain degree of possibility. If the material has a high degree of freedom (due to elasticity, for example), it is less predictable, since every circumstance can result in a large number of possible states. The higher the degree of freedom, the lower the inherent information. And vice versa: material constraints embody information.[4] Only by reducing the degree of freedom does a predictable action of the material become possible. This is – according to Franz Reuleaux – the fundamental principle of mechanical machines as chains of constrained elements [Reuleaux 1876, p. 46]. The gears made of stiff metal transform the energy input at the receiver into a specific work as the output. The mechanical constraints select predictable operations from the possible states. Thus, the constraints can be considered the programming and implemented information within the internal structure of the machine. The transfer or the input and its transformation into a constrained process reduce the possible states to the very precise mode of action prescribed by the machine.

In the case of a steam engine, this chain of constraints transforms steam energy into the intermittent movement of the piston, which is transformed in turn into the rotation of a wheel, which – in the case of a railroad – is constrained by the tracks to a linear movement, whereas, in the case of a car, an external coding by the steering wheel is required. Here, two modes of information processing become evident. First, the material 'programs' its activity through its intrinsic structure, comparable to mechanical gears in machines; second, the material is defined by a lot of possible states as less constrained and has to be controlled by means of outside information in order to generate predictable actions. Whereas the first mode integrates physical action and information processing in one and the same structure, the second mode separates the control operations from the mechanical device. This last version is the cybernetic and finally digital mode of separating information and physical work by isolating the control unit and its information processing from the material. Information then has to be supplied to the material from outside in order to be executed in the form of predictable work.

In the integrated version of an intrinsically programmed material, where mechanical work is simultaneously information processing, Boltzmann entropy and Shannon entropy can coincide, in the sense that the structural constraints decrease entropy

3 For Terrence Deacon, absence is a fundamental element of the informational dimension of matter [Deacon 2012].

4 See Chapter 12 "Information" and Chapter 13 "Significance" in: [Deacon 2012, pp. 371–420].

and increase information.[5] In this case, the small internal structures of the active material permit the fusion of information and work processes. Shannon's information theory, however, is modeled according to the second mode of separating information and working machine, since it is all about the transmission of information between sender and receiver. In the integrated mode, the receiver (in terms of Shannon's information theory and of Poncelet's machine theory) is not an empty black box – with a high degree of freedom – that has to be fed with information from outside, but a material that contains its intrinsic coded structure.

In nature (compared to artificial mechanical gears), much softer and more elastic materials are used for this integrative mode of information processing. Natural materials are based on ever-repeating constituents, such as proteins, polysaccharides, and minerals, with a hierarchical structure [Fratzl and Weinkamer 2007]. Due to this hierarchy of structures, materials can be adapted to a variety of sometimes conflicting functions [Weinkamer and Fratzl 2016], such as, simultaneously, optical, mechanical, and thermal functions, leading to multifunctional materials [Eder, Amini, and Fratzl 2018].

7 Information and Energy

The concept of a passive material is a convenient engineering concept. In such a picture, materials would react according to their set of material properties. Hence, the activity (of a robotic arm, for example) depends on an external input in the form of information (e.g., from digital processors) and in the form of energy (generated remotely) (see Fig. 7). Any intrinsic activity of the material is then considered a defect, an error, or a failure. Examples are beams bending under too much load, disrupting electrical cables, or plastic embrittlement under UV light – irrespective of the fact that in some systems these properties can become functional (in fuses, safety valves, etc.). The important conceptual difference is that an active material gains system properties, with information being processed directly in the material. An example of this is the complex curling movement that the *Erodium* awn performs when the air humidity changes (Fig. 6B). Active materials may also extract energy directly from the environment or even convert energy directly (e.g., the internal stresses that are released when certain seed pods explode).

5 See: [Deacon 2012, pp. 378–379]: "According to Shannon's analysis, the quantity of information conveyed at any point is the improbability of receiving a given transmitted signal, determined with respect to the probabilities of all possible signals that could have been sent. Because this measure of signal options is mathematically analogous to the measure of physical options in thermodynamic entropy, Shannon also called this measure the 'entropy' of the signal source. I will refer to this as Shannon entropy to distinguish it from thermodynamic entropy."

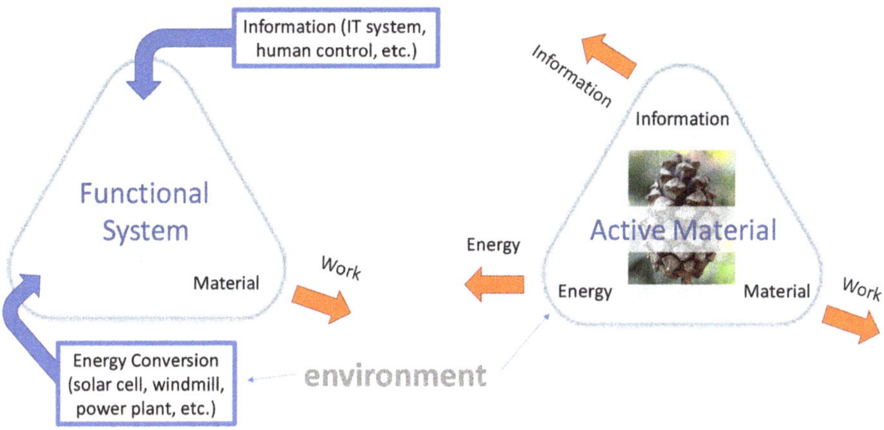

Fig. 7: Removing the dichotomy between information/energy and material in an active material. A functional system (left), such as a robotic arm, is based on a passive material (e.g., steel or aluminum) that is activated through the import of information from an IT system (e.g., a digital processor) and through the import of energy (previously converted from environmental or fossil sources). An active material (right) is a functional system in its own right. It is not just a passive material but carries information for the function (such as the cellulose fiber arrangement that encodes the opening movement of the pine cone) and takes the required energy directly from the environment (e.g., humidity changes to actuate the pine cone scales). Orange arrows symbolize potential outputs (e.g., work). Active materials may also convert environmental gradients into information (and thus work as sensors) or even into (electrical) energy, for example, by coupling to a piezo element. Reproduced from: [Eder et al. 2020, fig. 1]. © 2020 The Authors of [Eder et al. 2020]. Published by Wiley-VCH GmbH.

This difference between technical devices and biomaterials understood as machines is remarkable and corresponds to the two modes of operation mentioned above: of separating work and information processing and of integrating both. Furthermore, the way in which energy is supplied is different; whereas the technical arrangement of machines depends on an artificial environment that has to be established by larger amounts of external energy and information, biological material, in contrast, only uses the naturally existing environment, where sunlight, humidity, and gravitational forces are the basic requirements for its activity. The inner structure of a material – containing special functions – is the operator structure that becomes active through the extrinsic activity of, for example, water and temperature. In biological materials, this interaction with the environment combines information and energy for the mechanical work in one and the same process and structure. In this case, the material as an operator integrates several informational and mechanical activities, namely, acting as a sensor for the external conditions that trigger the coded action of the material, the processing and execution of information within the material structure, and finally the transformation of the activity into mechanical work as its output. For example, the internal structure of wood as an active material [Eder et al. 2020] contains an assemblage of different hydrophilic elements, whose geometry determines a certain material code and transforms the humidity gradient into mechanical movements. This integrated operator of sensor, structure, and mechanical execution can be seen as an analog code that, in contrast to digital coding, not only symbolically represents but also physically performs the action. In this sense, the geometric elements take on the role of an analog code that is the operative basis for the programmed material thus executing its intrinsic information.

The operator material has limited degrees of freedom and thus contains the necessary information that can be processed within the geometric structure, whereas the energy supply depends on the external environmental conditions. Nevertheless, storage and amplification of energy is also made possible by special structures such as spring mechanisms that can release stored energy in an explosive manner.

In terms of machinery, one can therefore distinguish three different types: (1) a cybernetic or digital machine, where the mechanical operation is fed by external artificial sources of energy and information; (2) an analog – or gear-controlled – machine, where the information is integrated or programmed as mechanical constraints within the gearbox, whereas the energy has to be added by an external artificial source; and (3) material as an operator that contains its information in its intrinsic geometric structures that are activated by the energy gradients of its natural environment.

8 Programmable and Self-Learning Materials

Recent developments in materials science have started to shatter the concept of immutable material properties as a paradigm for technical design. The concept of programmable materials seeks to modify the material property (that is, its operationality transforming an input into an output) depending on needs.[6] Programming steps can be of rather diverse types: the growth of the seed awn (Fig. 6B), for example, would involve a step in which microtubules in the living plant cell would control the orientation of the cellulose fibrils [Cosgrove 2016] that then provides the desired functionality after the death of this cell (so that only the woody cell wall remains). The internal structure would then be the code that defines the relationship between air humidity and the shape of the awn (Fig. 6B). Another possible type of programming is 3D fabrication used to generate internal structures that confer mechanical or optical properties to a material that it would not otherwise have. Properties could be programmed into a textile by different knitting or weaving procedures. Even thermomechanical processing is often used to adjust (program) the material properties of engineering alloys.

Perhaps, the simplest example of a programmable material is an electrical resistor that transforms voltage into current, the relevant material property P being its

6 See https://cpm.fraunhofer.de/en.html (accessed June 20, 2021); https://selfassemblylab.mit.edu/programmable-materials (accessed June 20, 2021).

resistivity. Since resistivities depend on temperature, a control of the environmental temperature would control the resistivity and thus allow the programming of the voltage–current relation. A more compelling example would be a material that changes its shape depending on a particular input, such as humidity or temperature. Going back to the awn of a plant seed (Fig. 6B), this material changes from a straight needle-like shape to a helix as a function of the air humidity in its environment. With the diagram in Fig. 8, the input is humidity and the output is shape. The corresponding material property *P* is complex and cannot be found in typical material property charts, but it fulfils the definition of relating the input (humidity) to the output (shape). *P* is complex because it depends on the internal fiber structure of this cellulosic material. With another fiber structure, the shape change would be different. By laying down a specific arrangement of cellulose fibrils, the plant programs the material for a specific behavior, relating humidity to desired shapes. Shape-changing objects have also been a strong focus in current research on programmable materials (see footnote 6). While it is easy to reprogram resistivity by changing the temperature, it is more difficult to reprogram the cellulosic seed awn, which was generated by the plant in a complex synthesis process. The cellulose fibril arrangement in the seed awn is therefore an analog code inscribed in the cell wall by a programming step, while the cell was still living and growing its cell wall. It then allows predictable shape changes even after the plant tissue is dead (in the sense that plant cells have lost their metabolic activity, leaving only the cellulosic cell walls behind).

Self-learning or adaptive materials possess even more exciting behaviors. As sketched in Fig. 8, the signal for the modification of the property *P* is derived directly from the output signal. In this way, the output signal has an influence on the material property *P*, which in turn influences how the input is converted into the output. Such

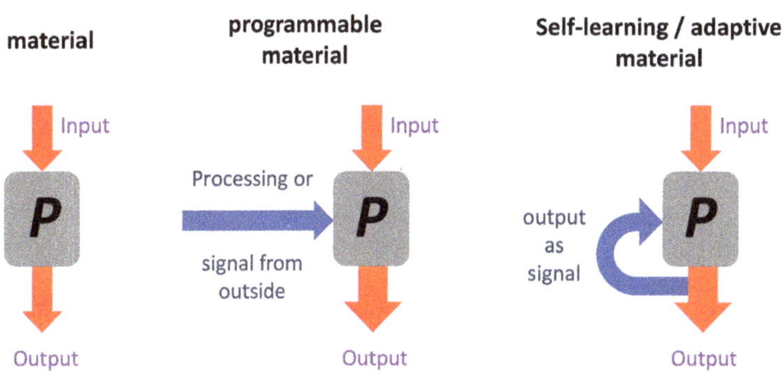

Fig. 8: Any material with property *P* is an operator that transforms the input from the environment into an output. Creating or modifying this property *P* by processing the material or through some outside signal changes the relation between input and output, allowing a programming of the material. Such materials are often also referred to as responsive. If the modification of the property *P* occurs in response to the level of output, a feedback loop is generated. Such a material would be adaptive or self-learning. © Graphic by the authors (P.F./W.S.).

feedback loops are a hallmark of living systems, and the combination of many such loops is known to lead to emergent lifelike behavior and is studied in the discipline of systems biology [Klipp, Liebermeister, and Wierling 2016; Alon 2019]. The muscles and even the bones in our bodies can be trained to become thicker and thus to respond better to challenges. If we train a muscle, the input is a mechanical load (generated by the weight to be lifted), and the output is the lift height. The mechanical contraction force of the muscle (relating input to output) depends on the muscle cross-section, which increases with training. Similarly, sustained loads on our bones increase the latter's thickness. Such materials are adaptive or self-learning. For an adaptive material, as sketched on the right in Fig. 8, the output might modify the material property in two principal ways. If an increase in the output leads to a modification of P so that the output for a given input is reduced, the system is stable and often called homeostatic. Many processes in living bodies are homeostatic, preserving the amount of material despite varying challenges from outside. If, on the contrary, an increase in the output leads to such a modification of P that the output for a given input increases further, then the situation is unstable. Instability may be problematic and lead to failure, but it may also be productive and lead to a new homeostatic equilibrium or induce growth, for example. The transfer of the concept of adaptive materials from biology to engineering is still in its infancy. It is not yet clear how adaptive materials can be fabricated based on nonliving components, and their behavior is complex and difficult to predict. The properties of adaptive (self-learning) materials depend on their history, and their combination may lead to emergent behaviors that cannot be inferred from the material properties of the individual components. In many ways, the challenges resemble those of artificial intelligence, except that the learning does not occur in formal networks but in analog and tangible materials.

9 Conclusion and Outlook: Logic, Code, and Material

Active materials change our classical understanding of material. The active structures of the material act as an operator that consists of an integrative system of material, energy, and code, which can be understood as a new type of hardware. This material hardware will no longer combine code and material by externally implementing symbolic algorithms in a passive material carrier but will embody their radical fusion. The understanding of materials as operators that are simultaneously their own material code is not limited to the study of biological materials; it provides a conceptual framework that fundamentally changes the way we conceive the relationship between material, information, and activity. This reconceptualization brings to light a downright revolutionary mode of material activity, one that incorporates an integrated version of code, working process, and building structure within its inner structure.

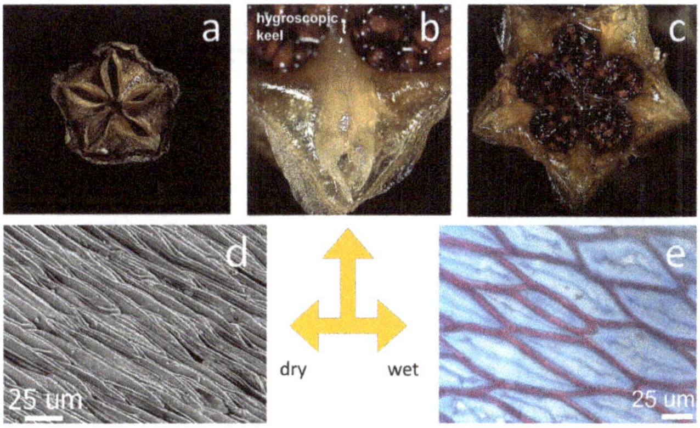

Fig. 9: The mechanical movement of the seed capsule of an ice plant (shown closed in (a) and open in (c)) is based on the intrinsic honey comb structure of the actuating keel shown in (b). The change from dry (d) to wet (e) makes the honeycomb structure swell and thus opens the closed capsule Adapted by permission from Springer Nature: [Harrington et al. 2011, figs. 1, 3].

The active material of the ice plant (see Fig. 9) is a switching structure that transforms the input into an output depending on a coded procedure. But it is simultaneously the unit that transforms the energy of the signal into the mechanical operation coded within the switching structure. The switching energy, however, is not electricity but water.

If we compare this water-fueled active structure with electronic switching elements, it is interesting to look back to the invention of the integrated circuit, which combined electronic elements such as transistors in a single wafer.

Digital hardware consists of an external activation of materials that obey the specific functions they represent within the preconceived operation activated by the externally added electricity. Figure 10A shows the elementary hands-on way the digital revolution started as integrated hardware 60 years ago. Instead of trying to minimize the single elements of switching circuits, integration here means incorporating all the elements into one and the same component.

Compared to this crude activation of material as a digital switching device, the active structures of the biomaterials we can identify today in nature exhibit a quite different level of sophistication. Therefore, it is clear that in the development of analog coding, biomaterials can be seen as highly promising active materials.

Our analysis of materials as operators and of their intrinsic material code shows that we have to reestablish the relationship between logic, code, and material. This includes – as its most important feature – the inversion of the classical idea of logic and code understood as an artificial intelligence implemented in our physical world, which is still essential for the digital world.

The emergence of the computer as a logical machine was due to the coupling of electrical engineering and logical operations. In analog electric circuits, the electrical

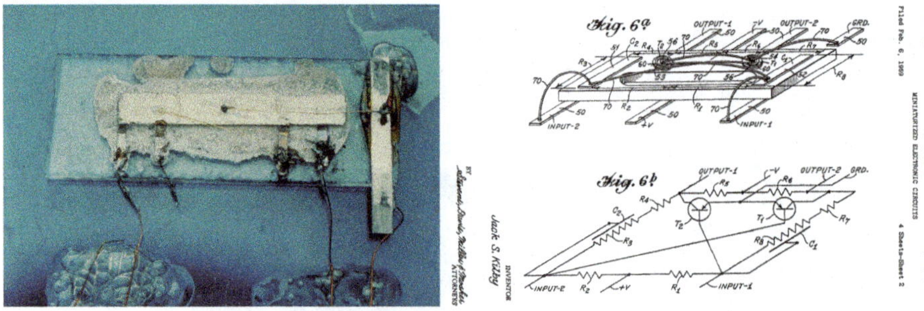

(A) (B)

Fig. 10: The first integrated circuit, developed by Jack Kilby at Texas Instruments in 1958 as a crude assemblage of wired materials. (A) A complete circuit of a phase-shift oscillator on a single crystal bar of germanium. (B) Kilby's patent drawing from 1964 showing the single body of the semiconductor material, which integrated all electronic components of the switching circuit [both from Jack Kilby, US Patent 3,138,743 Miniaturized Electronic Circuits. Patented June 23, 1964]. © Courtesy Texas Instruments.

flow is controlled by switches that open or close the circuit and thus make a machine run or a light burn. More complicated switching circuits – above all, in telephone networks or control units – require a large number of switching elements to produce the desired performance. This was Claude Shannon's starting point for optimizing switching circuits through a symbolic analysis [Shannon 1937]. Based on a binary logical calculus, Shannon could describe a mode in which logical operations could be 'interpreted' as switching circuits and thus transform a complex switching circuit into an algebraic expression (see Fig. 11).

This means inversely that logical operations such as addition and multiplication could also be performed by parallel and serial connections of switches. The electrical flow and its discrete sequential switching implement information transmission and processing as a sequential flow of signals that goes through matter but does not take into account matter's inner structure. Matter as digital hardware is some sort of material flowchart where the bistable switching elements perform logical operations. The very property of the material is reduced to its reliable and immediate reaction to the input commands. Thus, logical operations could be implemented in switching circuits that are fundamentally separated from the mechanical periphery they control.

This relation of logic, code, and material, however, has to be rethought within the context of adaptive materials. The coupling of logical operations and matter, which includes the fundamental separation between symbolic operations and mechanical work, has to be overcome. Instead of implementing logical operations within matter, it is necessary to conceive the 'logic' of the material structure itself. Instead of taking binary logical operations and looking for ways to find materialized modes of performance, one has to invert the procedure: the analysis of biological materials has to reveal their inner operational logic in terms of a basic code of the analogue.

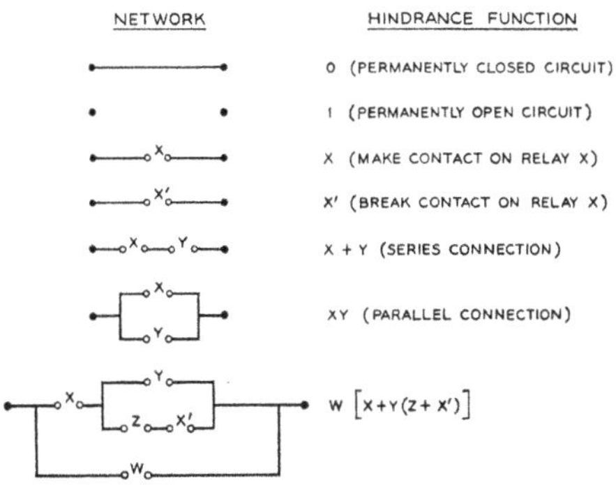

TABLE I

Analogue Between the Calculus of Propositions
and the Symbolic Relay Analysis

Symbol	Interpretation in relay circuits	Interpretation in the Calculus of Propositions
X	The circuit X.	The proposition X.
0	The circuit is closed.	The proposition is false.
1	The circuit is open.	The proposition is true.
X + Y	The series connection of circuits X and Y	The proposition which is true if either X or Y is true.
XY	The parallel connection of circuits X and Y	The proposition which is true if both X and Y are true.
X'	The circuit which is open when X is closed, and closed when X is open.	The contradictory of proposition X.
=	The circuits open and close simultaneously.	Each proposition implies the other.

Fig. 11: A symbolic analysis of relay and switching circuits. Left: [Shannon 1949, p. 60, Fig. 1]. Reused with permission of Nokia Corporation and AT&T Archives. Right: [Shannon 1937, p. 11, tab. 1]. © Massachusetts Institute of Technology.

A symbolic system that is not based on alphanumerically discrete elements such as letters or numbers, but on discrete and continous, symbolic, extended – and thus analogue – operations is the very classical realm of geometry. Conceiving in this sense geometric objects as operators, therefore, appears to be a basis for analyzing the operational character of materials. The materials science makes evident that a material's intrinsic operations are based on geometry, on interconnected hierarchical dynamic structures that perform mechanically. The operational character of this analog code is different from the sequential alphanumeric digital code, since it is a material code that performs the coded physical process at the same time. Thus, the approach of taking materials as operators raises the question of the intrinsic logic of the materials' geometry, which can be analyzed as a symbolic and material operation.

Bibliography

Abraham Y, Tamburu C, Klein E, et al. Tilted Cellulose Arrangement as a Novel Mechanism for Hygroscopic Coiling in the Stork's Bill Awn. Journal of the Royal Society Interface 2012, 9 (69), 640–647, DOI: 10.1098/rsif.2011.0395.

Alon U. An Introduction to Systems Biology: Design Principles of Biological Circuits. Boca Raton, CRC Press, 2019.

Ashby MF. Materials Selection in Mechanical Design. 2nd edn. Oxford et al., Butterworth-Heinemann, 1999.

Cosgrove DJ. Plant Cell Wall Extensibility: Connecting Plant Cell Growth with Cell Wall Structure, Mechanics, and the Action of Wall-Modifying Enzymes. Journal of Experimental Botany 2016, 67 (2), 463–476, DOI: 10.1093/jxb/erv511.

Deacon T. Incomplete Nature: How Mind Emerged from Matter. New York, London, W.W. Norton & Co., 2012.

Descartes R. Meditationes de prima philosophia. Paris, 1641 French, 1647.

Eder M, Amini S, Fratzl P. Biological Composites – Complex Structures for Functional Diversity. Science 2018, 362 (6414), 543–547, DOI: 10.1126/science.aat8297.

Eder M, Schäffner W, Burgert I, Fratzl P. Wood and the Activity of Dead Tissue. Advanced Materials 2020, 2001412, DOI: 10.1002/adma.202001412.

Fratzl P, Weinkamer W. Nature's Hierarchical Materials. Progress in Materials Science 2007, 52 (8), 1263–1334, DOI: 10.1016/j.pmatsci.2007.06.001.

Kant I. Critique of Judgement. Meredith JC, Ed. Trans. Walker N. Oxford, New York, Oxford University Press, 2007.

Keller EF. Active Matter, Then and Now. History and Philosophy of the Life Sciences 2016, 38, 11, DOI: 10.1007/s40656-016-0112-3.

Klipp E, Liebermeister W, Wierling C. Systems Biology: A Textbook. 2nd edn. Weinheim, Wiley-VHC, 2016.

Mach E. Internationale Wissenschaftliche Bibliothek. Vol. 59. Die Mechanik in ihrer Entwicklung. Historisch-kritisch dargestellt. 3rd edn. Leipzig, Brockhaus, 1897.

Meyer A. Eisenbauten: Ihre Geschichte und Ästhetik. Esslingen, Paul Neff Verlag (Max Schreiber), 1907.

Poncelet JV. Traité de mécanique industrielle: exposant les différentes méthodes pour déterminer et mesurer les forces motrices, ainsi que le travail mécanique des forces, Vol. 2. Liége, A. Leroux Fils Éditeur, 1845.

Reuleaux R. Kinematics of Machinery. Outlines of a Theory of Machines. Trans. and ed. Kennedy ABW. London, MacMillan & Co, 1876.

Shannon CE. A Symbolic Analysis of Relay and Switching Circuits. Master's thesis, Massachusetts Institute of Technology, 1937.

Shannon CE. The Synthesis of Two-Terminal Switching Circuits. Bell System Technical Journal 1949, 28 (1), 59–98.

Sitti M. Physical Intelligence as a New Paradigm. Extreme Mechanics Letters 2021, 46, 101340, DOI: 10.1016/j.eml.2021.101340.

Weinkamer R, Fratzl P. Solving Conflicting Functional Requirements by Hierarchical Structuring – Examples from Biological Materials. MRS Bulletin 2016, 41 (9), 667–671, DOI: 10.1557/mrs.2016.168.

Peter Fratzl is director at the Max Planck Institute of Colloids and Interfaces in Potsdam, Germany, heading the Department of Biomaterials. He is trained as an Engineer at the Ecole Polytechnique in Paris, France, and holds a doctorate in Physics from the University of Vienna, Austria. He has been pioneering the field of biological materials science and is one of the co-directors of the Cluster of Excellence 'Matters of Activity. Image Space Material' in Berlin.

Wolfgang Schäffner, a historian of science and media technologies, has been professor of the Cultural History of Knowledge at the Department of History and Theory of Culture at the Humboldt-Universität zu Berlin since 2009. He has been Director of the Hermann von Helmholtz Center for Cultural Techniques since 2003 and Director of the Cluster of Excellence 'Image Knowledge Gestaltung' at Humboldt-Universität zu Berin from 2012–2018. He has also been permanent guest full professor and director of the Walter Gropius Program at the Faculty of Architecture, Design and Urbanism at the Universidad de Buenos Aires since 2005 and head of the German-Argentinian Master-Program 'Open Design' of the Humboldt-Universität zu Berlin and Universidad de Buenos Aires. He is director of the Cluster of Excellence 'Matters of Activity. Image Space Material' which he heads together with Horst Bredekamp, Peter Fratzl and Claudia Mareis.

Interviews: Scientists on Active Matter

Michael Friedman, Karin Krauthausen, Thomas Speck

Interview with Thomas Speck: "You Don't Want to Build an Oak Tree – You Want to Invent It." Plants as Active Matter

KK: The research on active matter includes many different approaches, many different techniques and objects, and also defines and redefines many concepts. Before entering into the broader discussion on these changes, we would like to know how you would describe your particular path into this field called active matter. How would you consider this field? What kind of disciplines were and are important?

TS: I originally started in biology. I was trained as a biophysicist and my first work was way off materials research. I started with neurobiology [Speck et al. 1984], then switched to a totally unrelated subject: to fossil plants, doing my master's – at that time a diploma – and my PhD mainly on simulations of the mechanical properties of these plants [Speck and Vogellehner 1988; Rowe, Speck, and Galtier 1993; Speck, Spatz, and Vogellehner 1990]. To put it briefly, during the transfer from water to land the mechanical properties of the environment of these plants drastically change, and hence one cannot test fossil plants, since they are petrified. So one has to research fossil plants via recalculating mechanical properties from the thin ground sections one can make of them. Because of that my approach to mechanics was always very much triggered by functional morphology. In the process I realized that very little is known about growth habits in extant plants. We therefore started a 10-year project on lianas, climbers, and self-supporters in a tropical rainforest – you can call it basic research (see Fig. 1) [Rowe and Speck 1996; Speck and Rowe 1999]. The main reason for this research was to have better data for the recalculation of the properties of fossil plants. We realized then that, when considering living plants, very little is known about what makes a plant a successful climber. About the same time – 20 years ago more or less – I got an invitation to a biomimetics conference from Werner Nachtigall in Saarbrücken, one of the German 'godfathers' of biomimetics. When I told him that I did not work in biomimetics but rather on fossil plants and growth habits, he answered that what I was doing might be of interest for applications. This was more or less my first step into biomimetics, and from that point on I was really hooked on this subject [Spatz and Speck 1995; Speck, Rowe, and Spatz 1996]. I like doing basic research, analyzing how nature functions, abstracting, and then translating the results into technical materials and structures. Now it has

Acknowledgments: Michael Friedman and Karin Krauthausen acknowledge the support of the Cluster of Excellence 'Matters of Activity. Image Space Material' funded by the Deutsche Forschungsgemeinschaft (DFG, German Research Foundation) under Germany's Excellence Strategy – EXC 2025 – 390648296.

Open Access. © 2021 Michael Friedman et al., published by De Gruyter. BY-NC-ND This work is licensed under the Creative Commons Attribution-NonCommercial-NoDerivatives 4.0 International License.
https://doi.org/10.1515/9783110562064-003

(a) (b) (c) (d)

Fig. 1: (a) Lianas in the tropical lowland rainforest of French Guyana. (b) Slipped stem of the twining liana *Condylocarpon guianense*. (c) Cross section of an old stem of the liana *Condylocarpon guianense*. (d) Stem of the 'monkey ladder' liana *Bauhinia guianensis*. © Plant Biomechanics Group Freiburg.

been nearly 20 years since I have mainly been doing biomimetics. At the moment in my institute I would say 90% of the projects have links to biomimetics, especially to adaptive surfaces, adaptive materials, and so on.

KK: Before entering into a more detailed discussion on your research, it is worth emphasizing the role of technology. What kind of new technologies and devices have been used in the last 20 years in biomimetics and especially in the research on plants? Were these also adopted from other disciplines? Since the research on biomimetics and adaptive materials is highly interdisciplinary.

TS: I have to say that the last 20 years have been perfect for doing biomimetics, since many of the testing devices were developed and made accessible at a reasonable price to university institutes. At the moment we use various instruments from electron microscopy such as SEM, TEM, FIB TEM, and FIB SEMs,[1] but we also use microcomputed tomography (μCT) and magnetic resonance imaging (MRI) [Hesse et al. 2019a, 2019b]. The latter becomes important for us because we can test living plants: one can load and unload them, take them outside and let them grow under a specific load regime, and see how they react structurally to this regime. We therefore use a lot of imaging. We also have a very good high-speed camera lab for botanical research. This is very important for recording fast movements in plants. It is often said that plants are immobile, which is just wrong. We cannot see how they move. There are some which move very fast, like the traps of *Utricularia* and *Aldrovanda*,[2] or the Venus flytrap (see Fig. 2) [Poppinga et al.

1 SEM, scanning electron microscope; TEM, transmission electron microscope; FIB, focused ion beam.

2 *Utricularia* species are known as bladderworts, and *Aldrovanda vesiculosa* is known as the waterwheel plant.

2013, 2016, 2017; Westermeier et al. 2018; Sachse et al. 2020]. Or there are plants that move very slowly during growth processes. These kinds of movement, one could say, are outside our visible movement spectrum. Therefore, one may have the feeling, if there is no wind coming, that the plant is immobile, which is not true. We use time-lapse and high-speed cameras. Thirty years ago, it was extraordinarily complicated to do digital imaging with a high-speed camera. These technical improvements and developments have been beneficial for what I am interested in. At the moment, we can analyze everything from the molecule up to the organism on different hierarchical levels.

On the other hand, I do not do all of it personally but we collaborate a lot with modelers and colleagues from applied mathematics. Modeling is very important for us [Wolff-Vorbeck et al. 2019; Sachse et al. 2020]. And at the moment, we also collaborate with theoretical physicists to get a better understanding of self-repair mechanisms [Konrad et al. 2013; Klein et al. 2018]. Because of my training as a biophysicist I am not afraid to approach new methods. While a pure biologist might be hesitant due to the mathematics involved, I think that the phenomenon investigated might be complicated but that I can at least understand it at a level at which I can collaborate with somebody who is a real expert. We use a lot of different methods and a lot of theoretical approaches and simulations, mostly in collaboration. And this is perhaps the most important thing I have learned: this research cannot be done without an interdisciplinary group of people. Normally, in a collaborative project running in my institute, I am a real expert in 20–30%. The rest comes from colleagues. You need to understand each other, but you also have to have your own expertise. When everything comes together this might allow us to solve problems which nobody could have solved only with his or her own personal knowledge.

MF: You have touched on a lot of subjects which we are going to talk about, but I want to start with a topic which you have already mentioned implicitly: the relationship between biology and architecture [Gruber 2011]. You implied it when discussing mechanisms of plants but also when saying there is a need for interdisciplinary work, and it is obvious that this can only work when biologists understand this need. Looking at the recent work on biomimetics [Knippers, Schmid, and Speck 2019; Nachtigall and Pohl 2013], one can have the impression that there has been a radical change in how shapes and architecture in biology are considered. In the nineteenth century and at the beginning of the twentieth century one considered in architecture mainly inflexible, immobile structures, whereas in biology structures are seen as multifunctional and flexible. However, within the research on active matter, there is a change in the relations between biology and architecture. Can one say that a novel conception of structure for architecture emerges here?

TS: I think one could put it like that, but one has to see that there are totally different types of architects around. There are 'civil engineer architects,' who are basically engineers. They want to make functional things, which can be used for building. And there are 'artist architects,' who imagine what a building should look like. I have had the

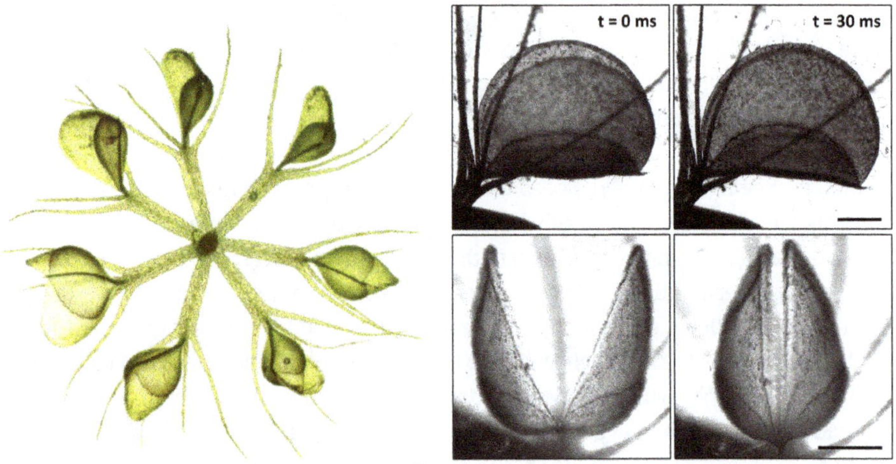

(a) (b)

Fig. 2: (a) Whirl with seven snap traps of the waterwheel plant *Aldrovanda vesiculosa*, a submerged carnivorous plant. (b) Open and closed snap traps of the waterwheel plant, which served as a model for the bioinspired façade shading system Flectofold. © Plant Biomechanics Group Freiburg.

advantage of collaborating with both in the last 10 years in a project in Vienna which was mainly driven by the design arts. It was called 'Growing as Building,'[3] and the approach was, one could say, a soft biomimetic one [Speck 2015]. Taking biomimetics into account, one had the idea of a building which was living, growing, and adapting. In the field of architecture closer to civil engineering, we collaborated with scientists from Stuttgart, from the Institute of Building Structures and Structural Design (ITKE)[4] and the Institute for Computational Design and Construction (ICD), who are also architects but mainly civil engineers or construction engineers. They come from, I would say, a hard biomimetic approach, wanting to improve the functions of a given thing, which we did successfully in a Collaborative Research Center SFB-TRR 141 "Biological Design and Integrative Structures – Analysis, Simulation and Implementation in Architecture,"[5]

3 The project ran between June 2013 and November 2015 in Vienna at the University of Applied Arts. The aim of the project was to "take[] growth patterns and dynamics from nature and appl[y] them to architecture with the goal of creating a new living architecture [...] [hence, to] develop architectural concepts for growing structures. Three main directions will be investigated: transfer of abstracted growth principles from nature to architecture, integration of biology into material systems and intervention of biological organisms and concepts with existing architecture. Key issues of investigation will be mechanisms of genetically-controlled and environmentally-informed, self-organised growth in organisms and the differentiation of tissues and materials." See: https://www.growingasbuilding.org/ (accessed May 23, 2020). See also the project publication [Imhof and Gruber 2016].

4 https://www.itke.uni-stuttgart.de/ (accessed June 26, 2021).

5 Collaborative Research Center SFB-TRR 141 "Biological Design and Integrative Structures – Analysis, Simulation and Implementation in Architecture" was funded by the Deutsche Forschungsgemeinschaft

and still do in other collaborative projects. These are totally different approaches, but I think both are capable of interacting really very well, because I think nature does both. Nature is stable when it has to be stable, and it is flexible and adaptive when it has to be flexible and adaptive. There is a vision of a living house which interacts with the environment like a tree or some other plant. It is really fascinating, but you also have to consider the needs of the people who live in this building. In the end, if you want to make a biomimetic approach successful, it has to be accepted by society, by the public. I think this is very important. Just as it is important to bring different cultures of architecture together. In the last 10 years, there are many examples of an increase in the communication between these different groups, but there is still a kind of gap, because you often hear, if you see one of these artist architects having an idea for a building, that the construction engineer says: 'It looks beautiful, but there's no way we can construct it.' On the other hand, when the construction engineer has a solution for something, then the artist architect says: 'But it looks totally different from my vision!' Bringing this together is extremely interesting. We have to tackle it from totally different angles. And in the end I think the most important thing is that we speak to each other and we understand each other's language. In my opinion one of the big problems we have in interdisciplinary work is that sometimes we just do not understand each other; we do not understand what one side wants, what the other side can deliver, and how to bring all this into a functional entity.

MF: Let me continue with a follow-up question which is somewhat general: how would you characterize the relation between biology and engineering? This relation is obviously not that engineering copies from biology ...

TS: Never never. In the end, the reason why biomimetics works at all is that we share the same physical environment. If the physical laws were different for living beings and technology, you could not do biomimetics. If you start from the biological side – what we call the bottom-up approach of biomimetics, or biology push – then, assume you have a new finding in biology, you see that this might be of interest; it shows an interesting function. You speak with your colleagues from the engineering department, the materials sciences department, and the architecture department, asking them whether this finding, this function which you observed in this specific plant, might be of interest for an application. If yes, then you start to analyze, to quantitatively understand the form-structure-function relationship of your biological role model. If you have done that, you start the abstraction, because you do not want to build an oak or a pine tree, you want to borrow some of the functions you have seen in the oak or in the pine tree. These abstractions are important. Once one has finished the abstraction, then

from 2014 to 2020 and included researchers from the Universities of Stuttgart, Freiburg, Tübingen, from the DITF Denkendorf and the Staatliche Museum für Naturkunde Stuttgart (SMNS) [Knippers, Nickel, and Speck 2016; Knippers, Schmid, and Speck 2019].

the engineers or materials scientists can start to produce their materials and structures, inspired from this. This is a (re)invention inspired by a natural role model. Very important during this process is that you often do simulations, mathematical modeling in different ways. You have to scale up or down for the application and techniques. You use different materials, and very often you learn a lot more about the structure-function relationship than we have seen in biology. Equipped with the knowledge of the transfer process, you can go back to your role model and better understand it. Then you can again feed new knowledge into an improved biomimetic development. In the end, if you are really lucky, then it is kind of a heuristic spiral. For all sides you have increase of knowledge. It does not always work but, when it does, this is an optimal and satisfying way to do science [Speck and Speck 2008; Masselter et al. 2012; Speck et al. 2012].

MF: Can you give an example of how a process like this works?

TS: We worked quite a lot on self-repairing materials; my wife Olga Speck, who is scientific coordinator at the Freiburg Center for Interactive Materials and Bioinspired Technologies (FIT) and a group leader in the Plant Biomechanics Group, works mostly on this subject.[6] We looked at a succulent plant – which does in fact have succulent leaves – from south Africa, the *Delosperma cooperi*, called pink carpet (see Fig. 3) [Konrad et al. 2013; Klein et al. 2018; Hesse et al. 2020]. If you take a leaf, you can cut it transversely, and the leaf makes a curvilinear motion; the cut is then closed. The same process takes place if you cut it longitudinally: also in this case the cut leads to a

Fig. 3: (a) Flowering plants of the pink carpet plant *Delosperma cooperi* in Freiburg Botanical Garden. (b) Fourteen days after the injury (arrow) the leaf has healed and shows a permanent though changed curvature at the site of the injury. Both images reprinted from [Konrad et al. 2013, p.48], with permission from Elsevier.

6 See the recent overview in: [Speck O and Speck T 2019]. See also: [Anandan et al. 2018].

curvilinear motion. This proves that there exist forces that close the cut. If you make a cut right around the leaf, then the leaf contracts.

This gives a lot of different ideas for applications. We did many measurements and analyzed them. We started working with theoretical physicists to do an analytical model, since we did not want to do a numerical model. We first tried to understand and predict what is possible, because a numerical model is primarily descriptive. With an analytical model you can make predictions and test them. Through this analytical model, we found out that the plant consists of five different tissue layers. Three of these layers are so dominant that we thought we could do the model with only three layers, which would make the research a lot easier. But we found out that this does not work. During the modeling we found out that we needed all five layers. This was the first feedback. The analytical model predicts that a lot of the movement is driven by elastic instability. In an intact plant, there are tissues under compression and others under tension. If you make a cut you disturb the equilibrium. The plant then has to come to a new equilibrium. We then started a numerical modeling [Klein et al. 2018]. We found out that in addition to the elastic instability there is also a hydraulic movement involved. This idea came into our mind because the behavior was different depending on the humidity of the air. By combining this numerical model with the analytical model we were able for the first time to ask the right questions and to do the right experiments to prove our hypothesis. We are not yet in the transfer process, but we have ideas for a bioinspired transfer with several layers of foils that are prestrained and prestressed. And even in the model-making process, which was mainly done for the abstraction, we had so much feedback that finally we had a much better understanding of the biological role model. Without trying to apply it, we would probably have stopped after the first experiment, because at that point we could already explain most of the movement process. That shows the power of transfer also for a better understanding of the biological role model in the process of 'reverse biomimetics' (see Fig. 4) [Speck and Speck 2008; Masselter et al. 2012]. It drives your questions further. If you want to transfer your finding to applications, you really have to understand what is going on in biology. Only studying the biological role model, you sometimes find an explanation that is good enough but does not tell the entire story, because the systems are so complex. That is the big advantage of a combination of basic and applied research. Because if you want to make a transfer to an applied material you really have to get your numbers right.

KK: What you were just discussing concerns the relation between geometry, form, and forces – a relation that was important in the biology of the nineteenth century, but its importance decreased in the twentieth century due to the rise of molecular biology and genetic explanations [Keller 2016]. How would you assess the situation today? What is the relationship between the different fields and orientations of biology? Is the role of molecular biology changing, maybe especially in the field of active matter research?

TS: The molecular biologists are so overwhelming in number, so we have stopped 'fighting them.' Right now we are at a level where we are starting to collaborate more again. A colleague of mine once said: 'The molecular biologists have the methods; the structural biologists have the questions.' Molecular biology should be a tool to understand organisms. And today they are very close to structural biologists even if the methods are very different. In the end we are all aiming to understand how plants, animals, and fungi interact, react, and live.

We have benefitted from what molecular biologists have found, and the aim is to bridge the gap between molecules and structures. In the end, if you look at all the biological entities, they are hierarchically structured. The question is where the (main) function is really embedded. Often it is on different hierarchical levels, that is, scale-overarching: sometimes on the more microscopic levels and sometimes on the macroscopic levels [Speck T and Speck O 2019; Fratzl, Speck, and Gorb 2016]. We have now reached a level of understanding where both aspects, structural and molecular, can interact very well. They can also interact to produce new biomimetic material structures and forms. In 10 or 15 years, we may see that the interaction is so intertwined that the question "are you doing molecular biology or are you doing organismic biology?" would be the wrong one. The question should be: what do you want to understand? And then you choose. I hope that there are still enough organismic biologists around at this time, because the number is decreasing.

It is interesting that structural organismic biologists collaborate with engineers and with architects more often than molecular biologists do. Probably because there is a strong link to functional and structural aspects in their scientific work. It was problematic that organismic biology had the reputation of being, let us say, a 'grandpa biology.' When I changed from neurobiology to paleobotany, some colleagues and friends asked if I was entirely mad. You go from an emerging field to a field where the scientists are as old as the fossils they are working with. But for me it was the best choice to make, because I started from the structure, and I am still taking advantage of this start. To summarize, both molecular and organismic are important, and they will interact more and more in the future.

KK: The genetic explanation describes a program from which everything develops. Whereas when you take the interaction of the living system with its environment into consideration – hence its embeddedness and dependence – then the story becomes more individual and complex. In the twentieth century, both perspectives have, unfortunately, often been seen as antagonistic, and the genetic explanation seemed to prevail [Keller 2000].

TS: That is definitely true. I think that the structures we are looking at are often seen as the 'frozen' results of a molecularly driven development – though describing it as frozen would be wrong; maybe one can say 'materialized results.' On the other hand, everything is material. If you start from the DNA, you go several levels up, and by self-organization you end up with proteins, or something that is the building block

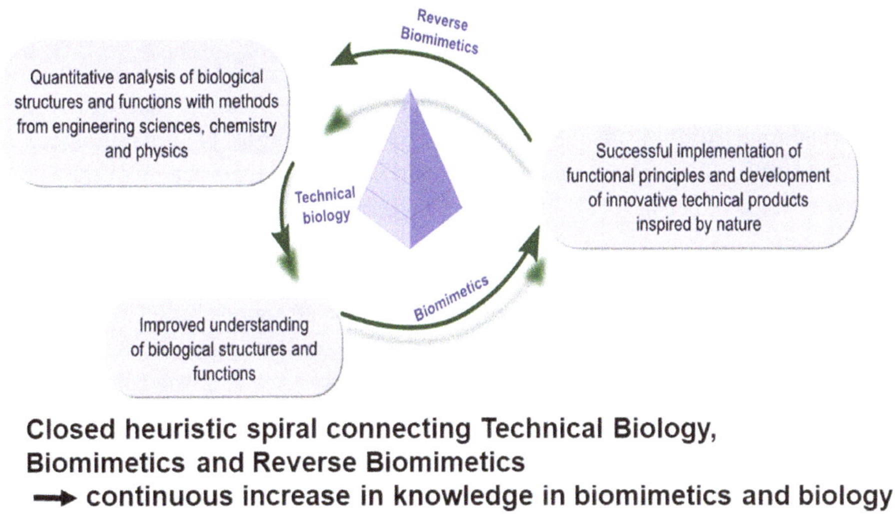

Fig. 4: Relationship between technical biology, biomimetics, and reverse biomimetics causing a continuous increase in knowledge in biology and biomimetics. Adapted from: [Masselter et al. 2012, p. 380]. © Taylor and Francis.

for the next hierarchical level. On the molecular side, it is like a building plan, as you said, of structures.

I think it is still a misunderstanding to look at 'frozen' states. We measure something in a given situation. But we have to keep in mind that the given situation in which the structure or organ was formed was dynamically influenced by the environment. If I measure, for example, the twig of a plant, it is important whether it is grown under a high mechanical load produced by the wind or whether it is very sheltered, whether it is in the shade or in the sun. This is what we have to keep in mind. We have to consider the reactiveness of the organisms to different environmental influences. This may be two sides of one coin. We see the result of a given situation a plant is growing in. That is why I always state whether our plants come from natural habitats or greenhouses. Because a greenhouse plant is something totally different. You can compare them, but you just have to keep in mind that there is a huge difference. Living beings are able to adapt their properties to their environment. For adaptive material systems, it is interesting to learn from this adaptability and to see how fast and how fine-tuned the reactiveness of plants is. We need molecular biology to properly understand this. Without it we can only describe such phenomena. When we have a molecular perspective, we can also describe the genesis, the development of the results we see and measure.

MF: The term 'frozen' certainly echoes older traditions in biology concerning the genetic code as a prewritten or predetermined 'program' or 'code' [Kay 2000]. In light of the new understanding that you just mentioned, should the concept of 'code' itself be redefined?

TS: It is hard to say. The flexibility of the genetic code is much higher than we thought for a long time. And it is also more resilient than we thought. We are still opening new doors. We still have just a glimpse. If you look at codes, there is a lot of activity in coding materials, such as algorithms that can predict material properties under changing environmental conditions. This has already been done, but there is also a lot of potential in it, that is, to structure materials through the coding of their properties to receive the wanted properties. This in my opinion is an approach that will become more and more important. At the moment it is still more descriptive. But in the future it might be like in pharmaceutics, where you can play a lot with the computer and find new pharmaceutical substances. This will also happen in the materials science in the near future, as several initiatives in which my institute is also involved show.

MF: So in a sense, the 'code' itself becomes more dynamic, more flexible, and more adaptive. You mentioned before the adaptiveness of plants, and I want to point out that in active matter research, and also in research on plants, it seems – also with this adaptiveness – that there is a crossing of the dead and living distinction. For example, in cellular and acellular self-healing [Harrington et al. 2015; Speck O and Speck T 2015, 2019; Hager et al. 2016], the latter obviously concerns organisms that are not alive. Or, to take another example, the mechanical behavior of climbing plants that still functions when the plant has died. Or the opening and closing of the pine cone: a mechanism that functions without the cells of the pine cone being alive.

TS: Well, if we touched before on one of the essential concepts for biology during the second half of the twentieth century, that is, 'code,' you now point to another essential concept, and the question is how one defines life. If you say life is biological life, then I would say it is easy to discern what structures and functions are active, even if the biological material is no longer alive in a biological sense. If you look at a 'living material,' then it is different. A living material can per definition be something that makes some adaptions that look life-like, that is, comprise some life-like features. But you do not need living processes in a biological sense to observe processes of adaptation. One basic question is where the energy comes from. If you take a pine cone, it is a hygroscopically driven movement of a structure that is dead in a biological sense but still working in fossils that are over 15 million years old, and can be nicely transferred into bioinspired structures applicable for building hulls, for example (see Fig. 5) [Correa et al. 2020; Poppinga et al. 2017, 2018]. Then the energy comes from the humidity and dryness. In the end, it is driven by the sun. Nearly all things we transfer to technical applications have to be done using materials that are dead in a biological sense. The pine cone is dead; the wood used for building purposes is dead. On the other hand, we want to see functions that we link to life: life-like functions. But for the bioinspired developments we do, for example, in our Cluster of Excellence *liv*MatS in Freiburg, there are never living cells in a biological sense involved. We want to 'smart up' dead materials systems so that they behave in a life-like way. This is a different approach to the one in which hybrid materials are made, where you can use living cells which can be held alive for a long time in a specific environment combined with artificial materials. With these hybrid structures, it is hard to say whether they are alive in a biological sense or not. I would say that everything we currently

Fig. 5: (a) Pine cones in closed state at high humidity. (b) Pine cones in open state at low humidity. (c) Longitudinal section through a pine cone showing the complex bilayer structure of the cone scales that is responsible for the hygroscopically driven opening and closing motion. (d) 4D printed biomimetic scales showing hygroscopically driven motion. All images republished with permission of The Royal Society (U.K.), from: [Correa et al. 2020, p. 3, fig. 1a–c, p. 10, fig. 5c]; permission conveyed through Copyright Clearance Center, Inc.

develop is at best life-like, or has at best life-like functions, but I would never say that it is living in a biological sense [Speck et al. 2021; Speck and Speck 2021].

MF: Maybe the whole definition of 'life' should be changed ...

TS: The biological definition is pretty clear. But I think you are right. The definition of life is extraordinarily complicated and has a lot of social, religious, and psychological baggage. There are discussions in parliaments concerning the question of when life starts. That is why I think the question of the definition of what life is one you cannot answer easily ... I like the term 'life-like' because it gives us an impression of what we are aiming for. We want to have materials, materials systems, which have properties that we think are pretty good in living systems, and we would like to transfer them to technical structures. This transfer is what we are aiming for.

KK: One might perhaps also consider J. Scott Turner's works, where he explains animal architecture as part of the physiology of the animal and therefore as 'living architecture.' If I may quote him: "By structurally modifying the environment, I will suggest, organisms manipulate and adaptively modify the ways energy and matter flow through the environment. In so doing, they modify the ways energy and matter flow through them" [Turner 2000, p. 3]. From this thermodynamic point of view life is in fact a non-equilibrium phenomenon and always about reaction, transformation, and transition.

TS: I think you are right – equilibrium is death. Everything that is in equilibrium is in one way or another dead. Not in a mechanical equilibrium but in a thermodynamic equilibrium. We have to accept that a new group of materials and materials systems which should have life-like functions has to be out of equilibrium [Walther 2020]. If not, it could not react. This is the basic change in the understanding of materials.

However, out-of-equilibrium materials systems are still rare. If you are out of equilibrium you need energy – a permanent energy supply in one way or the other. One of the big challenges is to make systems not only interactive or adaptive but also energy autonomous. Otherwise, you end up with something that has a lot of cables sticking out of it. This is not what we should aim for. At the moment, we often solve simple questions concerning the development of an adaptive material or structure with complex rather engineering-based answers. Because the digital approach is so cheap and so powerful, it is easier to solve the problem by putting some chips in and writing some program and using boring, dead materials. But it could be solved in a much cleverer way and much more easily with an adaptive, interactive material or materials system [Knippers and Speck 2012; Knippers, Schmid, and Speck 2019; Walther 2020]. This concerns the idea of coming (back) from the digital to the analogue. At the moment a lot of complex tasks that have to be done are done with inadequate materials. It is much like mathematics. As a (bio)physicist I get a mathematical solution, but this solution is neither elegant nor beautiful. And the mathematician aims to find a beautiful and elegant way. We often do not ask whether things could be done in a smoother, smarter way, with less energy. And this is where the new types of materials systems come in.

MF: To think again about nonequilibrium materials, can one say that these new materials have some aspects of instability, but at the same time possess mechanisms of survival and longevity, which function side by side?

TS: I think longevity is an important aspect. On the other hand, I would like to change 'longevity' to 'life duration.' Because in biology lots of organisms or parts of organisms are built more or less exactly for the time of functioning, then they stop functioning and decay in one way or the other. One of the examples I like the most is that I learned that in Europe a mobile phone typically is used for less than 2 years, but it takes decades to decay. Either you have to find a way to trigger decay at the end of use, which in biology is already built in – when the life processes stop, decay starts – or you have to build a phone that can be continuously upgraded and stay in use until structural decay starts. It is the same in architecture. We still build our houses with the same vision we use to build a cathedral which might last 800 years [Knippers, Nickel, and Speck 2016]. If you look at single family houses, they normally have a lifetime of about one and a half generations before they are either entirely rebuilt or taken down. We have to rethink how we use materials, how we build our environment, and what we can learn from living nature. There was a famous book from 1865 with the title *Homes Without Hands*, which is still worth reading today [Wood 1865]. It is about animals that build nests and other complex structures without, obviously, having hands to build them with. These 'homes without hands' are really fascinating because, when you look, if a bird's nest is to be used only one time, then the bird normally spends less energy building it than if it is to be used for several years. There you can see a kind of trade-off. They build it in a way that is energetically and material-wise 'meaningful.' But, on the other hand, we have to be careful. We often use terms and, even worse, ways of thinking that are very human centered. It is the same with beauty: we think that a lot of structures in biology are beautiful, or at least very aesthetic. I have problems understanding the evolutionary meaning of beauty. It is hard to map these anthropocentric terms onto biological and bioinspired terms and materials. In my opinion, a lot of beauty comes with functionality [Speck 2015; Speck T and Speck O 2019].

MF: I think the same problem occurs with the notion of life. Take, for example, what you mentioned about plants having dead parts which continue to function. Are plants considered to be alive by human standards? That brings me to the next two questions. The first refers to what you pointed out concerning houses. Can you say that one of the goals of not only architecture but also of biomimetic research is to construct materials that preserve or maintain their own 'house'? Like the bone that repairs itself [Fratzl and Weinkamer 2007]. Are there materials that prepare themselves beforehand to avoid damage?

TS: If you are suggesting a kind of training that materials can train themselves or can be trained to be adaptive to potential damage or high load situations, then this is definitely something interesting. One can think about interactive materials or materials systems in such a way that they can not only react to a stimulus, by flipping aside, for example, but also react to a stimulus in a trained way – the same way we train a muscle, for example. If you have a material that is put into an environment where there are no mechanical loads, then there is no need to use energy to strengthen the material. But if the material is put into a different environment where there are a lot of subcritical loads, it might be very advantageous if this material uses energy to strengthen itself through, for example, a (re)arrangement of substructures. There we come very close to a life-like behavior of technical materials and materials systems. That's what trees do when they grow in windy situations. In that case, they are typically bulkier, shorter, and wider. Why? Because in a wind-sheltered situation it is an advantage to be high and to have your leaves high up in the sun. However, if you face a lot of mechanical load, a tree 'can't afford that,' and then it is advantageous for the tree to stay shorter and bulkier. How can we use this idea of training? Up to now, it is very rarely thought in that way. Normally, one does not think about how to prevent failures in technical materials or structures by training. With the new types of materials and materials systems that are around we can think that way and perhaps start with the first developments along those lines.

MF: The second question in fact is connected to your answer. Regarding movement of plants, you said that we were not in a position to see the movement of plants. What are the types of mechanisms that are involved in this movement? Obviously, this research has a long history – for example, with Charles Darwin's book *The Power of Movement in Plants* from 1880. How do you consider movement in this array of mechanisms?

TS: First of all, one has to consider a major difference between plants and animals. You normally do not find localized hinges in plants. Hinges are often weak points. The same with doors: you have stiff elements and then you have the hinges. And what makes problems? The hinges, that is, the pivot joints, which are often prone to failure as they have a lot of gliding surfaces. They need maintenance and sometimes block. Plants typically do not have them. Sometimes they have ribbon-like joints as found in the staminal levers of the *Salvia* (see Fig. 6) [Reith, Claßen-Bockhoff, and Speck 2006; Poppinga, Masselter, and Speck 2013]. Already in that respect, movements in plants are very interesting. But first I have to say that there are totally different types of movements: there are nastic movements and tropistic movements. Tropistic movements go either in (e.g., in phototropism) or against (e.g., in skototropism) the direction of the stimulus; nastic movement always go in the same predetermined direction independent of the direction from which the stimulus acts. And what movements are interesting for us? We normally look at nastic movements with a predetermined movement pattern which are triggered by an external stimulus. We do not look a lot at growth processes, as growth is normally a very slow movement.

Fig. 6: Mode of functioning of the staminal lever in a flower of clary sage (*Salvia sclarea*) visited by a wood bee (genus *Xylocopa*). © Plant Biomechanics Group Freiburg.

This external stimulus is, for example, when pollinators visit plants or when a prey is trapped by a carnivorous plant. One specific example concerns the waterwheel plant, a sister of the Venus flytrap [Poppinga et al. 2013; Westermeier et al. 2018; Knippers, Schmid, and Speck 2019] but with a totally different movement pattern. The Venus flytrap makes a snap buckling movement with a curvature inversion [Sachse et al. 2020]. If you look from the outside, the trap, which is built by two leaf halves, has a concave shape. Then a triggering prey insect comes and, if this insect touches the trigger hairs twice in a given time window, then the trap snaps close [Poppinga et al. 2017]. The trap of the waterwheel plant always has convex halves (seen from the outside), and the movement comes about through the bending of the backbone, that is, the mid-vein of the leaf connecting the two trap halves. It is not a snapping by curvature inversion, as in the Venus flytrap, but a motion amplification. The backbone bends a tiny bit and, as a result of this bending, due to the hinged region found here, the trap halves close rapidly. So a totally different principle is found in two closely related plant species. Both principles lead to very fast active trapping.

Here, we find a possible transfer to architecture. The Venus flytrap is of limited interest because no one could image a façade shading system that gets triggered by heat or light and, as a result, snaps shut. It is open or closed and there is nothing in between, that is, it is either in one state or the other. The energetic landscape of the waterwheel plant is totally different. It is a very smooth and very fast continuous motion [Westermeier et al. 2018]. If it is triggered it completes the action, but with a smooth movement. Therefore, we have built, inspired by the waterwheel plant trap, a façade shading system which is stable in each opening or closing phase: the Flectofold (see Fig. 7) [Körner et al. 2018; Knippers, Schmid, and Speck 2019]. You can close it and then reopen it and keep it open or closed. This is one example where you really have to understand how actuation takes place and what the driving forces are. In the waterwheel plant, it is a combination of elastic stored instability and of hydraulics.

(a) (b)

Fig. 7: (a) Single element of bioinspired façade shading system Flectofold based on motion patterns of the snap trap of the waterwheel plant with a pneumatic actuation cushion bending the backbone. © Plant Biomechanics Group Freiburg & DITF Denkendorf. (b) Array of Flectofold elements as presented at an exhibition in the State Museum of Natural History in Stuttgart[7] showing the aesthetics of this system. © Plant Biomechanics Group Freiburg & ITKE Uni. Stuttgart.

KK: You noted two similar plants, each with a different structure, but only one has a structure with the potential of being transferred to architecture. In this case, one can say that the plant has a well-defined structure, and even if it goes through transformations it has clear functions. Maybe, as with the pink carpet mentioned before, there is a moment when a new structure emerges and becomes stabilized in the system. How would you describe this appearance of the new structure? As evolution? Emergence? Epigenesis?

TS: If a new structure evolves there is a nice saying that all living beings have to 'carry their evolutionary burden.' This is one reason, besides multifunctionality, for the sometimes surprising finding that – even after 3.8 billion years of biological evolution – you may find structures that would not be the first choice of an engineer for a specific (single functional) task. For example, the knee joint is not what you need

7 A virtual tour through the exhibition 'Baubionik – Biologie beflügelt Architektur (Biomimetics meets Architecture)' presented from 2017 to 2018 at the State Museum of Natural History in Stuttgart is possible at https://www.trr141.de/180409_Bionik/ (accessed February 20, 2021).

for a 90 kg, 1.80 m high European man who walks up- and downstairs. The reason is that it was inherited from much more lightweight, originally quadruped animals, where it functioned pretty well. Evolution could never put a sign up saying 'Closed for reconstruction.' It had always to function. Nevertheless, there are dramatic changes that might be caused by a single mutation. But in most cases you always carry with you what you inherit from your ancestors. This sometimes leads to evolutionary solutions that are different from straightforward technical solutions for a specific task but often offer a totally new approach to a question. This is a big advantage in biomimetic research. All the evolutionary processes are random; nevertheless, they lead to a highly adaptive and efficient solution in a typically multifunction situation. Often a solution that is 'good enough' but 'cheap' with respect to energy and/or material demand is the one favored by evolutionary processes [Vincent 2002]. The question that is hard to answer is the number of generations needed. We see that life started about 3.8 billion years ago, and then you count the number of attempts that were taking place. A lot of mutational changes are inferior to what existed before. But some may be better for the existing situation or at least not worse, and then may be kept. As soon as the environment changes, these mutations may have a big advantage and therefore can establish themselves more and more. We have to see this change in a continuous fashion. But, as mentioned above, sometimes you also have drastic changes caused by single mutations.

KK: But could not these changes come from the environment?

TS: In this sense, epigenetics is important. It was vastly misjudged until about 20 or 30 years ago. The environment has a much bigger impact on development and phenotype than we thought before. Changes that came from the environment can be inherited by the offspring. Some may even be permanent. Some others are found in the first generation and then vanish. We are just starting to understand how much environment can influence the genetic background of the organism and how this can be passed on to the next generation.

MF: The discussion on changes and the two different types of plant movement brings me back to the beginning of our conversation: how to model this movement. What is the role of modeling with respect to what can be predicted? You use computer programs, but what do they actually model?

TS: What we often model with movements is their kinetics and kinematics. We try to understand how changes in structure and material influence the movement. This can be done very well with finite element methods where you can play around relatively easily with material properties. More complex situations are performed when you change shape or movement pattern because then you have to do a remodeling.

Numerical models are well established. They help to understand in one way the biological role model and to transfer the abstraction to a technical application. Personally, I like analytical models. Because there you can do predictions, which is very hard to do with numerical modeling [Konrad et al. 2013; Wolff-Vorbeck et al. 2019].

Numerical modeling is typically more or less descriptive. You can change parameters within the numerical model but with analytical models we can really predict what would change if we change this or that part of the model. This is important because then we can design new experiments and test the predictions. With fossil plants we could nearly only use analytical models, because the state of preservation was often so bad that it was impossible to do a numerical modeling. What we need to analyze with the growth habit of a fossil plant is the recalculation of the different mechanical properties of the different ontogenetic stages of the stem and twigs [Speck 1994; Speck and Rowe 2003]. This means that different ontogenetic stages must be preserved, which is rare. But, if you have, for example, three or four out of the six you would normally need, then, through an analytical model, you could remodel the missing stages. This would be much more complicated and unreliable with numerical modeling. For me analytical models are much more helpful than the numerical ones, which are good for transfer and for understanding the structures at a given moment, but not so much for predictions.

KK: You have just presented two types of modeling, and this brings me to the next question. In active matter research, several methods of description and prediction are involved: numerical and analytical modeling as well as all kinds of visualizations. All these tools have their own possibilities and constraints. I am wondering what it will mean for materials science when these different approaches come together.

TS: That is very hard to answer. We normally use statistics to give our measurements and results a given reliability. This is a test to see how reliable our data and the interpretations derived from this data are. For us statistics is just a tool; one that states, for example, that these values are highly significantly different, significantly different, or not different at all. This allows us to make interpretations about the functional differences. We do not do real probabilistic modeling in a way this is used to make predictions. What we use are the numerical and analytical models. As I said, for me the numerical models are descriptive. But for the analytical models we really try to get information about the next questions we have to ask. It is a way to look at the things that should be done next. Where are the next open and unanswered questions? How can we tackle them? From a scientific point of view, I like analytical models very much. From the point of view of technical applications, these numerical models are the language to speak with engineers and materials scientists, because it is the language they are used to. As to various types of visualization, we use all necessary and available approaches, from light microscopy through SEM and TEM to confocal laser microscopy, μCT, and MRI, and combine these studies with (micro)mechanical tests to analyze quantitatively the form–structure–function relationship on all hierarchical levels necessary [Hesse et al. 2019a, 2019b].

KK: The last question is a more general one. How would you describe the bigger picture of active matter, and where do you see yourself in this picture? What are the challenges that active matter research faces, and what could be a future horizon?

TS: The bigger picture certainly consists of soft robotics and interactive materials systems and structures with embedded energy and embedded intelligence; this is in my opinion the future of the discipline. We still have an overload of solutions using microchips and external control systems, which is driven by the fact that these systems have become very fast and very powerful with regard to data management. But often it would be much better to have the decision-making and energy source directly embedded in the material or materials system. This is what we have to learn. I have to admit that I am not a friend of big data and machine learning. It has many advantages but it often does not answer the central question of understanding the underlying form–structure–function relationship. We just get a pattern. But if you ask, for example, where a failure comes from – we have no idea. It does not help us with a broader understanding. It helps us to keep things going. What my vision would be – and this is where biology comes in – is to have clever filters with which we learn to understand the really important information. Because – as also in our private lives – we are overflooded with information, we need filters that filter out only the relevant environmental information which comes to the material, and then we need to develop materials systems that can respond quickly and locally, without a central control unit, that is, without a brain so to speak. This would speed up these materials systems that can be used for soft machines and would make them less complex and perhaps even smarter, as well as less prone to failure. Because then you could also use redundancy. And what is more, our daily lives would also benefit from having adaptive materials. But what we really have to keep in mind is the acceptance. One certainly accepts and is accustomed to materials that do not change shape and behavior. But if you had, for example, a mug that would fold around your hand for ensuring a safe grip, then you really have to see whether the public would accept that. If you have no acceptance, even good ideas can die very quickly.

Where do I see myself? Taking inspiration from biology, in order to understand it and to abstract it. That is the process I like most, the abstraction. If you have five different levels which all contribute to a given function, to understand which of the five are really necessary is challenging. I hope that bioinspired materials and materials systems, adaptive matter, get the attention they deserve, since the potential is huge.

Bibliography

Anandan S, Rudolph A, Speck T, Speck O. Comparative Morphological and Anatomical Study of Self-Repair in Succulent Cylindrical Plant Organs. Flora 2018, 241, 1–7, DOI: 10.1016/j.flora.2018.02.008.

Correa D, Poppinga S, Mylo M, Westermeier A, Bruchmann B, Menges A, Speck T. 4D Pine Scale: Biomimetic 4D Printed Autonomous Scale and Flap Structures Capable of Multi-Phase Movement. Philosophical Transactions of the Royal Society London A 2020, 378 (2167), 20190445, DOI: 10.1098/rsta.2019.0445.

Fratzl P, Speck T, Gorb S. Function by Internal Structure – Preface to the Special Issue on Bioinspired Hierarchical Materials. Bioinspiration and Biomimetics 2016, 11, 6, DOI: 10.1088/1748-3190/11/6/060301.

Fratzl P, Weinkamer R. Hierarchical Structure and Repair of Bone: Deformation, Remodelling, Healing. In: Van Der ZS, ed. Self-Healing Materials. Dordrecht, Springer, 2007, 323–335.

Gruber P. Biomimetics in Architecture: Architecture of Life and Buildings. Vienna, Springer, 2011.

Hager MD, Van Der ZS, Schubert US, eds. Self-Healing Materials. Springer, 2016.

Harrington MJ, Speck O, Speck T, Wagner S, Weinkamer R. Biological Archetypes for Self-Healing Materials. In: Hager M, Van Der ZS, Schubert U, eds. Self-Healing Materials. Advances in Polymer Science, Vol. 273. Cham, Springer, 2015, 307–344.

Hesse L, Bunk K, Leupold L, Speck T, Masselter T. Structural and Functional Imaging of Large and Opaque Plant Specimen. Journal of Experimental Botany 2019a, 70 (14), 3659–3678, DOI: 10.1093/jxb/erz186.

Hesse L, Kampowski T, Leupold J, Caliaro S, Speck T, Speck O. Comparative Analyses of the Self-Sealing Mechanisms in Leaves of Delosperma Cooperi and Delosperma Ecklonis (Aizoaceae). International Journal of Molecular Sciences 2020, 21 (16), 5768, DOI: 10.3390/ijms21165768.

Hesse L, Leupold J, Poppinga S, et al. Resolving Form-Structure-Function Relationships in Plants with MRI for Biomimetics Transfer. Integrative and Comparative Biology 2019b, 59 (6), 1713–1726, DOI: 10.1093/icb/icz051.

Imhof B, Gruber P, eds. Built to Grow – Blending Architecture and Biology. Berlin, Basel, Birkhäuser, 2016.

Kay LE. Who Wrote the Book of Life? Stanford. Stanford University Press, 2000.

Keller EF. The Century of the Gene. Cambridge MA, Harvard University Press, 2000.

Keller EF. Active Matter, Then and Now. History and Philosophy of the Life Sciences 2016, 38 (11), 1–11, DOI: 10.1007/s40656-016-0112-3.

Klein H, Hesse L, Boljen M, et al. Finite Element Modelling of Complex Movements During Self-Sealing of Ring Incisions in Leaves of Delosperma Cooperi. Journal of Theoretical Biology 2018, 458, 184–206, DOI: 10.1016/j.jtbi.2018.08.023.

Knippers J, Nickel KG, Speck T, eds. Biomimetic Research for Architecture and Building Construction: Biological Design and Integrative Structures. Biologically-Inspired Systems, Vol. 9. Cham, Springer, 2016.

Knippers J, Schmid U, Speck T, eds. Biomimetics for Architecture: Learning from Nature. Basel, Birkhäuser, 2019.

Knippers J, Speck T. Design and Construction Principles in Nature and Architecture. Bioinspiration and Biomimetics 2012, 7 (1), 015002, DOI: 10.1088/1748-3182/7/1/015002.

Konrad W, Flues F, Schmich F, Speck T, Speck O. An Analytic Model of the Self-Sealing Mechanism of the Succulent Plant Delosperma Cooperi. Journal of Theoretical Biology 2013, 336, 96–109, DOI: 10.1016/j.jtbi.2013.07.013.

Körner A, Born L, Mader A, et al. Flectofold – A Biomimetic Compliant Shading Device for Complex Free Form Facades. Smart Materials and Structures 2018, 27 (1), 017001, DOI: 10.1088/1361-665X/aa9c2f.

Masselter T, Barthlott W, Bauer G, et al. Biomimetic Products. In: Bar-Cohen Y, ed. Biomimetics: Nature-Based Innovation. Boca Raton, CRC Press, 2012, 377–429.

Nachtigall W, Pohl G. Bau-Bionik: Natur – Analogien – Technik. Berlin, Heidelberg, Springer, 2013.

Poppinga S, Bauer U, Speck T. Volkov AG: Motile Traps. In: Ellison AM, Adamec L, eds. Carnivorous Plants: Physiology, Ecology, and Evolution. Oxford, Oxford University Press, 2017, 180–193.

Poppinga S, Masselter T, Speck T. Faster than their Prey: New Insights into the Rapid Movements of Active Carnivorous Plants Traps. Bioessays 2013, 35 (7), 649–657, DOI: 10.1002/bies.201200175.

Poppinga S, Nestle N, Šandor A, et al. Hygroscopic Motions of Fossil Conifer Cones. Scientific Reports 2017, 7, 40302, DOI: 10.1038/srep40302.

Poppinga S, Weisskopf C, Westermeier A, Masselter T, Speck T. Fastest Predators in Plant Kingdom: Functional Morphology and Biomechanics of Suction Traps Found in Utricularia, the Largest Genus of Carnivorous Plants. AoB PLANTS 2016, 8, plv140, DOI: 10.1093/aobpla/plv140.

Poppinga S, Zollfrank C, Prucker O, et al. Toward a New Generation of Smart Biomimetic Actuators for Architecture. Advanced Materials 2018, 30 (19), 1703653, DOI: 10.1002/adma.201703653.

Reith M, Claßen-Bockhoff R, Speck T. Biomechanics of Salvia Flowers: The Role of Lever and Flower Tube in Specialization on Pollinators. In: Herrel A, Speck T, Rowe NP, eds. Ecology and Biomechanics: A Mechanical Approach to the Ecology of Animals and Plants. Boca Raton, CRC/Taylor & Francis Group, 2006, 123–146.

Rowe NP, Speck T. Biomechanical Characteristics of the Ontogeny and Growth Habit of the Tropical Liana Condylocarpon guianense (Apocynaceae). International Journal of Plant Science 1996, 157, 406–417, DOI: 10.1086/297357.

Rowe NP, Speck T, Galtier J. Biomechanical Analysis of a Palaeozoic Gymnosperm Stem. Proceedings of the Royal Society London B 1993, 252, 19–28, DOI: 10.1098/rspb.1993.0041.

Sachse R, Westermeier A, Mylo M, Nadasdi J, Bischoff M, Speck T, Poppinga S. Snapping Mechanics of the Venus Flytrap (Dionaea muscipula). Proceedings National Academy of Sciences 2020, 117 (27), 16035–16042, DOI: 10.1073/pnas.2002707117.

Spatz H-C, Speck T. Mechanische Eigenschaften von Hohlrohren am Beispiel von Gräsern. In: Nachtigall W, ed. Technische Biologie und Bionik 2: 2. Bionik-Kongress, Saarbrücken 1994, Mainz, Akad. Wiss. u. Lit. Stuttgart, Fischer, 1995, 91–132.

Speck O, Speck T. Selbstreparatur in Natur und Technik. Biologie in Unserer Zeit 2015, 45, 44–51, DOI: 10.1002/biuz.201510558.

Speck O, Speck T. An Overview of Bioinspired and Biomimetic Self-Repairing Materials. Biomimetics 2019, 4 (1), 26, DOI: 10.3390/biomimetics4010026.

Speck O, Speck T. Functional morphology of plants – a key to biomimetic applications. New Phytologist 2021, 231, 950–956, DOI.org/10.1111/nph.17396.

Speck T. A Biomechanical Method to Distinguish Between Self-Supporting and non Self-Supporting Fossil Plants. Review of Palaeobotany and Palynology 1994, 81 (1), 65–82, DOI: 10.1016/0034-6667(94)90127-9.

Speck T. Approaches to Bio-Inspiration in Novel Architecture. In: Imhof B, Gruber P, eds. Built to Grow – Blending Architecture and Biology. Basel, Birkhäuser, 2015, 145–149.

Speck T, Mutz M, Ohl S, Spatz H-C. The Electroretinogram of Drosophila to Flickering Light: Assesment of the Contributions of Receptor Cells R7 and R8. Journal of Insect Physiology 1984, 30 (6), 437–440, DOI: 10.1016/0022-1910(84)90022-2.

Speck T, Poppinga S, Speck O, Tauber F. Bio-Inspired Life-Like Motile Materials Systems: Changing the Boundaries Between Living and Technical Systems in the Anthropocene?. The Anthropocene Review 2021, 1039275, DOI: 10.1177/20530-962-1039275.

Speck T, Rowe NP. A Quantitative Approach for Analytically Defining Size, Form and Habit in Living and Fossil Plants. In: Kurmann MA, Hemsley AR, eds. The Evolution of Plant Architecture. Kew, Royal Botanic Gardens, 1999, 447–479.

Speck T, Rowe NP. Modelling Primary and Secondary Growth Processes in Plants: A Summary of the Methodology and New Data for the Early Lignophyte. Philosophical Transactions of the Royal Society London B 2003, 358, 1473–1485, DOI: 10.1098/rstb.2003.1347.

Speck T, Rowe NP, Spatz H-C. Pflanzliche Achsen, Hochkomplexe Verbundmaterialien mit erstaunlichen mechanischen Eigenschaften. In: Nachtigall W, Wisser A, eds. Technische Biologie und Bionik 3: 3. Bionik-Kongress Mannheim 1996, Mainz, Akad. Wiss. u. Lit. Stuttgart, Fischer, 1996, 101–131.

Speck T, Spatz H-C, Vogellehner D. Contributions to the Biomechanics of Plants: I. Stabilities of Plant Stems with Strengthening Elements of Different Cross-Sections Against Weight and Wind Forces. Botanica Acta 1990, 103, 111–122.
Speck T, Speck O. Process Sequences in Biomimetic Research. In: Brebbia CA, ed. Design and Nature IV. Southampton, WIT Press, 2008, 3–11.
Speck T, Speck O. Emergence in Biomimetic Materials Systems. In: Wegner LH, Lüttge U, eds. Emergence and Modularity in Life Sciences. Cham, Springer Nature Switzerland, 2019, 97–115.
Speck T, Speck O, Neinhuis C, Bargel H. Bionik – Faszinierende Lösungen der Natur für die Technik der Zukunft. Freiburg, Lavori-Verlag, 2012.
Speck T, Vogellehner D. Biophysical Examinations of the Bending Stability of Various Stele Types and the Upright Axes of Early 'Vascular' Land Plants. Botanica Acta 1988, 101, 262–268, DOI: 10.1111/j.1438-8677.1988.tb00042.x.
Turner JS. The Extended Organism: The Physiology of Animal-Built Structures. Cambridge MA, London, Harvard University Press, 2000.
Vincent JFV. Survival of the Cheapest. Materials Today 2002, 5 (12), 28–41, DOI: 10.1016/S1369-7021(02)01237-3.
Walther A. Viewpoint: From Responsive to Adaptive and Interactive Materials and Materials Systems: A Roadmap. Advanced Materials 2020, 32 (20), 1905111, DOI: 10.1002/adma.201905111.
Westermeier A, Sachse R, Poppinga S, et al. How the Carnivorous Waterwheel Plant (Aldrovanda Vesiculosa) Snaps. Proceedings of the Royal Society London B Biological Sciences 2018, 285 (1878), 20180012, DOI: 10.1098/rspb.2018.0012.
Wolff-Vorbeck S, Langer M, Speck O, Speck T, Dondl PW. Twist-To-Bend Ratio: An Important Selective Factor for Many Rod-Shaped Biological Structures. Scientific Reports 2019, 9, 17182, DOI: 10.1038/s41598-019-52878-z.
Wood JG. Homes Without Hands: Being a Description of the Habitations of Animals, Classed According to Their Principle of Construction. London, Longmans, Green, 1865.

Thomas Speck studied biology at the University of Freiburg (PhD in 1990) and in 1996 received the venia legendi for botany and biophysics. After a visiting professorship at the University of Vienna, he was offered professorships at the Humboldt Universität zu Berlin and at the University of Freiburg, where from 2002 to 2006 he was an associate professor for botany and director of the Botanic Garden. After declining the offer of a full professorship and the directorship of the Botanic Garden at the Freie Universität Berlin, in 2006 he became full professor for botany: functional morphology and biomimetics in Freiburg.

He is deputy managing director of the Freiburg Center for Interactive Materials and Bioinspired Technologies (FIT), a scientific member of the Freiburg Materials Research Center (FMF), and a spokesperson of the Cluster of Excellence 'Living, Adaptive, and Energy-autonomous Materials Systems' (*liv*MatS). Thomas Speck is also spokesperson of the Competence Network Biomimetics, and vice chair of the Society for Technical Biology and Bionics.

He has received several scientific awards, is the (co)editor of several scientific books and journals, and has published more than 300 scientific articles in peer-reviewed journals and books in the fields of functional morphology, biomechanics, biomimetics, evolutionary biology, and paleobotany. He has over 10 patents for bioinspired products.

Joanna Aizenberg, Michael Friedman, Karin Krauthausen

Interview with Joanna Aizenberg: On Responsive and Adaptive Materials

MF: The research field of active matter is a very heterogeneous field dealing with several types of materials: active materials, smart materials, and hybrid materials among others. Before entering a discussion on these various distinctions, we would like to begin with your own background in this field. Can you describe how you see the beginnings of this field of research and how your research started and is situated in it?

JA: I entered the field through my PhD research on bioinspired materials, in which I examined the skeletons of marine organisms to try to uncover the interesting material features of these skeletons and of biological structures in general [Aizenberg et al. 1997; Aizenberg et al. 2001; Sundar et al. 2003; Aizenberg et al. 2005]. I became fascinated by the way biological crystalline materials were organized in radically different ways than conventional human-made materials – many combined inorganic and organic components structured in intricate detail from the molecular to the macroscale – and, even more interestingly, I was able to show how the active, bottom-up processes the organisms use to make them can be mimicked in synthetic systems. This research made clear the amazing potential of self-organizing materials. Around the same time, I discovered that the brittle star exerts yet another level of dynamic control over its skeleton: in response to light, it pumps fluid through its porous skeleton to cover and uncover the surface with a protective layer of fluidic 'sunglasses,' and the deep-sea sponge evolves a hierarchically structured, fiber-optically illuminated glass 'house' inhabited by a pair of shrimp (see Fig. 1). As I studied these and other organisms – like echinoderms that use dynamic spiny surfaces to keep themselves clean – what became obvious to me was that there was one fundamental feature of the biological world which would be absolutely critical to mimic, but which at that time was almost completely absent from nonliving materials. That feature was the ability to respond to the environment, to be active, to reconfigure, to change properties, and adapt to meet changing functional needs. Twenty years ago, our materials were mostly optimized to do one thing and never change, but it has become clear that this would not be enough if we are to design sustainable, energy-efficient buildings, safe and durable infrastructure, or targeted medical interventions. The living world is brewing with an infinite variety of dynamic materials that form and adapt in response to everything from light

Acknowledgment: Joanna Aizenberg is acknowledging the contribution of Dr. Alison Grinthal.
Michael Friedman and Karin Krauthausen acknowledge the support of the Cluster of Excellence 'Matters of Activity. Image Space Material' funded by the Deutsche Forschungsgemeinschaft (DFG, German Research Foundation) under Germany's Excellence Strategy – EXC 2025 –390648296.

Open Access. © 2021 Joanna Aizenberg et al., published by De Gruyter. BY-NC-ND This work is licensed under the Creative Commons Attribution-NonCommercial-NoDerivatives 4.0 International License.
https://doi.org/10.1515/9783110562064-004

to heat to water to chemical species. My group's research now bridges fields as diverse as optics, mechanics, self-assembly, fluid dynamics, and catalysis, but is united by an underlying passion: to understand and creatively build on the fundamental principles of such living systems to develop not just new materials but entirely new ways of envisioning how we design and interact with materials.

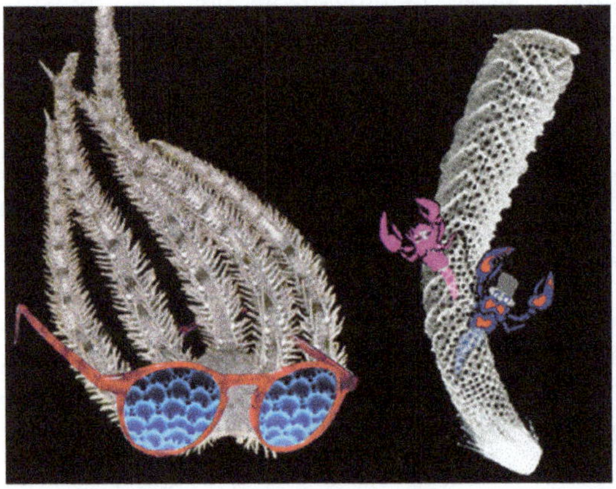

Fig. 1: Left: Marine brittle stars dynamically regulate their light exposure by pumping pigment through tiny pores in their lens-covered skeleton to 'put upon themselves' unique fluidic transition 'sunglasses' (close-up seen in sunglass lenses). Right: A marine sponge known as the Venus's flower basket makes a remarkably sturdy house for a pair of shrimp despite being made of glass, thanks to an intricately woven hierarchical architecture. As a bonus, the house also serves as a waveguide for light. © The Aizenberg lab's.

MF: Your answer brings us to one of the key concepts of active matter research, one which you also often discuss: the concept of architecture, often in the sense of an internally and hierarchically structured 3D shape.[1] Can one say that, with the realization of the potential benefits of dynamic man-made materials, akin to the ones existing in living matter, the concept of architecture changes, that architecture itself becomes dynamic and responsive, in other words reacting, harnessing and transforming energy, and adapting?

JA: Yes, absolutely. Biology uses the concept of a highly hierarchical assembly of features to put together an architecture that can give rise to complex synergies and feedback between components and across scales. In nature, materials are built from the molecular scale to the nanoscale through the microscale and all the way to the macroscopic level, and this hierarchical complexity of the resulting structure

1 See [Aizenberg et al. 2005, p. 275]: "Structural materials in nature exhibit remarkable designs with building blocks, often hierarchically arranged from the nanometer to the macroscopic length scales." See also [Noorduin et al. 2013].

introduces ways to, for example, channel mechanical force into a chemical reaction, or finely control the transport of heat or fluids or light within the material. The new properties that emerge from such architectures and from their precisely controlled combinations of different materials can far exceed the possibilities of any of the components alone.

KK: In your work you often refer to the dynamic between chemistry and the mechanical component,[2] which characterizes living systems on the micro- to macrolevel. Living systems seem to be continuously transforming or developing, and this development is triggered through the interaction between chemical and mechanical processes, which are also intrinsically linked. Could you describe this dynamic in more detail?

JA: By training I am a physical chemist, but when I started looking into biological structures, whether at their architectural or crystal growth or optical properties, I began to appreciate that it is often not possible to understand how they form and interact with the environment without taking a more interdisciplinary approach that encompasses the coupling of chemistry and mechanics. Even from my earliest work on bioinspired crystal growth, I could see how feedback between chemical reactions and mechanical stresses that develop within the growing material may lead to 'self-healing' behavior that corrects structural anomalies during growth – and when manipulated, can be used to precisely control pattern formation. This constant interplay between chemistry and mechanics becomes even more prominent in biological and synthetic materials that actively move and reconfigure. We have developed hybrid materials where a chemical change within a gel drives the up-and-down motion of microscale hairs or fins [Sidorenko et al. 2007], and have even taken these systems further to design homeostatic materials that actively maintain a steady state through chemo-mechano-chemical feedback (see Fig. 2) [He et al. 2012]. In the latter case, when the material is brought out of a steady state, such as a temperature set point, it is capable of bringing itself back: a chemical change in the gel mechanically drives the hairs to stand up, which in turn triggers a chemical reaction that restores the temperature – and ultimately leads to another chemical change in the gel that mechanically drives the hairs back down, to await another cycle.

Some of the most interesting dynamic processes occur, where mechanics and chemistry are further coupled to other energy domains, so that other triggers such as light-induced or magnetic-induced fields can be used to control changes in material

2 See: [Grinthal and Aizenberg 2013, p. 7072]: "A living organism is a bundle of dynamic, integrated adaptive processes: not only does it continuously respond to constant changes in temperature, sunlight, nutrients, and other features of its environment, but it does so by coordinating hierarchies of feedback among cells, tissues, organs, and networks all continuously adapting to each other. At the root of it all is one of the most fundamental adaptive processes: the constant tug of war between chemistry and mechanics that interweaves chemical signals with endless reconfigurations of macromolecules, fibers, meshworks, and membrane."

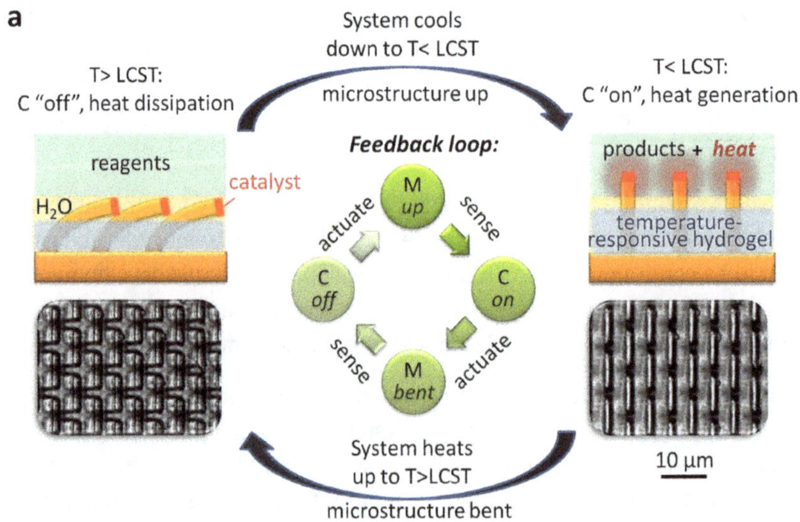

Fig. 2: A schematic representation of chemo-mechano-chemical feedback loop, in which mechanical action of temperature-responsive gel is coupled with an exothermic reaction. Material from: [He et al. 2012, fig. 3], Springer Nature.

properties. These feedback processes also extend beyond the solid state to include fluid dynamics and transport, as seen everywhere – from transport of water and gases in trees to secretion and healing processes in fluid membranes.

KK: In your research, you deal with the organism as a hierarchically organized bundle of feedback processes, since all levels of the living system – from cell to tissue to organ – are continuously adapting to each other and to the environment of the whole system. This dynamic is connected to the flows of energy between chemical processes and physical action, as you said. From a thermodynamic point of view, living systems are 'open' systems because they are incessantly dependent on the flows of energy and material passing through them. In this understanding, a living system is not in a state of equilibrium. How does research in the field of chemistry describe these systems?

JA: This question gets down to the very basic definitions of chemistry, that is, the fact that chemistry as a science was designed to describe equilibrium.[3] Almost everything one does in chemistry, all the equations we know as chemists – for example, concepts of chemical kinetics and thermodynamics and how reactions take place – all these by now are very well-described concepts related to chemical equilibrium. So, often, when we try to dissect a system that is out of equilibrium, we still understand it in

3 One of the important discoveries in chemistry during the 1950s–1960s was the Belousov–Zhabotinsky reaction as an example of a nonequilibrium thermodynamics not dominated by equilibrium thermodynamic behavior.

relation to local states, where there is a local equilibrium. But as the boundaries between chemistry, physics, and biology start to blur, some chemists are moving beyond this view and integrating ideas from reaction-diffusion systems, statistical physics, nonlinear dynamics, chaos, and complexity, with the aim of describing the emergence of life, pattern formation, self-replication, and adaptation in terms of dynamic, out-of-equilibrium processes. I think that, as more researchers are trained in interdisciplinary programs and become intrigued by the beauty of these systems, even basic chemistry will start to encompass such complex behaviors. The move to do so is coming from other fields as well, especially the study of earth systems, atmospheres, and oceans, which also tie chemistry to highly complex, out-of-equilibrium feedback processes and pattern generation.

Chemists are also recognizing the need to rethink the experimental assumptions used to understand living systems. Much of biological chemistry has been based on standard solution chemistry, with kinetics and thermodynamics analyzed as if everything were floating in a dilute solution in a tube. But of course this picture is nothing like the hierarchically organized bundle of feedback processes, where chemical processes may have little chance of ever reaching equilibrium. As with theory, many of the essential experimental techniques are being developed by reaching across disciplinary boundaries – both by focusing on a smaller scale, such as pushing and pulling on individual biomolecules to understand how their enzymatic activity might respond to an environment of mechanical forces, and by looking at the bigger context, such as by developing a variety of probes and sensors to interrogate chemical behavior along with, for example, mechanical changes on larger length- and timescales. Data from such experiments will no doubt further inform the development of out-of-equilibrium theory.

MF: Related to this issue, you have noted also that responsive behavior in synthetic materials is mainly driven by a two-state switch between properties.[4] You mention that a way to overcome or escape from this deadlock is to think about dynamics or dynamic fluids in a way that transforms our conception of static materials. If we consider fluids as something that points toward an essentially changing geometrical structure of materials, can one think about geometry as something that will prompt such changes? Can a new concept of geometry and its role in how materials are shaped help us bring forward new mathematical tools?

JA: The essential question here is how one might design a material that can undergo a continuum of responses – so that, rather than switching between two stable states, one can rationally program finely graded morphological changes. In traditional solid

4 See: [Aizenberg Biomineralization and Biomimetics Lab]: "Materials that adapt dynamically to environmental changes are generally limited to two-state switching of single properties, and only a small number of strategies that may lead to materials with continuously adjustable characteristics have been reported."

materials, responses generally rely on programmed switches in elastic force balance such as buckling instabilities, but living materials widely integrate liquid, liquid crystalline, and other fluidic components along with solids. These introduce another class of forces – surface tension, capillarity – that can potentially influence morphology in a more graded manner by driving local flows or 'sculpting' the liquid's surface. Although a liquid might seem too dynamic to even be considered a material, an answer is seen everywhere in biology: liquids coat surfaces from our eyes to our stomach lining to ants' feet, and, in nearly every case, are kept from escaping by an underlying porous solid that holds the liquid in place through capillary forces within the pores [Wong et al. 2011]. In the synthetic liquid-infused materials our group has designed, we have shown how manipulating the pore geometries by stretching or bending the material can lead to a continuum of liquid surface morphologies [Yao et al. 2013]. In other cases, the pore geometry remains fixed but guides local capillary flows of an infused ferrofluid under magnetic force, again leading to highly predictable graded surface morphologies.

Yet, even as we began to demonstrate these possibilities, one of my students was actually challenged at a conference with 'But a liquid isn't really a material.' This reinforced to us how profoundly two components of a hybrid material can influence each other's behavior. Of course, we would not wear clothes or live in houses made of liquid, but combining it with a structured solid gives us entirely different ways to think about both the solid and the liquid. Together they make the solid's geometry dynamic and the liquid's surface both stable and sculptable through selective control over tiny local flows, leading to the emergence of dynamic or adaptive material properties far different from those of either component alone (see Fig. 3).

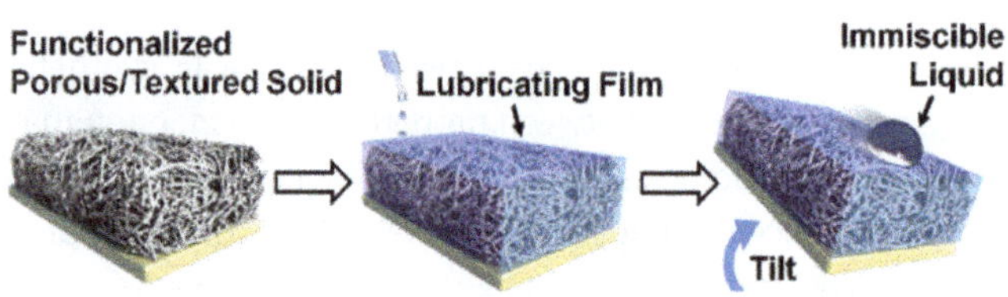

Fig. 3: Schematics showing the fabrication of a SLIPS (Slippery Liquid-Infused Porous Surface) by infiltrating a functionalized porous/textured solid with a low-surface energy, chemically inert liquid to form a physically smooth and chemically homogeneous lubricating film on the surface of the substrate. Material from: [Wong et al. 2011, p. 444, fig. 1], Springer Nature.

For us an essential part of designing these or any material systems is not just producing a particular structure or properties, but fundamentally understanding the scientific processes underlying how they form and behave under different conditions, so we and others can rationally design and further imagine yet more possibilities. Especially in the hierarchical and multiphase materials we have been discussing, this often means developing new comprehensive theoretical models to connect previously disparate aspects of physics, mechanics, fluid dynamics, chemistry, and so on, even if we do not always need to reinvent the basic math of these fields.

Probably, the best example of these principles in the biological system is bone [Currey 2002]. This is a material that has structural hierarchy from the molecular scale to the nanolevel to the macrolevel. There are at least 10 levels of structural hierarchy until it arrives at the final bone. It has channels for bodily fluids to pass through, it has collagen (organic material), and it has inorganic material (apatite). Each of these exists in an intimate relationship with the other.

MF: To focus on the plurality of materials or hierarchies in the same 'material,' can one also claim that one of the characteristics that enables these pluralities, one might say, is always a change in these hierarchies? These are never stable, in the sense, as you mentioned, that they neither reach nor attain a fixed equilibrium. There is always a change in this hierarchy – the hierarchies themselves or this plurality are never fixed.

JA: Not necessarily always, but certainly in many areas of biology, this is exactly the case. Especially given the integration of solids, liquids, liquid crystals, gels, and glasses, the different components constantly synthesize, dissolve, or metabolize each other, nucleate and trigger phase changes and phase separations, redefine boundaries and flip inside/outside topologies, so that not only the organization but even the existence of hybrids can change on different length- and timescales. All of these activities have been harnessed in the service of making the system inducible and adaptive, by providing access to a range of geometries and hybrid combinations that allow constant transformations and a response to environmental pressures necessary for survival.

KK: How would you then define active materials? According to their responsiveness? Or to their adaptiveness?

JA: My colleagues and I have had a long discussion on what are responsive, what are adaptive, and what are homeostatic materials.[5] All these types of dynamics certainly overlap, but to me they differ in the complexity of what goes on inside the material and, consequently, of how it interacts with stimuli.

'Responsive' is the most basic term and means something changes in the material when it is exposed to a stimulus. Although people like to point out that shattering when smashed with a hammer is a response, we take 'responsive' to mean that a specific stimulus, such as light, pH, or a magnetic field, evokes a specific, defined change in a material property or behavior, such as color or adhesiveness –

5 On the recent discussion on homeostasis, see: [Lerch, Grinthal, and Aizenberg 2020]. See also all the papers in Volume 32 (20) of *Advanced Materials* (2020) called "From Responsive Materials to Interactive Materials."

straightforward input–output. Although most often this implies that the change is reversible, that the material will return to its original state once the stimulus is removed, this is not always true, as in food packaging designed to report tampering by permanently changing the color.

Adaptive materials add a level of complexity – rather than just a simple direct input–output, the stimulus-induced change is met with a competing force (e.g., a chemical change competes with larger scale elastic resistance), and the output results from balancing the two via mutual feedback. In this sense, the response is 'adaptive' even at the level of its core mechanism. The internal feedback then enables more complex and more diverse stimulus–response behavior that can potentially produce better adapted materials, as small changes in the stimulus can alter the balance in a variety of non-straightforward/nonlinear/more nuanced ways.

Homeostatic or self-regulating material behavior incorporates internal feedback cycles into a larger 'meta-feedback' with the stimulus. We have recently discussed how the concept of homeostasis can operate within the material at the most basic mechanistic level, by incorporating what we call 'mini-homeostatic modules' – paired processes in which the stimulus-triggered response is opposed by a second process that drives it back toward the original state. The result is not simply a new static compromise, but a dynamic system that continuously evolves as, for example, it gets another kick from the stimulus on its way back to the initial state. While the material will eventually return to the original state once the stimulus is removed and the show is over, in the meantime it can produce an infinite variety of response trajectories depending on how and when in the cycle the stimulus is presented.

This latter case can also be considered the most active, since – as the 'self' in 'self-regulated' implies – its own internal dynamics enable it to actively interact with the stimulus rather than passively obeying it.

KK: The biological and the physical description of living beings has often influenced the philosophical answer to the question: what is life? On the other hand, research in physics and in biology has always also been inspired by what at a certain time and in a more general sense was understood under the notion of life. In light of your research and your experience, how do you consider the relation between natural biological materials and artificial synthetic materials, even if the latter are inspired by nature?

JA: First, inspiring as life is, I do not put it on a pedestal or over-romanticize its 'perfection' compared to nonliving materials. There are many examples where the structures nature produces or the materials it uses are absolutely not optimal. Nature has a very limited range of materials to utilize – it just evolves a certain structure in a way that works at the time for survival and reproduction purposes, using the available material components.

What nature does do in a mind-bogglingly effective way is couple – reactions, forces, materials, timescales and length scales, instabilities, defects, mistakes – whatever it finds or whatever arises can be plugged into the churning gears via the highly structured, hierarchical architecture. In the process, it creates highly complex multilevel physical and chemical networks that not only drive all sorts of otherwise energy-demanding reactions and movements but also provide redundancy to preserve robust, non-failing function – not only compensating for mistakes but turning them into opportunities.

In our own work, we do not aim to mimic life, but nature helps us look at our own synthetic materials from often vastly different and counterintuitive perspectives to see possibilities we did not see before. For all of life's wild and crazy complexity, it is often the more fundamental principles that lead us to new ideas – much of our work stems from discovering how complex coupling, feedback, hierarchy, and redundancy can emerge even in simple, boring materials.

MF: As you said, not only are there a lot of redundancies, but nature also selects for example from a very limited variety of materials,[6] whereas in the domain of synthetic materials there is a larger variety of materials that one can produce in the laboratory. To focus on redundancy, can you describe how synthetic materials stand in relation to biological ones?

JA: When we design a material, we do not have to worry about a whole range of things that nature has to optimize for. Let us say I want to make the strongest material possible. There are some interesting inspirational examples in biology, but living creatures also have to reproduce. In that respect, I have a somewhat easier task than biology because we can ignore functions that are not critical *and* use a range of extremely strong synthetic materials, such as steel or metals, for example, that might be out of the question for biology. If I want to make synthetic materials that have structural color similar to that used by butterflies or beetles – which rely on organic materials – we have access to a whole range of potentially superior inorganic materials. Thus, to design a synthetic optical system inspired by nature, I can choose to follow nature's structural design principles but reformulate them in completely different materials such as metal oxides, for example, which can give us additional optical properties that might augment the structure in unique ways. This may create a novel bioinspired material with optical properties optimized for the application in mind.

At the same time, we may want to think twice before dismissing certain features of the biological materials as irrelevant to our needs. Even if we are not interested in designing buildings that can lay eggs that hatch and grow into new buildings, we may well want a sustainable building that can perform multiple, often dynamic

6 On this selection with regard to optical phenomena see, for example: [England and Aizenberg 2018].

functions, such as changing its critical functional characteristics to adapt to the environment, for example, absorbing or repelling moisture depending on the weather, harvesting energy, collecting light, using dynamic windows that adapt their transparency or reflectance in response to outside temperature, and even self-healing damage. At this point, most building components are designed to optimize one function at a time, but in the future we need to think about multifunctional materials capable of performing a range of functions. Biology gives us numerous examples of how what initially looks like a trade-off may provide built-in redundancy, adaptiveness, or other desired properties that could enhance our needs in unexpected ways, whether or not the organism had the same use for them.

KK: I would like to come back to this antagonistic principle between chemistry and mechanics and to the 'complex' systems that can be built out of the relation between chemical processes and mechanical properties. In one of your articles you once called this dynamic relation a "dense, interconnected jungle".[7] That is to say that in living systems we are not dealing with a totally controlled or determined architecture but with one that is perpetually in a process of becoming because it deals with instabilities. What is the significance of instability and complexity for the research on active materials?

JA: As I discussed earlier with respect to complexity, feedback – within a material and with a stimulus – is at the core of sophisticated adaptive and self-regulated behavior. But, despite what we might think from looking at biology, complexity can emerge from very simple systems. Dynamic feedback between even two simple elements can lead to the emergence of properties that are not trivial and not easily explainable by either of these components. A beautiful example of complexity from simplicity that does not even rely on mechanics is the gardens of intricate microscale flowers we have grown by combining two simple competing chemical precipitation reactions in a beaker (see Fig. 4) [Noorduin et al. 2013]. We can sculpt vases, grow the flowers in them, even grow snails on the leaves, just by modulating air flow into the beaker or adjusting the temperature to shift the balance between the two reactions – highly interactive meta-feedback behavior that we have been able to theoretically model and rationally control based on the simple starting components. Local instabilities arise all the time. By watching the structure's evolution in real time and using our model, we can study how instabilities may be suppressed by the feedback, leaving characteristic local signatures in the structure, or be

7 See: [Grinthal and Aizenberg 2013, p. 7072]: "By sequestering chemical species in confined microarchitectures, living materials set the stage for energetically coupled reaction cascades that would likely be unfavorable anywhere else. But the reaction components do far more than simply mingle inside a passive architecture. Rather, they are the architecture: reactants and catalysts are strung together in long chains, twisted into filaments, woven into networks, and packed into bilayers, with all of them wrapped around each other in a dense, interconnected jungle."

Fig. 4: Gardens of microflowers, complete with leaves, blossoms, and even a few snails, can be sculpted simply by combining two competing chemical precipitation reactions in a beaker and taking advantage of unique reaction-diffusion pathways that result. The size of each microstructure is ~ 100 times smaller than the width of a human hair. © The Aizenberg lab's.

amplified and radically alter the global pattern evolution. Guided by our model we can even predict what conditions may drive the system into the 'Valley of Death,' a regime where feedback stops entirely.

MF: To continue the discussion on instability, should these instabilities be resolved? As you note in your work, these instabilities may also be harnessed [Hu, Kim, and Aizenberg 2017]. This employment of instabilities does not set out to resolve them but to reach some sort of static, stable state.

JA: Much of our work would not even exist without instabilities – the patterns, motions, and adaptive behaviors in many of the systems we study rely on tiny instabilities as triggers. For example, in one of these systems, surface arrays of nanoscale hairs, the hairs are designed to stand straight up, but in an evaporating solvent they tend to flop over due to local imperfections that break the symmetry of a hair's response to capillary forces [Pokroy et al. 2009]. This was understandably dismissed as failure by many people, since the hairs then stick to each other and get tangled up. But we discovered that by fully understanding and modeling the competing forces in the system – capillary, elastic, and adhesion – we could harness this symmetry-breaking to make the hairs spiral around each other into hierarchical swirls, and even program hierarchical stages of unspiraling (see Fig. 5). The chiralities are random, but, if we directly program an instability into each hair by introducing just a slight tilt, we can control which way the symmetry breaks so the hairs spiral with either all left- or all right-handed chirality.

We also harness instabilities to entirely transform the topology of microscale cellular surfaces – lattices of interconnected triangles, circles, or other shapes (see Fig. 6)

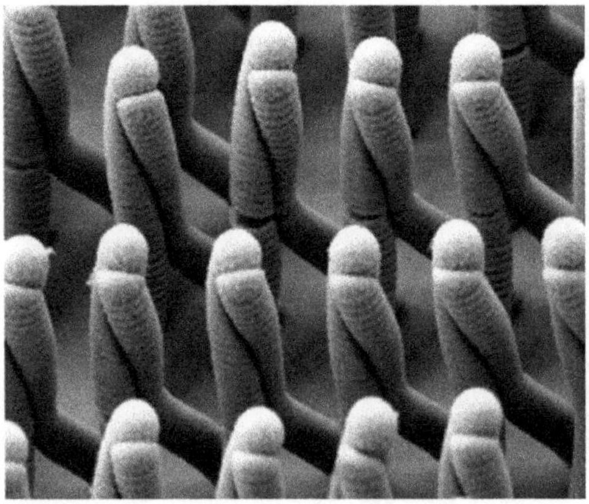

Fig. 5: Initially, vertical nanoscale hairs spiral around each other in an evaporating liquid, drawn together by meniscus forces, held by surface adhesion, and kissing with uniform chirality if given a slight initial tilt. © The Aizenberg lab's.

[Li et al. 2021]. In this case, we temporarily swell the structure with liquid to soften it and then use capillary forces to zip the walls together, switching, for example, a triangular topology to a hexagonal, with half as many compartments and double the nodes. Due to the symmetry of the initial system none of this would happen without tiny local instabilities that cause the swelled walls to buckle in one direction or another. Again, guided by a theoretical model, we are able to program the buckling direction by introducing tiny angular imperfections at each node. The programmed instabilities can even be used for encryption, since the imperfections are invisible and the encrypted patterns appear only upon transformation.

MF: Alongside this classification of instabilities, you also introduce a differentiation between structural instability and material instability. Could you explain what do you mean by this? Can one really sharply distinguish between the two types? How do you consider the relation between material and structure?

JA: Although there is certainly some ambiguity and overlap depending on what field you are in, we usually use structure to refer to geometry or configuration, generally at the nanoscale or above, and material to mean the composition or phase of the substance it is made of. For example, in the work by Hu et al. you mentioned earlier (and all our hybrid systems based on it), where a micropillar is embedded in a gel, we have both types of instability working together. The structural instability consists of the pillar abruptly buckling under mechanical force; the material instability refers to a discontinuous phase change due to the gel's polymer strands mixing with water in response to temperature. Differentiating matters most in cases like this, where the two types of instability operate in the same system but on different timescales and length scales, which must be explicitly taken into account in a theoretical

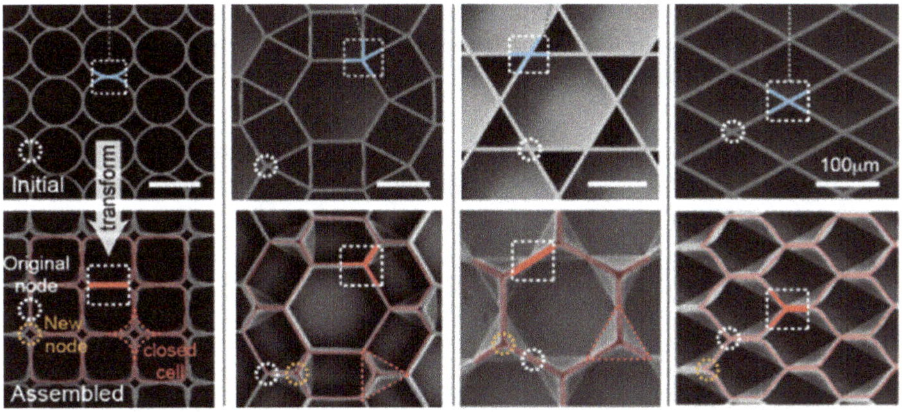

Fig. 6: The generalized lattice consists of nodes with different angles. The edges forming the smallest angle will undergo zipping for angle-guided capillary-driven transformations from circular to square (i), rhombitrihexagonal to hybrid hexagonal (ii), kagome to hexagonal (iii), and rhombic to hexagonal (iv). We note that, compared to the triangular-transformed and kagome-transformed hexagonal lattices, which are comprised of nodes connecting three equal double walls, nodes in the hexagonal lattice assembled from the diamond structure comprise two single walls and one double wall, resulting in additional anisotropy along the horizontal direction. Material from: [Li et al. 2021, p. 389, fig. 3], Springer Nature.

model to predict the system's response behavior and – importantly – can be controlled modularly by distinct handles to influence the response.

KK: I would like to return to the theme with which we started the discussion on instabilities, which concerns the relation between natural biological materials and synthetic ones. In the research on smart or active materials, one finds a variety of objects and research approaches. These include soft robotics and the implementation of computing in soft robotics. I wonder what your relation is to robotics, as I know that some of your objects were intended for and also find applications in the soft robotics context [Kumar et al. 2017].

JA: I approach this topic from a materials point of view, since I am not coming from pure robotics. The majority of things that I have seen so far are robots that are externally controlled, such as, for example, pneumatically or electrically controlled systems. Increasingly these are made of soft materials and are easier to manipulate because you can inflate them, you can swell them, you can bend them, but you still have to use a pneumatic or electric cord to send a signal to, say, move the arm. The direction soft robotics is starting to head in now is one in which the components are smart materials, so that the robot is not externally controlled, but the material itself can transduce signals from the environment into specific responses that enable the robot to navigate in the right direction or catch, release, secrete, or sort as appropriate.[8] Many lines of research in our group – such as those we have been discussing – investigate the fundamental principles required to design, control, and predict such materials, and it is becoming possible to envision not only robots but buildings, medical therapies, and more being constructed from autonomous, programmable matter – where it is the material itself that operates, responds to environment, harvests energy from the environment, and operates accordingly.

Bibliography

Advanced Materials. Special Issue: From Responsive Materials to Interactive Materials. 2020, 32 (20), DOI: 10.1002/adma.202070153.

Aizenberg J, Hanson J, Koetzle TF, Weiner S, Addadi L. Control of Macromolecule Distribution within Synthetic and Biogenic Single Calcite Crystals. Journal of the American Chemical Society 1997, 119 (5), 881–886, DOI: 10.1021/ja9628821.

Aizenberg J, Tkachenko A, Weiner S, Addadi L, Hendler G. Calcitic Microlenses as Part of the Photoreceptor System in Brittlestars. Nature 2001, 412 (6849), 819–822, DOI: 10.1038/35090573.

Aizenberg J, Weaver JC, Thanawala MS, Sundar VC, Morse DE, Fratzl P. Skeleton of Euplectella sp.: Structural Hierarchy from the Nanoscale to the Macroscale. Science 2005, 309 (5732), 275–278, DOI: 10.1126/science.1112255.

Aizenberg Biomineralization and Biomimetics Lab. Dynamic, Adaptive Fluidic Surfaces for Multifunctional Tunability. Cambridge, MA: The President and Fellows of Harvard College, (2021. Accessed January 26, 2021, at https://aizenberglab.seas.harvard.edu/dynamic-adaptive-fluidic-surfaces-multifunctional-tunability-0.).

Currey JD. Bones – Structure and Mechanics. Princeton, Princeton University Press, 2002.

England GT, Aizenberg J. Emerging Optical Properties from the Combination of Simple Optical Effects. Reports on Progress in Physics 2018, 81 (1), 016402, DOI: 10.1088/1361-6633/aa8372.

Grinthal A, Aizenberg J. Adaptive all the Way Down: Building Responsive Materials from Hierarchies of Chemomechanical Feedback. Chemical Society Reviews 2013, 7 (42), 17, 7072-7085, DOI: 10.1039/C3CS60045A.

He XM, Aizenberg M, Kuksenok O, Zarzar LD, Shastri A, Balazs AC, Aizenberg J. Synthetic Homeostatic Materials with Chemo-Mechano-Chemical Self-Regulation. Nature 2012, 487 (7406), 214–218, DOI: 10.1038/nature11223.

Hu Y, Kim P, Aizenberg J. Harnessing Structural Instability and Material Instability in the Hydrogel-Actuated Integrated Responsive Structures (HAIRS). Extreme Mechanics Letters 2017, 13, 84–90, DOI: 10.1016/j.eml.2017.02.003.

Kumar K, Liu J, Christianson C, et al. A Biologically Inspired, Functionally Graded End Effector for Soft Robotics Applications. Soft Robotics 2017, 4 (4), 317–323, DOI: 10.1089/soro.2017.0002.

8 See the recent discussion on interactive materials: [Advanced Materials 2020], as well as the abstract to the volume: "Interactive materials mimicking the embodied intelligence of biological systems lead to a paradigm shift from materials that merely respond to external signals to materials that sense their environment, process the information received, and respond or adapt autonomously."

Lerch MM, Grinthal A, Aizenberg J. Viewpoint: Homeostasis as Inspiration – Toward Interactive Materials. Advanced Matter 2020, 32 (20), 1905554, DOI: 10.1002/adma.201905554.

Li S, Deng B, Grinthal A, et al. Liquid-Induced Topological Transformations of Cellular Microstructures. Nature 2021, 592, 386–391, DOI: 10.1038/s41586-021-03404-7.

Noorduin WL, Grinthal A, Mahadevan L, Aizenberg J. Rationally Designed Complex, Hierarchical Microarchitectures. Science 2013, 17, 832–837, DOI: 10.1126/science.1234621.

Pokroy B, Kang SH, Mahadevan L, Aizenberg J. Self-Organization of a Mesoscale Bristle into Ordered, Hierarchical Helical Assemblies. Science 2009, 323 (5911), 237–240, DOI: 10.1126/science.1165607.

Sidorenko A, Krupenkin T, Taylor A, Fratzl P, Aizenberg J. Reversible Switching of Hydrogel-Actuated Nanostructures into Complex Micropatterns. Science 2007, 315 (5811), 487–490, DOI: 10.1126/science.1135516.

Sundar VC, Yablon AD, Grazul JL, Ilan M, Aizenberg J. Fibre-Optical Features of a Glass Sponge – Some Superior Technological Secrets Have Come to Light from a Deep-Sea Organism. Nature 2003, 424 (6951), 899–900, DOI: 10.4161/org.5843.

Wong TS, Kang SH, Tang SKY, Smythe EJ, Hatton BD, Grinthal A, Aizenberg J. Bioinspired Self-Repairing Slippery Surfaces with Pressure-Stable Omniphobicity. Nature 2011, 477 (7365), 443–447, DOI: 10.1038/nature10447.

Yao X, Hu Y, Grinthal A, Wong T-S, Mahadevan L, Aizenberg J. Adaptive Fluid-Infused Porous Films with Tunable Transparency and Wettability. Nature Materials 2013, 12, 529–534, DOI: 10.1038/nmat3598.

Joanna Aizenberg received her BS in chemistry from Moscow State University (MSU) and her PhD in structural biology from the Weizmann Institute of Science (WIS). After spending nearly a decade at Bell Labs, Joanna Aizenberg joined Harvard University, where she is the Amy Smith Berylson Professor of Materials Science and Professor of Chemistry and Chemical Biology.

The Aizenberg lab's research is aimed at understanding some of the basic principles of biological architecture and the economy with which nature solves complex problems in the design of multifunctional, adaptive materials. These biological principles are then used as guidance in developing new, bioinspired synthetic routes and nanofabrication strategies that lead to advanced materials and devices with broad implications in fields ranging from architecture to energy efficiency to medicine. Research topics of interest include biomimetics, smart materials, wetting phenomena, bio-nano interfaces, self-assembly, crystal engineering, surface chemistry, structural color, and biomineralization.

John Dunlop, Karin Krauthausen

Interview with John Dunlop: Shape-Changing Materials

KK: You are working on the activity of 'living' as well as 'dead' materials, especially with phenomena of shape changing and the possibility of a bioinspired transfer from shape-changing mechanisms in living systems to synthetic systems. Many of your studies belong to the field of active matter research, although you do not explicitly refer to this concept in your publications. May I ask what your understanding is of active matter? And how as a scientist did you find your way into this relatively new field of research?

JD: I suppose I should start with the way I got into active matter. It started from the perspective of my PhD, where I was interested in processes of changing morphology of the internal structure of metals.[1] In the project, we were trying to understand how crystals grew and changed shape due to different energy inputs. For example, if you deform a material, you can store lots of energy from the deformation process, and this is then transformed, if you supply enough heat to the surroundings, into a process by which crystals will change size, orientation, shape, and so on. We were trying to understand these processes and develop simple physically based models to describe their behavior. The idea was that if we could determine a relationship between structure and mechanical function, and if we could understand how structure changes with temperature, for example, we could then predict how mechanical properties of the material would evolve. Subsequently, in February 2006, I started working in the Department of Biomaterials at the Max Planck Institute of Colloids and Interfaces (MPICI) in Potsdam.[2] It took a little bit of time but I began to become interested in processes of describing structural shape changes initially in bones, but shortly afterward also in plants. These two processes of shape change have a commonality with what I was looking at before in that we were trying to come up with very simple descriptions of how in the architecture of trabecular bone, for example, shape and orientation changes in response to the external environment [Dunlop and Fratzl 2013;

1 John Dunlop did his PhD on internal variable modeling of creep and recrystallization in zirconium alloys in the years 2002–2005 at the Laboratoire de Thermodynamique et de Physico-chimie des Matériaux of the Institut national polytechnique de Grenoble in France [Dunlop, Bréchet, and Legras 2004; Zurob, Bréchet, and Dunlop 2006].

2 In 2006, Dunlop continued his research at the MPICI in Potsdam (Germany), first as a postdoctoral scientist and in 2007 also as Alexander von Humboldt Research Fellow. From 2008 to 2017, he led his own research group 'Biomimetic Actuation and Tissue Growth' at the MPICI [Dunlop 2021].

Acknowledgment: Karin Krauthausen acknowledges the support of the Cluster of Excellence 'Matters of Activity. Image Space Material' funded by the Deutsche Forschungsgemeinschaft (DFG, German Research Foundation) under Germany's Excellence Strategy – EXC 2025 – 390648296.

Open Access. © 2021 John Dunlop, Karin Krauthausen, published by De Gruyter. BY-NC-ND This work is licensed under the Creative Commons Attribution-NonCommercial-NoDerivatives 4.0 International License.
https://doi.org/10.1515/9783110562064-005

Gamsjäger et al. 2013; Fratzl and Weinkamer 2007]. In terms of plant tissues, this was a work done together with the research group of Ingo Burgert at the time.[3] That was an attempt to describe how objects such as the pine cone or the ice plant change shape in accordance with changes in the environment, changes in humidity or water content, for example (see Fig. 1) [Harrington et al. 2011; Burgert and Dunlop 2011].

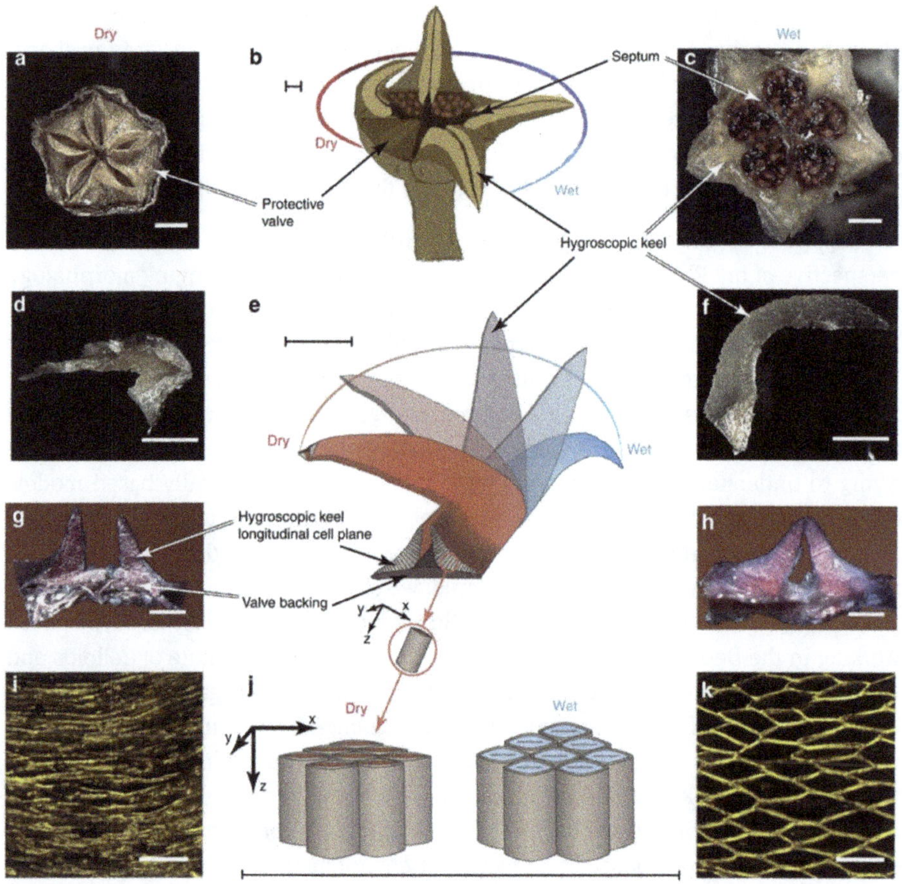

Fig. 1: Hygroscopic opening of the ice plant seed capsule is controlled by tissue microstructure. Illustration of the different levels of architectural complexity in the ice plant seed capsule at different length scales (*a–k*). The images on the left-hand side show the different tissue structures in the dry state and on the right-hand side in the wet state. The central column gives schematic illustrations of the different tissues. Scale bars are defined as: *a* and *c* = 2 mm; *b*, *e*, and *j* ≈ 1 mm; *d* and *f* = 1 mm; *g* and *h* = 0.5 mm; *i* and *k* = 0.1 mm. Reprinted by permission from Springer Nature: [Harrington et al. 2011, fig. 1].

3 Until 2011 Ingo Burgert led the research group 'Plant Biomechanics and Biomimetics' at the MPICI in Potsdam (Germany) (https://www.mpikg.mpg.de/5788791/plant-biomechanics-and-biomimetics, accessed May 5, 2021).

Through that I got into this field of active matter. This continued during my time at the MPICI and is also being carried out here in my research group 'MorphoPhysics' at the Paris Lodron University of Salzburg.[4] Returning to the question of what I consider active matter to be in this context, I judge it as being a class of materials that respond to internal energy (coming from cells acting in response to their environment), or to energy that in principle comes from the external environment. These materials use that energy to change their structure, their arrangement. This could be an internal or an external structure. I see it as being a fairly broad topic that would encompass the kinds of shape changes you get in objects that can swell and change form in their environment. I could also see it as being in processes where cells act locally in response to their environment. You get some sort of collective behavior. You can also observe this sort of collective behavior in general. This is what you have in the descriptions of swarm behavior, the flocking of birds or insects [Vicsek et al. 1995; Grégoire and Chaté 2004]. All of these would also fit in this perspective. My view in this field is really about thinking of materials as such, and how they change their structural shape in response to their environment.

KK: If you look back at the development of your research on shape-change phenomena (and in this sense on active matter), which influences, which disciplines, and which tools were important for you and helped you on the way?

JD: My background is varied in that I studied materials engineering and chemistry. The materials engineering or materials science background is one where you take concepts from physics or chemistry, and then you apply them to understand the object you are interested in as an engineer. Through the materials science background, you have quite a good overview of the different fields. Chemistry gives you a good molecular understanding of what is going on. But the most important part would be the physics of biological systems. It is there that these concepts of emergent phenomena have become more mainstream over the last 10 to 15 to 20 years.[5] The descriptions of these phenomena and the tools used have become very helpful for research, particularly in the field of active matter.

From another standpoint, you could say that there are basic tools of engineering, for instance the simulation tools that we use. Having an understanding of mechanics and of basic physical phenomena such as diffusion and similar sorts of things is also pretty important. And having the opportunity to talk to people from other disciplines

4 Since 2017, Dunlop is professor of biological physics at the Department of the Chemistry and Physics of Materials at Paris Lodron University of Salzburg, where he leads the research group 'MorphoPhysics' (https://www.morphophysics.com/john-dunlop, accessed May 5, 2021). For the most recent research in the fields of actuation in plants and tissue growth, see: [Dunlop et al. 2020; Ehrig et al. 2019].

5 For an excellent introduction into models of self-organization, see: [Camazine et al. 2003].

is crucial, like talking to biologists in order to understand the important features that you observe in a cell culture or in an organism.

Perhaps the most central point for me would be the training I acquired in materials science where you describe, for example, the phenomena you see and then try and come up with relatively simple models to represent the physical principles that are governing any phenomena that you are observing. Even if the models themselves are not fully successful in describing everything, you still learn a lot – if you keep their limitations in mind. I think the materials sciences are quite strong in this respect, because they really do bring these simplified models from chemistry and physics together in order to describe applied problems. This is relatively easy to transfer to the description of biological systems, at least from a physical viewpoint.

KK: The research field of active matter has developed strongly during the past 20 to 30 years.[6] It is supported by different disciplines and probably expresses a broader interest among the sciences in the activity of materials. Today, the research on active matter seems to be quite diversified. For a scientist working in this field, is it still possible to get the bigger picture?

JD: Getting an overview of the entire field is probably very difficult or unrealistic, especially if you want it to be detailed. But what you can certainly do is to get a relatively broad overview of key types of phenomena that I would say fit into the field of active material or matter research. By calling it 'materials' here, my own background is coloring my discourse about it. You can divide the field into subtopics or sub-foci. You can go in the direction of emergent phenomena or patterning – that is already a large field in itself [Camazine et al. 2003]. But it has been shown in quite a few biological systems that you can describe patterning of pigmentation in the skin or patterning of coloring of shells and things like that by relatively simple reaction-diffusion equations [Nüsslein-Volhard and Singh 2017; Kondo 2002; Turing 1952; Malacincski and Bryant 1984; Meinhardt and Klingler 1987]. To give you an idea, we can consider a chemical system in which we have a pigment whose production is stimulated by the presence of component A (this species is autocatalytic, meaning it stimulates its own production) but is inhibited by the presence of component B. Under situations where these components are well mixed, the system will tend to a stable steady state, meaning the system will present either a pigment or not. When the system is not well mixed, the spatial diffusion of species A and B can give rise to spatially inhomogeneous patterns, leading to complex pigmentation patterns in the system. It has been shown that these types of models can also apply to biological systems at the level of the cell up to the organism.

From the perspective of an overview, it is also important to highlight the role of mechanics in active matter, in that much of what we describe as being active matter

6 For a good introduction in the different field of active matter research, see: [Gompper et al. 2020].

consists of the interaction of objects and materials through forces [Ambrosi et al. 2011; Trepat and Sahai 2018]. As I see it, that is a key point. The description of shape changes, or the description of how one goes about changing the shape of an object or transforming its internal structure, implies that you need to give a detailed account of forces, of how mathematically you can describe a shape. You need to be able to describe how a surface changes its geometry, how its internal pattern changes with time or in response to a signal [Kollmannsberger et al. 2011; Guiducci et al. 2015; Ehrig et al. 2019]. I believe that is another component that is very important. And then perhaps there is the applied direction, which would be examples of these sorts of changes, patterning, and how that fits with the first two.

If you can nail down what you mean by active matter, then you can get an overview. But of course, by going in-depth with each example, you can very quickly get lost in the details. And also, I guess, the other problem is that it is such a broad topic that you can find many examples in the biological, physical, or mathematical literature. It can be very difficult to get into active matter, especially when you come from a different field. Being a newbie in the field is somewhat tough to digest. It is possible, however, to get an overview of the important themes.

KK: Would you say that active matter works as an umbrella term or migrating concept that is able to 'bridge' between different disciplines?

JD: For sure. For example, from the experience of the research I have been conducting on shape changing, you end up using these different tools in widely different disciplines. For example, when we explored macroscopic shape changes of the ice plant or the *Banksia* seed pods, we used the same computational tools that we use to describe the behavior of synthetic materials [Guiducci et al. 2015; Harrington et al. 2011, footnote 6; Huss et al. 2019]. Furthermore, the description of the reaction-diffusion equations mentioned earlier works well in chemistry, but it can also be used in biology. The physics of this is also very interesting and can also be characterized in this respect [Halatek and Frey 2018]. There is certainly a lot of commonality there. To view it as a bridge makes a lot of sense.

KK: You mentioned the role of energy in active materials. Gautam I. Menon in his 'tentative' (as he called it) definition of active matter also referred to energy from an internal source or as a trigger from the environment for explaining activity.[7] In Menon's account, active matter is understood as a far-from-equilibrium system that can show emergent behavior. Hence, he refers often to living systems whose behavior is

7 See [Menon 2010, p. 194]: "Active matter is a term which describes a material (either in the continuum or naturally decomposable into discrete units), which is driven out of equilibrium through the transduction of energy derived from an internal energy depot or ambient medium into work performed on the environment. Such systems are generically capable of emergent behaviour at large scales."

explained by thermodynamic theories and statistical physics. In your own research on active materials, you often use mathematical descriptions, for instance the geometry of an internal or external structure, in order to explain either the growth of cells or the actuation in plant materials. Could you tell me more about the importance of geometry for the explanation of activity?

JD: In order to answer this question, it is necessary to describe the two directions that interest me. On the one hand, I am into changes in environmental conditions – and I suppose in this respect this is inspired by these structures that you can see in plants that actuate, like a pine cone (see Fig. 2), wheat awns, the ice plant, or the *Banksia* seed pods [Dunlop, Weinkamer, and Fratzl 2011; Harrington et al. 2011; Huss et al. 2019].

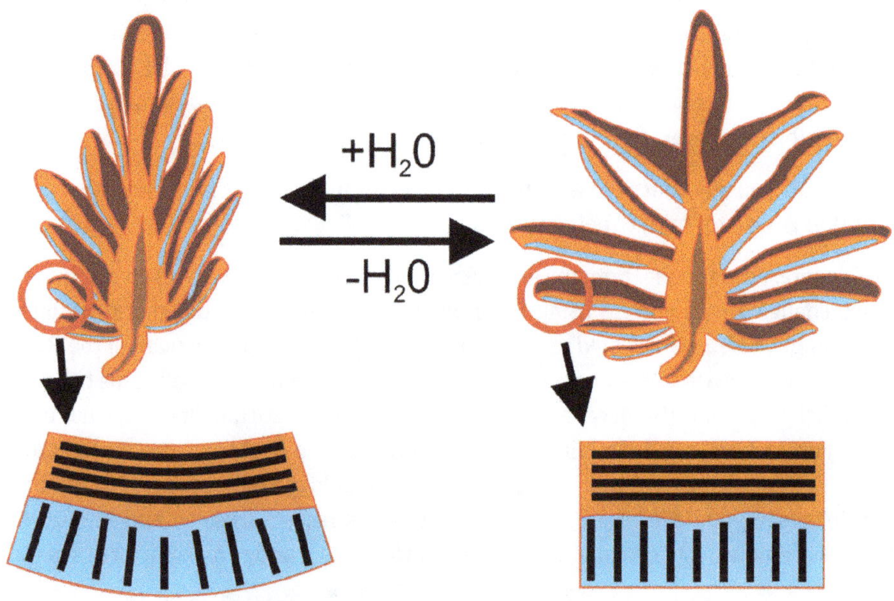

Fig. 2: Hygroscopic opening of pine cones. These cones are closed when wet and open when dry for seed release. Sketches of how different orientations of the cellulose microfibrils inside different layers gives rise to bending motion upon changes in moisture content. From: [Dunlop, Weinkammer, and Fratzl 2011, p. 76, fig. 7]. License: CC BY-NC-ND 3.0 (https://creativecommons.org/licenses/by-nc-nd/3.0/).

These actuation processes occur in a dead material, so there is no metabolism going on anymore in these objects. The seeds of course are still living, but the material surrounding them is dead. Somehow, the activity or potential for activity is in-built into the structure, and when a stimulus comes along certain components will swell or change their volume. They do it within the constraints of the surrounding tissues that maybe do not swell so much or maybe more. Depending on how these are arranged in space, you can get very different types of 3D movements and shape changes. What interests me is to sort of step back a little bit. Hence, on the one hand to work together with experimentalists who study these sorts of plant-based materials or observed structures and to assist them in sort of analyzing the process of movement and actuation that can be observed, but on the other hand to take a step back and ask how the geometric arrangement of say swellable and non-swellable components controls the macroscopic shape change that you observe. And, of course, how you can use that in a design process. For example, if you say you want to have a certain shape change, is there a unique way of achieving it? In principle, there are different ways of achieving macroscopic deformation. What is interesting is that it turns out that there is no unique solution for a particular shape change [Turcaud et al. 2011; Reyssat and Mahadevan 2009; Fratzl and Barth 2009]. Thus, you can have the classic example of a bilayer where you can have a sharp interface between material A and material B, and that will give a certain curvature. But you can also introduce a gradient to your material properties going from material A to material B with no sharp interface, and it can give you the same curvature as well. You simply have to play with the swellability of the different components in the right way. I guess trying to explore these is all one thing. What is very important for understanding this is a description of the mechanical properties of a material, because you need the elastic modulus to describe how stress relates to strain or force relates to displacement in the material. And then you need a description of how the material changes its volume with respect to some external stimulus, whether the latter is temperature or swellability due to changes in humidity or something of the sort. Hence, you have something that depends on the stimulus and you have your mechanical properties, and then you can in principle take tools that you use for engineering – such as the finite element method, a numerical method for solving partial differential equations in two or three space variables [Zienkiewicz et al. 2005] – and then stimulate your objects and see how they change shape and form. The geometry comes in the context of boundary conditions. Not only do you have shape change of some composite materials consisting of different components but you also have the fact that, if you constrain an object inside some physical environment, you will then influence the sorts of shape changes that are possible – and you will also push the shape changes to go in certain directions versus in others. In some of the work that we were doing together with Leonid Ionov from the Leibniz Institute of Polymer Research in Dresden (Germany), we were observing that by spatially controlling the speed at which an environmental signal comes into effect – in this case the diffusion of water through its bilayers – you could control the process by which your actuation occurs and you can push your actuation in certain directions, which would not necessarily be the lowest energy configuration (see Fig. 3) [Stoychev et al. 2012; Stoychev et al. 2013].

In this way, by playing with boundary conditions – which by definition means playing with the geometry of the surroundings – you can then influence your shape changes. This is also a very interesting direction to go. In this sense, you study materials which

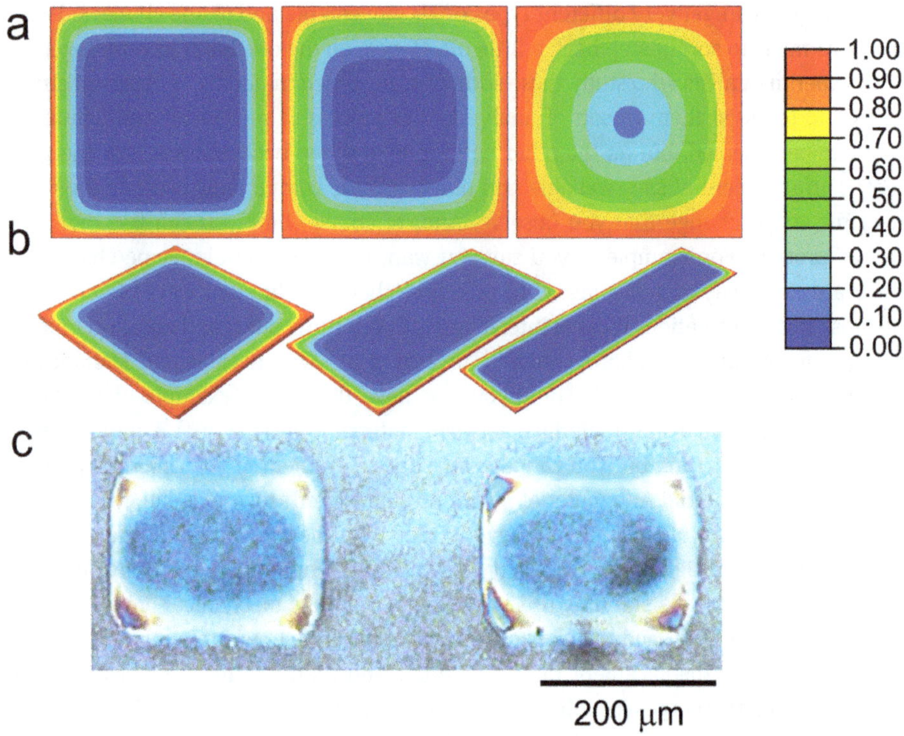

Fig. 3: Color map of the calculated swelling (from 0 to 1) controlled by water diffusion in the active monolayer with a lateral constant boundary condition (blue is non-swollen) dependent on (a) time and (b) shape obtained by finite element simulations as well as (c) experimentally obtained microscopy snapshot of swollen P(NIPAM-BA) – PMMA bilayer after a few seconds of swelling. From: [Stoychev et al. 2012, p. 3930, fig. 6]. © 2012, American Chemical Society.

have the potential for activity built in, and it is the arrangement of the materials that determine in what direction shape changes will occur.

KK: That is one focus of your research. And the other focus?

JD: The other focus goes in the direction of living systems. The goal is to understand how the shape of the environment controls processes of growth, processes of cell migration, processes of cell patterning. When I started off as a postdoc in Peter Fratzl's Department of Biomaterials at the MPICI in Potsdam (Germany), I met Monika Rumpler from the Ludwig Boltzmann Institute of Osteology in Vienna (Austria), who was using 3D printing to create scaffolds for tissue engineering. From a certain perspective, this was a very applied topic as the ultimate goal would be to build a scaffold that you can implant in a fracture gap in bone and then hopefully create an ideal environment to stimulate cells to move in and produce new bone. What Monika did was that she first started looking at rather complex scaffolds with multiple interconnections, with high porosity. The research group could show that the cells would grow inside these scaffolds, but it was difficult to really observe individual holes or pores inside these scaffolds and what was going on there. They then had the great idea to use rapid prototyping or 3D printing to create single pores or single holes where you can explore the role of shape and the shape of this pore on how cells grow and pattern [Rumpler et al. 2008]. What got me excited when I was working with them was that the structures that appeared in these pores seemed describable with very simple models – either models which you normally use to describe, for instance, the shape of fluids on surfaces, or models that you use to describe crystal growth and similar things coming from the materials sciences or physics (see Fig. 4) [Dunlop et al. 2010; Bidan et al. 2012; Gamsjäger et al. 2013; Fischer et al. 2015; Lecuit and Lenne 2007]. I guess it was one of these strange eureka-style moments where you sit down and start doing some calculations and you realize we get simulated shapes that match very well the shapes of the experiments. This is how I got into this topic. I began to become interested firstly in developing theoretical models to understand how cells interact with surfaces and shapes where the shapes are much bigger than the size of individual cells. And then with time I got more and more into designing experiments to test models, and then there was this nice feedback process between experimenting, simulating, and modeling.

When one thinks about the role of geometry, I guess the topic or our interest was how do the external boundary conditions of a growing tissue, hence a growing set of cells – how does this shape influence the collective behavior of cells, inside or on these objects? Can you come up with models that give you an average behavior of cells and tissues at a variety of different length scales that you can then travel back and link to experiments? There is a lot to be done related to understanding how these geometric boundary conditions influence cells. So, what is the signaling process? Is it that cells are attached to the substrate and can then somehow sense local geometric features like local curvature? When we talk about length scales that are much larger than the single cell, the cells in principle would not respond to this. This is something that you can only get when many cells operate together, and then you arrive at this whole field of collective behavior. I think that is where these models of swarming behaviors could be very interesting.

KK: This mathematical modeling of growth processes reminds me of D'Arcy Wentworth Thompson's *On Growth and Form* (1917/1942), where he wanted to unify biology, physics, and mathematics in order to find universal laws for the morphology and structural development of living beings. Of course, with today's mathematical tools and high-tech instruments like scanning electron microscopes, this is a different situation. Could your research on 'MorphoPhysics' (as well as the research being carried out by other scientists in this field) possibly lead to a geometry of growth processes in living materials?

JD: In some sense, I suppose. There are several questions that would be behind it. What we saw, at least in our first experiments, is that you could very well describe

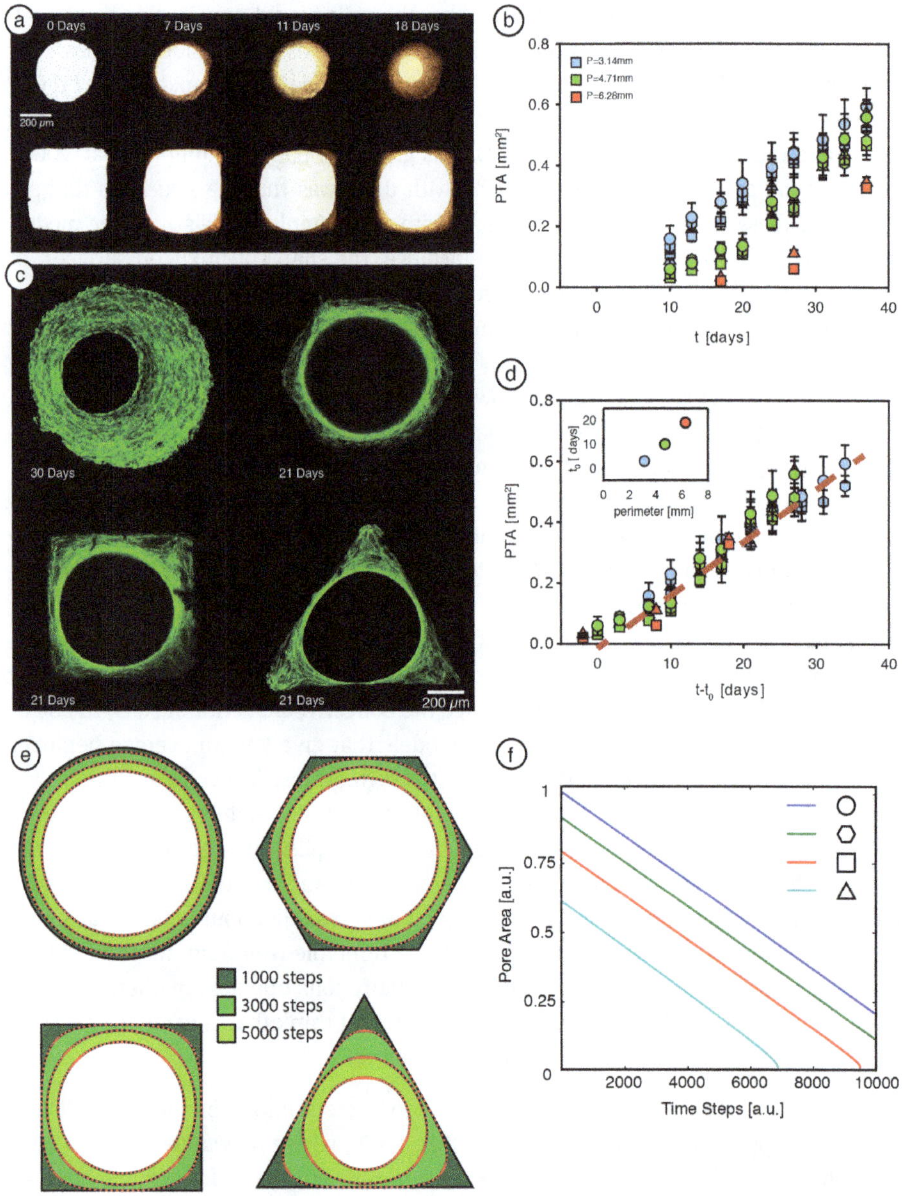

Fig. 4: Experiments and simulations of tissue growth in convex pores. (a) Phase contrast microscopy images of two pores within hydroxyapatite (HA) scaffolds during MC3T3-E1 cell culture experiments. Tissue appears brown, scaffold is black, and medium appears white (images produced by C. Bidan). (b) Projected tissue area as a function of culture time for four different pore cross sections (circular, hexagonal, square, and triangular) and three different pore perimeters (3.14, 4.71, and 6.28 mm). Error bars indicate standard error ($n = 10$) (data from [Rumpler et al. 2008]). (c) Laser scanning confocal microscopy images of four different HA pores containing tissue

the shape changes with models that are essentially geometric descriptions. Thus, this curvature-driven growth model that we had is a model that takes an interface, which can have a varying curvature, and you would allow that shape to change or move as a function of that local curvature [Rumpler et al. 2008; Bidan et al. 2012; Dunlop et al. 2010]. The simplest version of such a model is that a point on an interface moves at a rate directly proportional to its local curvature. You then calculate the motion for all points on the interfaces and then move them. This then gives you a new shape, which would subsequently change iteratively. This works quite well up to a certain point, at least in the system we were working with or the systems we have looked at. The question then is: what is actually controlling it or why is it that in the majority of our experiments after about 20 or 25 days of cell culture the growth process starts slowing down? What is going on? Is this a process by which the cells are differentiating, gaining new characteristics so that they are not behaving in the same way they were earlier on in the culture, or is it a process by which the geometry has changed so much that there is this feedback effect, the cells have created a geometry where they do not want to grow anymore? And then there is also the question of what is responsible for this geometry sensing. How are the cells actually doing it? Are they doing it by means of collective behavior, because they are attached to each other, glued to each other by these adhesive molecules? Why do the cells then start exerting tension on their surroundings? Is this due to their cytoskeleton? They start pulling on each other. The underlying causes are to be understood in the direction of mechanics or forces between interacting objects in a constrained environment.

KK: In the description of growth processes or actuation phenomena, there are different explanations involved that come from different disciplines like physics, biology, and chemistry, as well as mathematics. I wonder how these different understandings can come together. Joanna Aizenberg and Alison Grinthal from Harvard University, in a paper on responsive materials from 2013, wrote that the microarchitecture of living materials should not be conceived as a static configuration or a passive scaffold but more like a "complex multi-scale feedback dance" [Grinthal and Aizenberg 2013, p. 7073]. In their article, they also call it a "dense interconnected jungle" of

Fig. 4 (continued)
produced by MC3T3-E1 cells in which actin has been stained (green) with phalloidin-FITC. The image of the circular pore contains tissue grown for 30 days; the other pores show images after 21 days of culture. (d) The data from (b) is replotted as a function of culture time minus lag time, which is the time taken for linear growth to start. The red dotted line indicates the linear growth of tissues on hydroxyapatite scaffolds in convex-shaped pores. (e) Predicted tissue geometries according to the curvature-driven growth model [Rumpler et al. 2008] at three different time points. Each shape has the same starting perimeter. (f) Calculated remaining pore area (white) as a function of time for the simulations shown in (e) From: [Dunlop 2015, p. 20, figs. 2–6]. © John Dunlop.

"reactants and catalysts" – to which one could maybe add: of forces and environmental conditions – hence very different signaling.[8]

JD: I like the imagery of that. There are quite a few groups that are working in this direction, especially in the context of regenerative medicine or in cell cultures – one could mention Anja Geitmann, Christopher Chen, Celeste Nelson, Xavier Trepat, Dennis Discher, Carl-Philipp Heisenberg, Viola Vogel, to name just a few. These groups are trying to understand what is the role of mechanics and what is the role of chemistry or biochemistry on cell behavior or tissue behavior or a regenerating organ. If you have a broken joint or bone then you have by definition mechanical signals; your muscles contract and there will be some sort of mechanical response. One knows that cells respond to that. But they also respond to various growth factors. What is not entirely clear is how these things work. For example, your mechanical signaling would tend to force a cell or an organ to go down process A, and chemical signaling forces it to go down process B. At what point do you switch from one to the other? These sorts of things are difficult to isolate. Indeed it is quite likely that, if you make the right combination, you might get a tissue to go down process C, which goes in a completely different direction.

It is interesting that even with static materials – although you could also say that they are in some sense active – there is this point where you can do experiments on a material and calculate the stress to rupture. What is the loading the material can handle? You can design your object for this type of loading. You can then do the same sort of experiments for a corrosive environment for example. So we need to be able to measure the concentration of acid that a material can withstand before it begins to be eaten away. Following on this, you can – as an engineer – put in your design conditions based on these two criteria and build your components. In some fairly well-known examples, you get to the point where your materials start breaking at stresses and concentrations much lower than what you planned for because you have these combined phenomena [Hänninen 2003]. This process is called stress corrosion cracking, which has been found to be responsible for some pipeline and bridge failures. It means that you must develop whole new scenario criteria to describe this process.

This is what is being done in engineering and the same is required for understanding this in biology. It is known that mechanics can influence cell behavior. But it is also known that mechanics can influence chemistry. There is this great work from Viola Vogel's Laboratory of Applied Mechanobiology at the ETH Zürich where they look at the unfolding of protein complexes when you load them with force.[9] You can see changes in the chemical activity of these complexes due to the presence

8 See [Grinthal and Aizenberg 2013, p. 7072]; see also the interview with Joanna Aizenberg in this volume.

9 For the research of the Laboratory of Applied Mechanobiology, see online: https://appliedmechanobio.ethz.ch/ (accessed May 5, 2021). See also: [Vogel 2006; Vogel and Sheetz 2009].

of external forces. This gives rise to a molecular mechanism whereby mechanical signaling can change local chemistry and mechanics. Then you have this very interesting feedback effect (see Fig. 5) in which cell response to a physical environment changes that environment.

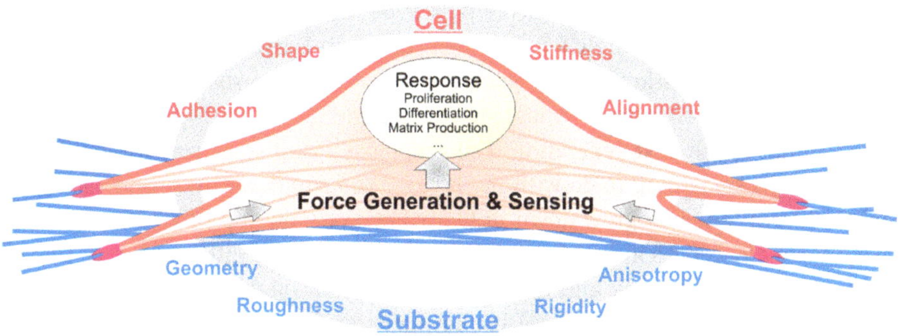

Fig. 5: Cells attached to a substrate respond to its physical properties via the generation of forces applied by the cytoskeleton to the focal adhesions. These mechanical signals give rise to responses in cell behavior, for example with changes in proliferation, differentiation, and extracellular matrix production. From: [Kollmansberger et al. 2011, fig. 1]. Reproduced by permission of The Royal Society of Chemistry.

Whether this feedback process is deterministic or not, I think one could say that the process by which you load up a molecule and describe force displacement or the unfolding process is one thing, and the process by which the chemical reactions would occur is another thing, but in the end you have to describe both together in one system to be able to be predictive.

I think that there is still a lot of interesting work to be carried out. This is also true in the applied direction, where people look at these so-called 'self-healing materials' – which in some sense you could also describe as being a component of active materials – that will respond to the presence of a defect by allowing new material to polymerize and reseal the crack. These filler materials can fill the gap either by surface tension, or by being actively pumped into the fracture gap of an engineering object [Hager, Zwaag, and Schubert 2016; Thomas and Surendran 2020]. But they have also very interesting feedback effects, because the healing material is never going to be the same as the original material that was there. Due to the fact that in your first cycle you can crack and heal, it is quite likely that in the second cycle the behavior will be different. Looking at this over a long period of time goes in the same direction. You need to consider both. Not only the chemistry of the healing process but also the mechanics of the cracking process and how all of these fit together. That would be a fascinating addition to this concept of active matter, which I had not really thought about before. In this respect, healing would be part of that as well.

KK: When you are dealing with living systems it is not only the relation of chemistry and mechanics that is relevant but also the role of genetics. What happens to the feedback dance if you manipulate the genetics of cells? How does this affect the chemical signaling, geometric signaling, or mechanical signaling?

JD: The classical, neo-Darwinistic style of describing living systems would say that everything is written in the genetic code. Maybe that viewpoint of how shape appears in biology is not entirely fair with respect to neo-Darwinism, in the sense that there it is more a question that the genetic code gives rise to some sort of organism structure, which then is more or less successful, and if it survives it can pass these genes further to the next generation. The shape itself is in some way irrelevant unless it has got something to do with the fitness. You could imagine two genetically identical trees: one you grow down in the valley floor, one up in a crack in a cliff face. Due to differences in environment, they will create very different forms. Due to the constraints of the rock and high winds, one plant might be highly deformed, leaning to one side, while the tree in the plains or protected environment grows high and straight. What is clear is that the environmental constraints can clearly influence the shape. From that perspective, you could say that genetics gives an organism the tools whereby the various processes can occur in reaction to the various environmental constraints, which can be immense in size.

I think one key point is related here to what Thompson in his book *On Growth and Form* wrote on living organisms: they operate in a physical world so there are physical constraints and as such these constrains will put limitations on what is actually possible.[10] That being said, the growing organism will be acting against that. Through the influx of energy into a system, through the influx of molecules from the surroundings, an organism will be able to create shapes. But, if you have a floppy membrane surrounding a single cell, the membrane has a certain bending energy, a certain stretching energy. This will have an influence on the sorts of shapes that it can then achieve. The same goes for a growing organism: there are certain processes that will occur, but these processes are constrained inside this physical world. The key point in general is to find out to what extent these constraints influence the sorts of processes that occur. If you go back to the swarming models of Tamás Vicsek and others [Vicsek 1995, footnote 7], if you allow swarming in confined spaces, the fact that you have physical walls or you can constrain objects to a curved surface will influence the sort of patterns in self-organization that you can achieve. This is again

10 See [Thompson 1945, p. 10]: "Cell and tissue, shell and bone, leaf and flower, are so many portions of matter, and it is in obedience to the laws of physics that their particles have been moved, moulded and conformed. [...] Their problems of form are in the first instance mathematical problems, their problems of growth are essentially physical problems, and the morphologist is, *ipso facto*, a student of physical science."

something you will observe with cells, bacteria, or the patterning of living interacting objects in space.

I do not know if it is the environment influencing the genetics, although there are studies that show that in principle this is possible – in epigenetics say. The view of how organisms can change with time is undergoing expansion. But what we are in principle focusing on is how this given organism responds in terms of its shape to the physical environment. The driving force, if you like, for shape changes, which would consist of how cells interact with their environment. If you provide constraints, this can then influence on how the cells organize. If you give an anisotropic environment where cells are aligned on a fiber-like structure, they will also align, but, if they then start producing extra cells in a matrix, such materials will also have an influence [Bidan et al. 2016; Ehrig et al. 2019]. They will be influenced by the orientation of the cells, which will then feedback on the next layer of cells that will feel again this anisotropic environment. You have this intense feedback process occurring between environmental conditions and behavior of cells. Environmental conditions are the local conditions around cells and growing tissues, which will then change the local physical environment once again. This type of feedback process is very interesting. If you start pushing cells in a certain direction, you can try to get them aligned in a particular way. Your genetic description would say that they would try to switch back to their desired or idealized direction. To what extent would this mechanical or geometrical signal then override what is occurring at the genetic level? I am not sure that this has been really understood.

KK: I would like to come back to the relation of living matter to its environment in a different way. Living entities are described in energetic terms as open systems and, moreover, as systems far from equilibrium, because they are incessantly dependent on energy and material inflows from their environment and therefore have to come to terms with a wide variety of environments. Living systems adapt either by changing their internal structures or by building external structures to regulate the influx – just as the beaver constructs a burrow and thus adapts the environment for itself, creating its own partially stable niche. The biofilm of bacteria would be another example. J. Scott Turner has written on this kind of 'niche construction,' and he describes it as a "[p]hysiology of the [e]nvironment" [Turner 2000, p. 7].[11] What I am heading at is: because living systems are open systems, they have the necessity and the possibility to actively design their boundaries, but they cannot fully close these boundaries – and this means that they always have to deal with instabilities, both in a positive sense (they evolve by adapting to changing conditions) and in a negative sense (too much instability can destroy the system). I wonder how this productive role of instability in structural development and environmental relations is reflected in the research on active materials?

11 For an application of Turner's ideas on biofilm (as an architecture that balances the instabilities that appear because of changes in the environment), see: [Hengge 2020].

JD: There is wonderful work on instabilities done in and around the group of Lakshminarayanan Mahadevan from the Department of Physics at Harvard University.[12] In his Soft Math Lab, they looked at patterning – say of the villi in the lining of the gut or stomach (see Fig. 6). What they could show was that the sort of complex oscillations could be genetically determined, but the immediate physical cause was a buckling instability [Savin et al. 2011].[13] Picking up your word instability here, this is a mechanical instability, which is then coupled back to growth processes and other similar things.

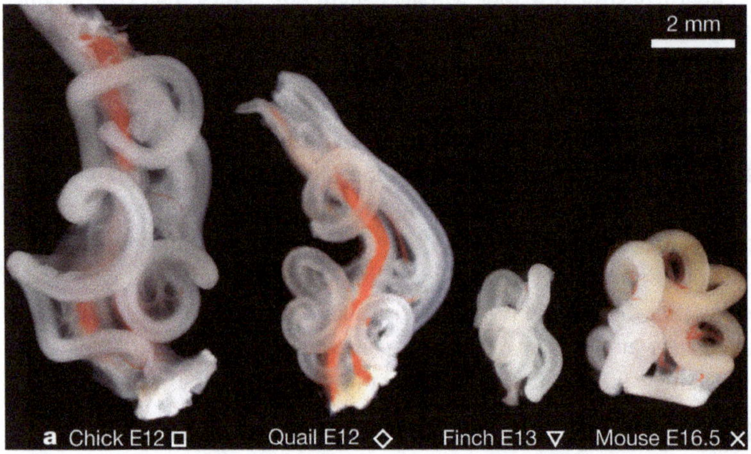

Fig. 6: Gut looping patterns in the chick, quail, finch, and mouse (to scale). The comparison shows qualitative similarities in the shape of the loops. Reprinted by permission from Springer Nature: [Savin et al. 2011, fig. 5a].

From the perspective of physics or mathematics, what is nice about these kinds of concepts is that you do not need that much information to create a very complex structure. You have this very complex patterning that you see, highly convoluted surfaces, but in principle, it comes from relatively simple phenomena. The information that an organism needs is what thickness of tissue needs to be grown on what other thickness of other tissue – and then due to differential growth you come up with this pattern formation. And it comes automatically out. That is a very efficient way of coding for structure. You do it in the properties of the materials that are stuck together and how they are arranged in space and that gives you an extra dimension to work with. This feedback effect between your external structure and your structure

12 https://softmath.seas.harvard.edu/ (accessed May 5, 2021).

13 For an application of this research to the brain, see also: [Tallinen et al. 2016].

which is influenced by your organism. The examples of Scott Turner are something that is fascinating to explore, because the example of structures such as nests and burrows built or in some way “grown” by animals can then give rise to very interesting pattern formation at much larger length scales than what one would expect if it was just deterministic growth alone. It becomes much more efficient in terms of the information storage that is required, because it is built into the physics which are already there in the environment or in our world that we are living in.

KK: My last question aims at the future horizon of the research on active materials. What are the next challenges in this field of research?

JD: A big challenge that I would at least like to be involved with is the question of how extracellular matrix components organize overlarge length scales, giving rise to certain types of structures, function, and form. Again, this is about internal structures of tissues. What is fairly well understood is how a certain arrangement of component materials gives rise to a certain function. That you can do because it is a static situation. But then how does this actually come to be? That is a big question. Liquid crystal modeling here is very interesting because it seems to suggest at least that there is a certain self-organization process that is occurring in controlling that. An alternative is to use agent-based models that can explicitly deal with the extracellular matrix that is produced when tissue grows. This would be for example collagen or fibronectin when you are talking about bone or skin. In plants, you have cellulose microfibrils, chitin in arthropods, all of which are further complicated by the mineralization process that may occur in the environment created by cells. How these extracellular matrix components are organized in space will have an influence on the properties of the environment for future generations of cells. So again, you have this feedback effect. Describing this at local level is incredibly difficult. Finding a way of accomplishing that is certainly part of the research of the future. It is a question of coming up with suitable experiments where you can really test these ideas in 3D. I do not see much that is being done or fully understood in this direction. One other way of describing the patterns that are observed in these extracellular matrix components is to make analogies with other sorts of similar patterns that are being observed in synthetic material, for instance in liquid crystals [Gompper et al. 2020]. There is the work of Pierre-Gilles de Gennes and Jacques Prost, who have pushed forward this idea, but observing these processes live would be a key step in understanding [De Gennes and Prost 1993]. Why is that important? It is important in terms of basic biology and also in terms of medicine and in terms of the understanding of certain types of diseases or regeneration processes. Ideally, you want tissues to regenerate in the patterning they had beforehand. If they do not you sort of end up with a scar tissue with different properties, and that can be problematic. If one has an understanding of how these sorts of organizational processes during growth or during remodeling of tissues actually occur, then you can maybe apply that to artificial systems of course.[14] That could be an ultimate dream.

What one can always win from studying active matter – and in particular if you go in the direction of understanding how the physical constraints surrounding this complex interacting system can have a huge effect on how growth occurs – is how patterning occurs at different length scales and different levels. Studies in this direction can have very wide implications. If, for instance, you look at the geometric constraints here in Salzburg, where you have mountains that surround the area, and regions where you cannot build, that means the town is built around certain types of constraint, which have an effect on things like transport and how the town can grow. It is not just a deterministic process where the growth of a town is in the hands of city planners. The surroundings have an implicit effect on how growth can occur. And this is something that is generalizable over many length scales and many different fields. Probably in the social sciences you could also have these sorts of concepts where constraints due to certain topics that are not able to be discussed in a particular environment might force certain ideas to come out preferentially. For example, constraints of media: the constraint of the way we read, the constraint of a printed book versus a digital object – these constraints change how that media can actually be presented – the constraints of film, where in principle you will not rewind to see things again, because typically it goes from start to finish without any or much interaction. These sorts of things are all linked to the studying of active matter.

Bibliography

Ambrosi D, Ateshian GA, Arruda EM et al. Perspectives on Biological Growth and Remodeling. Journal of the Mechanics and Physics of Solids 2011, 59 (4), 863–883. DOI: 10.1016/j.jmps.2010.12.011.

Bidan CM, Kollmannsberger P, Gering V et al. Gradual Conversion of Cellular Stress Patterns into Pre-Stressed Matrix Architecture During in Vitro Tissue Growth. Journal of the Royal Society Interface 2016, 13 (118), 20160136, DOI: 10.1098/rsif.2016.0136.

Bidan CM, Kommareddy KP, Rumpler M et al. How Linear Tension Converts to Curvature: Geometric Control of Bone Tissue Growth. Plos One 2012, 7 (5), e36336. DOI: 10.1371/journal.pone.0036336.

Burgert I. Plant Biomechanics and Biomimetics. Munich: Max Planck Institute of Colloids and Interfaces, (2021. Accessed May 5, 2021, at https://www.mpikg.mpg.de/5788791/plant-biomechanics-and-biomimetics.)

Burgert I, Dunlop JWC. Micromechanics of Cell Walls. In: Wojtaszek P, ed. Mechanical Integration of Plant Cells and Plants. Series: Signaling and Communication in Plants. Vol. 9. Berlin, Heidelberg, Springer, 2011, 27–52, DOI: 10.1007/978-3-642-19091-9_2.

14 Ideas of this sort can be seen in the work of André Studart (ETH Zürich), who tries to mimic biological structures to create new functional materials [Ferrand et al. 2015].

Camazine S, Deneubourg JL, Franks NR, Sneyd J, Theraula G, Bonabeau G. Self-Organization in Biological Systems. Princeton, Princeton University Press, 2003.

Dunlop JWC. The Physics of Shape Changes in Biology. Habilitationsschrift zur Erlangung der Lehrbefähigung für das Fach Physik. University of Potsdam, October 2015. (Accessed May 5, 2021, at: https://publishup.uni-potsdam.de/opus4-ubp/frontdoor/deliver/index/docId/9655/file/dunlop_habil.pdf.)

Dunlop JWC. Biomimetic Actuation and Tissue Growth. Munich: Max Planck Institute of Colloids and Interfaces, (2021. Accessed May 3, 2021, at https://www.mpikg.mpg.de/5896344/Biomimetic-Actuation-and-Tissue-Growth.)

Dunlop JWC, Bréchet Y, Legras L. A Coupled Recovery/Recrystallisation Model for Zirconium Alloys. Influence of Hydrogen. Materials Science Forum 2004, 467-470, 629–634. DOI: 10.4028/www.scientific.net/MSF.467-470.629.

Dunlop JWC, Fischer FD, Gamsjäger E, Fratzl P. A Theoretical Model for Tissue Growth in Confined Geometries. Journal of Mechanics and Physics of Solids 2010, 58 (8), 1073–1087. DOI: 10.1016/j.jmps.2010.04.008.

Dunlop JWC, Fratzl P. Multilevel Architectures in Natural Materials. Scripta Materialia 2013, 68 (1), 8–12, DOI: 10.1016/j.scriptamat.2012.05.045.

Dunlop JWC, Ionov L. Shape-Programmed Folding of Stimuli-Responsive Polymer Bilayers. ACS Nano 2012, 6 (5), 3925–3934, DOI: https:/doi.org/10.1021/nn300079f.

Dunlop JWC, Weinkamer R, Fratzl P. Artful Interfaces within Biological Materials. Materials Today 2011, 14 (3), 70–78, DOI:10.1016/S1369-7021(11)70056-6.

Dunlop JWC, Zickler GA, Weinkamer R, Fischer FD, Fratzl P. The Emergence of Complexity from a Simple Model for Tissue Growth. Journal of Statistical Physics 2020, 180, 459–473, DOI: 10.1007/s10955-019-02461-7.

Ehrig S, Schamberger B, Bidan CM et al. Surface Tension Determines Tissue Shape and Growth Kinetics. Science Advances 2019, 5 (9), eaav9394, DOI: 10.1126/sciadv.aav9394.

Ferrand HL, Bouville F, Niebel TP, Studart AR. Magnetically Assisted Slip Casting of Bioinspired Heterogeneous Composites. Nature Materials 2015, 14 (11), 1172–1179.

Fischer FD, Zickler GA, Dunlop JWC, Fratzl P. Tissue Growth Controlled by Geometric Boundary Conditions: A Simple Model Recapitulating Aspects of Callus Formation and Bone Healing. Journal of the Royal Society Interface 2015, 12 (107), 20150108, DOI: 10.1098/rsif.2015.0108.

Fratzl P, Barth FG. Biomaterial Systems for Mechanosensing and Actuation. Nature 2009, 462, 442–448, DOI: 10.1038/nature08603.

Fratzl P, Weinkamer R. Nature's Hierarchical Materials. Progress in Materials Science 2007, 52 (8), 1263–1334, DOI: 10.1016/j.pmatsci.2007.06.001.

Gamsjäger E, Bidan CM, Fischer FD, Fratzl P, Dunlop JWC. Modelling the Role of Surface Stress on the Kinetics of Tissue Growth in Confined Geometries. Acta Biomaterialia 2013, 9 (3), 5531–5543, DOI: 10.1016/j.actio.2012.10.020.

Gennes P-G de, Prost J. The Physics of Liquid Crystals, Oxford, Clarendon Press, 1993.

Gompper G, Winkler RG, Speck T et al. The 2020 Motile Active Matter Roadmap. Journal of Physics: Condensed Matter 2020, 32, 193001, DOI: 10.1088/1361-648X/ab6348.

Grégoire G, Chaté H. Onset of Collective and Cohesive Motion. Physical Review Letters 2004, 92 (2), 025702, DOI: 10.1103/PhysRevLett.92.025702.

Grinthal A, Aizenberg J. Adaptive all the Way Down: Building Responsive Materials from Hierarchies of Chemomechanical Feedback. Chemical Society Reviews 2013, 42 (17), 7072–7085, DOI: 10.1039/C3CS60045A.

Guiducci L, Weaver JC, Bréchet YJM, Fratzl P, Dunlop JWC. The Geometric Design and Fabrication of Actuating Cellular Structures. Advanced Materials Interfaces 2015, 2 (11), 1500011, DOI: 10.1002/admi.201500011.

Hager MD, Zwaag S van der, Schubert US. Self-Healing Materials. Cham, Springer, 2016.
Halatek J, Frey E. Rethinking Pattern Formation in Reaction-Diffusion Systems. Nature Physics 2018, 14, 507–514, DOI: 10.1038/s41567-017-0040-5.
Hänninen HE. Stress Corrosion Cracking. In: Milne I, Karihaloo B, Ritchie RO, eds. Comprehensive Structural Integrity. Vol. 6. Environmentally Assisted Fatigue. Ed. Petit J, Scott PM. Amsterdam et al., Elsevier, 2003, 1–29.
Harrington MJ, Razghandi K, Ditsch F et al. Origami-Like Unfolding of Hydro-Actuated Ice Plant Seed Capsules. Nature Communications 2011, 2, 337, DOI: 10.1038/ncomms1336.
Hengge R. Linking Bacterial Growth, Survival, and Multicellularity – Small Signaling Molecules as Triggers and Drivers. Current Opinion in Microbiology 2020, 55, 57–66, DOI: 10.1016/j.mib.2020.02.007.
Huss JC, Fratzl P, Dunlop JWC, Merritt DJ, Miller BP, Eder M. Protecting Offspring Against Fire: Lessons from Banksia Seed Pods. Frontiers in Plant Science 2019, 10, 283, DOI: https:/doi.org/10.3389/fpls.2019.00283.
Khan A, Jawaid M, Raveendran SN, Asiri AMA, eds. Self-Healing Composite Materials. From Design to Applications. Duxford et al., Woodhead Publishing, Elsevier, 2019/2020, DOI: 10.1016/C2018-0-02118-0.
Kollmannsberger P, Bidan CM, Dunlop JWC, Fratzl P. The Physics of Tissue and Extracellular Matrix Organisation: How Cells Join Forces. Soft Matter 2011, 7 (20), 9549–9560, DOI: 10.1039/C1SM05588G.
Kondo S. The Reaction-Diffusion System: A Mechanism for Autonomous Pattern Formation in the Animal Skin. Genes to Cells 2002, 7 (6), 535–541. DOI: 10.1046/j.1365-2443.2002.00543.x.
Lecuit T, Lenne PF. Cell Surface Mechanics and the Control of Cell Shape, Tissue Patterns and Morphogenesis. Nature Reviews Molecular Cell Biology 2007, 8, 633–644. DOI: 10.1038/nrm2222.
Malacincski GM, Bryant SV, eds. Pattern Formation. A Primer in Developmental Biology. New York, London, Macmillan Publishing Company, Collier Macmillan Publishers, 1984.
Meinhardt H, Klingler M. A Model for Pattern Formation on the Shells of Molluscs. Journal of theoretical Biology 1987, 126 (1), 63–89, DOI:10.1016/S0022-5193(87)80101-7.
Menon GI. Active Matter. In: Krishnan JM, Deshpande AP, Kumar PB, eds. Rheology of Complex Fluids. New York, Springer, 2010, 193–218.
Nüsslein-Volhard C, Singh AP. How Fish Color their Skin: A Paradigm for Development and Evolution of Adult Patterns. BioEssays 2017, 39 (3), 1600231, DOI: 10.1002/bies.201600231.
Reyssat E, Mahadevan L. Hygromorphs: From Pine Cones to Biomimetic Bilayers. Journal of The Royal Society Interface 2009, 6 (39), 951–957, DOI: 10.1098/rsif.2009.0184.
Rumpler M, Woesz A, Dunlop JWC, Dongen JT van, Fratzl P. The Effect of Geometry on Three-Dimensional Tissue Growth. Journal of the Royal Society Interface 2008, 5 (27), 1173–1180, DOI: 10.1098/rsif.2008.0064.
Savin T, Kurpios NA, Shyer AE, Florescu P, Liang H, Mahadevan L, Tabin CJ. On the Growth and Form of the Gut. Nature 2011, 476, 57–62, DOI: 10.1038/nature10277.
Stoychev G, Zakharchenko S, Turcaud S, Dunlop JWC, Ionov L. Shape-Programmed Folding of Stimuli-Responsive Polymer Bilayers. ACS Nano 2012, 6 (5), 3925–3934, DOI: 10.1021/nn300079f.
Stoychev G, Turcaud S, Dunlop JWC, Ionov L. Hierarchical Multi-Step Folding of Polymer Bilayers. Advanced Functional Materials 2013, 23 (18), 2295–2300, DOI: 10.1002/adfm.201203245.
Tallinen T, Chung JY, Rousseau F, Girard N, Lefèvre J, Mahadevan L. On the Growth and Form of Cortical Convolutions. Nature Physics 2016, 12, 588–593. DOI: 10.1038/nphys3632.
Thomas S, Surendran A, eds. Self-Healing Polymer-Based Systems. Amsterdam, Oxford, Cambridge MA, Elsevier, 2020, DOI: 10.1016/C2018-0-03002-9.

Thompson DW. On Growth and Form. Cambridge, New York, Cambridge University Press, The Macmillan Company, 1945.
Trepat X, Sahai E. Mesoscale Physical Principles of Collective Cell Organization. Nature Physics 2018, 14 (7), 671–682, DOI: 10.1038/s41567-018-0194-9.
Turcaud S, Guiducci L, Fratzl P, Bréchet YJM, Dunlop JWC. An Excursion into the Design Space of Biomimetic Architectured Biphasic Actuators. International Journal of Materials Research 2011, 102 (6), 607–612, DOI: 10.3139/146.110517.
Turing AM. The Chemical Basis of Morphogenesis. Philosophical Transactions of the Royal Society B 1952, 237 (641), 37–72, DOI: 10.1098/rstb.1952.0012.
Turner JS. The Extended Organism: The Physiology of Animal-Built Structures. Cambridge MA, London, Harvard University Press, 2000.
Vicsek T, Czirók A, Ben-Jacob E, Cohen I, Shochet O. Novel Type of Phase Transition in a System of Self-Driven Particles. Physical Review Letters 1995, 75 (6), 1226–1229, DOI: 10.1103/PhysRevLett.75.1226.
Vogel V. Mechanotransduction Involving Multimodular Proteins: Converting Force into Biochemical Signals. Annual Review of Biophysics and Biomolecular Structure 2006, 35 (1), 459–488, DOI: 10.1146/annurev.biophys.35.040405.102013.
Vogel V, Sheetz MP. Cell Fate Regulation by Coupling Mechanical Cycles to Biochemical Signaling Pathways. Current Opinion in Cell Biology 2009, 21 (1), 38–46, DOI: 10.1016/j.ceb.2009.01.002.
Zienkiewicz OC, Taylor RL, Zhu JZ. The Finite Element Method: Its Basis and Fundamentals. 6th ed. Amsterdam et al., Elsevier Butterworth-Heinemann, 2005.
Zurob HS, Bréchet Y, Dunlop J. Quantitative Criterion for Recrystallization Nucleation in Single-phase Alloys: Prediction of Critical Strains and Incubation Times. Acta Materialia 2006, 54 (15), 3983–3990. DOI: 10.1016/j,actamat.2006.04.028.

John Dunlop studied chemistry and materials engineering at the University of Western Australia, before he moved to Grenoble, France, to start a PhD in Metal Physics with Yves Bréchet. There he studied the microstructure of and the physical changes that could occur in spent nuclear fuel rods during transport. After his PhD, he moved to the MPICI in Potsdam, Germany, to the Department of Biomaterials of Peter Fratzl. After the end of his Humboldt fellowship, he stayed on in Potsdam for almost 10 years working as a group leader researching issues related to understanding the physics of shape changes in living and dead materials. Since 2017, John Dunlop is professor for 'Biological Physics' at the University of Salzburg and heads the 'MorphoPhysics' group there, continuing to work on a variety of topics related to active materials.

Michael Friedman, Jean-François Joanny, Karin Krauthausen

Interview with Jean-François Joanny: Activity, Instabilities, and Defects

MF: The emerging research on active matter offers a new conception of what matter is, what materiality is. But it did not just arrive out of the blue. Professor Joanny, you come from a background of soft matter [Marchetti et al. 2013]. Other people in the field of active matter have other backgrounds, in physics, chemistry, biology, and so on. It could be suggested that this field is an intersection of these domains, or even an effect of these domains reflecting on each other. Could you tell us how you started working in the field of active matter? What were the initial research directions?

J-FJ: As you mention, I had been working since my PhD on soft condensed matter, polymer physics, colloids, wetting phenomena. In 2003, I decided to reorient my research to study the applications of soft matter theory to physical questions raised by biology and I joined the physics department of Institut Curie, where I started working on the actomyosin cytoskeleton. There, in collaboration with Jacques Prost, Frank Jülicher, and Karsten Kruse, we started to make a hydrodynamic theory of what we called active gels [Kruse et al. 2005]. We also rapidly realized the connections with the work of Sriram Ramaswamy, who was working on active liquid crystals, and the fact that the results that we were obtaining for the cytoskeleton were very similar to those described for swimmers. This was indeed a theory of active matter, but the name itself came later. We were talking of active materials. For me, Hugues Chaté is the one who first used the term 'active matter.'

KK: Even before the term 'active matter' was explicitly mentioned in your research, you had already focused on 'active processes' in soft matter materials like gels and red blood cells [Betz et al. 2009; Turlier et al. 2016]. You showed that the flickering of these cells is not related to mere thermal agitation but is a true activity of the cell and linked to the protein-driven nonequilibrium state of the system. Could you explain the characteristics of these active processes? Is this kind of activity a sign of a living system?

J-FJ: Usually the starting point of my research is a biological question. If we focus on the research of red blood cells this is particularly interesting. There were several groups of people claiming that red blood cells were in the process of dying or already dead – in other words that they were like a system in equilibrium. This appeared quite odd to me. There was a young researcher at Institut Curie, Timo Betz, who had discovered a very elaborate way of measuring the fluctuations in the position of the membrane of the red blood cell. He had carried out several beautiful

Acknowledgment: Michael Friedman and Karin Krauthausen acknowledge the support of the Cluster of Excellence 'Matters of Activity. Image Space Material' funded by the Deutsche Forschungsgemeinschaft (DFG, German Research Foundation) under Germany's Excellence Strategy – EXC 2025 – 390648296.

Open Access. © 2021 Michael Friedman et al., published by De Gruyter. BY-NC-ND This work is licensed under the Creative Commons Attribution-NonCommercial-NoDerivatives 4.0 International License.
https://doi.org/10.1515/9783110562064-006

experiments on vesicles, and I suggested that he check with these experiments and the results obtained to ascertain whether red blood cells are alive. He then measured the fluctuation of the cell membrane and the response to any external force, and we summarized the results as follows:

> [The] data sets demonstrate[] the validity of the FDT [fluctuation-dissipation theorem][1] for a purely passive RBC [red blood cell], as expected. In strong contrast, fresh RBCs [...] show a clear violation of the FDT in the low-frequency regime $f < 10$ *Hz* [...]. This result is a direct and conclusive demonstration that an active mechanical process contributes to the flickering of the RBC membrane at timescales above 100 ms. [Turlier et al. 2016, p. 514]

In short, given that the fluctuation was too large, it meant that the red blood cell was active. At the time, I did not imagine that it would be passive. But to show the experiment properly was very difficult. Once we had these results, we sat down and attempted to make a model showing why the red blood cell was out of equilibrium. The way T. Betz performed this was to look at all of the proteins in a red blood cell that would consume energy. Though not many were found, one of these seemed to us more important than the others. That was the protein linking the cytoskeleton of the red blood cell to the membrane. This protein consumes energy, which meant that the chemical reaction was out of equilibrium. Hence, we went back to the theories of the red blood cell but added this fact that the chemical reaction was out of equilibrium. Because the geometry of the red blood cell is rather complicated, this was not very quantitative. It has a shape (see Fig. 1) which mathematically is quite involved. Gerhard Gompper's group in Jülich did some highly complex numerical simulations of the real shape, but by replacing the real shape with a spherical shape, we had all of the other arguments showing that the red blood cell was a system in nonequilibrium, which to me means that it is alive.

KK: This is particularly interesting because one may understand from your research on blood cells that the instability, which is part of this state of nonequilibrium, is not something that the living system wants to avoid at all costs, but rather something that the living system can handle and even benefit from. Could one claim that instability is necessary for the entire functioning of the living system?[2]

1 According to this theorem, which is one of the most important theorems in equilibrium statistical physics, when there is a process that dissipates energy, turning it into heat (e.g., friction), there is a response process related to thermal fluctuations. Ryogo Kubo describes it as follows: the theorem "states that the linear response of a given system to an external perturbation is expressed in terms of fluctuation properties of the system in thermal equilibrium. This theorem may be represented by a stochastic equation describing the fluctuation [...]" [Kubo 1966, p. 255].

2 For example, in another subfield of the research on active matter, that is, in the framework of the research on the dynamics of suspensions of self-propelled particles, Jean-François Joanny and Sriram Ramaswamy note that "[i]nstability and self-generated flow are conspicuous features of active fluids. Indeed, the idea of hydrodynamic instability in living matter, and even the term active stress, goes back at least to the work of Finlayson & Scriven [1969]" [Joanny and Ramaswamy 2012, p. 47].

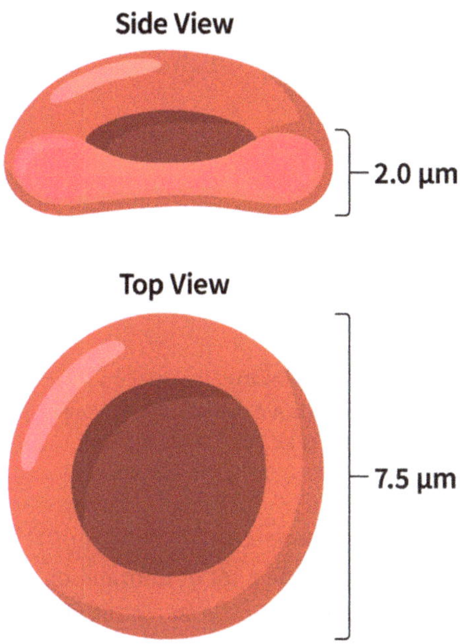

Fig. 1: The shape of a red blood cell. © Graphic drawn by M.F.

J-FJ: One can certainly claim so. With respect to instability, active systems are rather counterintuitive in some cases. Something we did at the very beginning was to look at thin active films. Imagine a surface covered with a film of active fluid and an orientation of the molecules of this fluid which is parallel to the surface. If one were to make a rather naive guess, one might conclude that the vector field for the orientation of the molecules remains constant throughout the film thickness. If one puts this state back into the equation, it works. Thus, there is a solution in which the orientation remains constant. But one has to worry about whether this solution is stable – and it is not. If the film is very thin, this solution remains stable; however, there is a critical thickness, and above this critical thickness, instead of remaining parallel, the molecules in the film tilt in the center, creating a gradient of stress that throws the system out of equilibrium and drives a fluid flow – even in the absence of external pressure. Once we calculated that, this was indeed the case. Numerical simulations made by other groups confirmed this result. But there was still no experiment. Very recently, one of my colleagues, Pascal Silberzan, did experiments with cells in two dimensions, plating cells on a solid surface. The cells were elongated, and we were looking at defects, counting them. Here, topological defects are singular points of the cell orientation field.[3] Such

3 For example, when the vector field V is a velocity field, a singular point is often called a rest point. When V is a field of forces, the singular point is called an equilibrium point.

defects (see Fig. 2) annihilate each other in pairs of opposite signs, and hence we were counting them as a function of time.[4]

Fig. 2: An example of (integral) curves along a vector field. The various defects are outlined with colored circles. From: [Duclos et al. 2017, p. 59, fig. 1(b)]. Republished with permission by the authors of [Duclos et al. 2017].

We came to the conclusion that these defects could provide a lot of information on the behavior of the cells. My colleague decided to plate the cells in a confined region, putting the cells inside a circular region, that is, in a disk-like region. However, nothing remarkable results if you put cells inside a circle, because there is friction of the cell on the substrate, so all of the energy dissipates, and hence the cells behaved like a passive system. So then, we tried to put the defects inside a band and, following this idea, a few weeks later we noted that the band spontaneously shears; without being pushed in any way, on the right-hand side, it was going up and on the left-hand side, it was going down. We realized that this was the same instability we had predicted before – although it was not exactly the same thing because there was friction on the substrate. Our theoretical reflections were ideas we had had more than 10 years before, whereas the experiment itself only took place recently. That said, it fits almost perfectly with the theory.

MF: If we concentrate on the defects, these can be considered and described as purely mathematical objects. This brings us to the relations between mathematics, simulation, and experimentation. As you have just mentioned, some experiments arrive only much later, after the simulation and the theory.

J-FJ: Though sometimes this also happens much earlier. Moreover, I would not claim that what I do is purely mathematics, because we use the general laws of

4 See [Duclos et al. 2017, p. 58]: "On the theoretical side, several recent works have addressed the behaviours of defects in active nematics by adding active terms to the hydrodynamic equations of liquid crystals. [These approaches] […] all converge in showing that active systems differ from passive nematics by [among other things] […] the creation/annihilation of pairs of defects of opposite signs […]."

physics, we use the rules of mechanics, we use the laws of statistical physics, we use the laws of hydrodynamics, and we elaborate on those laws using mathematics. I have a colleague who is an applied mathematician who says that anybody who does not have a lab is an applied mathematician. There is some truth to that, but my interest is not in doing the mathematics but rather in finding explanations that, as one of my colleagues mentioned, could be explained to my grandmother – to explain why a given system behaves in a certain way.

MF: Can one say that the mathematical tools one uses are not sufficient? That other explanations must also be taken into account?

J-FJ: About mathematics, I do not know, but I can certainly say that physics is not sufficient. You have to understand the biology of the system. Chemistry plays a role as well. I would say that I do not think that anything in a biological system would violate the laws of physics, but you have to couple it to chemistry. If you take cells for example, one should also consider their division, but that does not derive from the laws of physics, except when you have a theory at the very microscopic level. I do not believe in the theory of everything.

KK: Could the research on active matter or, more generally, on the activity of materials eventually bridge the gap between the 'life sciences' and physics?

J-FJ: That is indeed an interesting observation, though a physicist becoming interested in biology is not something new. There are famous examples such as Max Delbrück[5] or Erwin Schrödinger. Look at Schrödinger's book *What Is Life?* [Schrödinger 1992 (1944)]; [6] there he speculates among other things about the "hereditary code-script," what we call the genetic code [ibid., p. 20ff.]. Hence, there have always been people interested in explaining biological facts using physical ideas. And such individuals came from different backgrounds. There are also people interested in experimenting in physics from a biological perspective, utilizing optics or even fast optics to look at proteins, spectroscopy, or similar things.[7] Such interests have always existed. The new fact for me is that people like myself believe that what they know and learn from soft matter can be applied to biology. Thus all of these ideas – that you should find simple variables, that you eliminate the ones that are not relevant, that you should consider the symmetries to make things simple – are ideas that come from soft matter. There are a few things that do not exist in the soft matter I was studying previously. One of them is that biological molecules (proteins) have very specific interactions. In many

5 Max Delbrück (1906–1981) was a German-American biophysicist who helped develop the molecular biology research program in the late 1930s.

6 A gene for Schrödinger is a "material carrier of a definite hereditary feature" [Schrödinger 1992 (1944), p. 29].

7 For the biophysical research on optogenetics and the importance of this new field, cf. the commentary in: [Deisseroth 2011]. For the biological perspective, see: [Krueger et al. 2019].

cases, a protein of a given type interacts only with a precise, specific type of proteins and not with all the other ones. Biologists talk about lock and key mechanisms – just as the key must match the lock, the two proteins must also match in order to interact. The second specific feature, which also exists to a lesser extent in soft matter, is the nonequilibrium aspect of active matter. This endows them with many new and very rich properties such as self-assembly or spontaneous motion that does not exist in classical soft matter.

KK: When considering the research on nonequilibrium processes in living and non-living systems, would you say that there has been a big leap in the last ten to twenty years?

J-FJ: There has been progress. There are general laws that can be applied to systems in a nonequilibrium state. I was speaking about the fluctuation–dissipation theorem (in relation to blood cells); our generalization of that works for systems in a nonequilibrium state, is pretty surprising. People could perhaps have discovered this in the nineteenth or early twentieth century, because the ideas are very simple, but they did not.

KK: To continue discussing the theme of nonequilibrium, in your articles you understand active gels as complex materials which have an internal fuel consumption that works as a sort of means of dissipation. In one of your papers, you have described such materials as belonging to a "new class." What is meant here by 'new class'?[8]

J-FJ: There are two meanings. The first concerns the difference between active matter or active materials and what I would call more generally materials out of equilibrium. The second relates to the question of whether such active materials can be used to make anything useful. For example, there are all of these aspects of what is called microfluidics that concern the movement or flow of liquids in very small channels. Some people used an array of rotating biological motors that can drive the flow. Thus – though this is not my expertise – in a sense people attempt to make materials with new properties by using ideas that already exist in biological systems (they mimic biological systems), or they use biological objects such as the motors to make new types of materials.

8 [Kruse et al. 2005, p. 5] (on eukaryotic cytoskeletal gels as 'complex materials'): "We call these gels and more generally all gels working in the presence of a permanent energy consumption 'active' gels." And on p. 14: "We have introduced in this manuscript equations which describe the long wavelength and low frequency behavior of active gels. Although we have written the equations specifically in the case where the activity is due to motor proteins, they should apply, in their principles, to all gels in which a permanent source of dissipation is at work. Such gels define a new class of materials. For instance, a conventional 'physical' polar gel, absorbing a high frequency ultrasonic wave should obey, in the low frequency, long wavelength limit, the set of equations proposed here."

KK: Nonequilibrium systems with sudden generation of flows and stresses can be difficult to describe and explain, either in a quantitative or in a qualitative way. Is this considered an obstacle or, on the contrary, a motivation to continue researching?

J-FJ: The thing that attracted me to active matter is the fact that you find behavior that is rather unexpected. If you work on fluids, you know that shear waves do not propagate. And this is precisely due to viscous dissipation in the fluid. In active fluids, they do this because such dissipation is compensated by injecting energy, and then waves can propagate. To me, this is kind of unusual or something rather unexpected, and researchers are always attracted to such things. For example, active materials or biological systems fluctuate a lot (they jiggle a lot under a microscope) – what certain people call giant fluctuations.[9] When one hears this for the first time, one is quite surprised: why does it violate things I have always believed in? But there is a reasonable explanation for everything. What I mean is that qualitatively there is a reasonable explanation; quantitatively it is more difficult to work out.

MF: A description which is often employed in the discourse on active matter for these kinds of structures is 'smart materials.' Considering the structures appearing in the framework of the research on active matter, can one claim that the questions concerning these structures no longer concentrate on unstable systems behaving randomly, irrespective of any structure whatsoever, but focus instead on systems which are structured in a dynamic way?

J-FJ: I agree that there are structures that appear in dynamical systems out of equilibrium. For example, a property that one wants to study in biological systems is self-assembly. It is much more efficient if you carry this out in systems out of equilibrium, because it can be undone and redone if you do not get it right. In this way, one can find a much richer number of states than those found in systems at equilibrium. And, of course, one of the areas one wants to be interested in is morphogenesis: why do animals look like they do? People have had general ideas for a long time with respect to such things. My claim would be that the research on active matter should concentrate [also] on these ideas.

9 See, for example, the paper from 2006, which associates active phenomena with out-of-equilibrium systems and giant fluctuations [Chaté, Ginelli, and Montagne 2006, p. 1]: "Over the last decade or so, physicists have been looking for common, possibly universal, features of the collective motion of animals, bacteria, cells, molecular motors [...]. Among the emergent properties of these groups of 'active' or self-propelled particles (SPP), distinctively out-of-equilibrium features have been found, such as the existence of long-range orientational order in two-dimensional 'ferromagnetic' flocks of polar SPP [...]. Another set of striking intrinsically nonequilibrium properties have recently been predicted by Ramaswamy and co-workers [...]. They considered, in particular, the case of apolar but oriented SPP and argued that such 'active nematics' should differ dramatically from the usual (equilibrium) case. In particular, their approach [...] predicts that giant density fluctuations arise in the ordered phase of such media."

One can take as another example the task of how to explain the shape of bones. This is a question of tissue growth and development. There is research that is attempting to calculate the shape of a bone with ideas very close to those relating to what we call active matter [Willie, Duda, and Weinkamer 2013].

KK: Is it a question of 'dynamic structures' – of structures that emerge, evolve, and change, and can therefore go through very different processes?

J-FJ: Calling it a dynamic structure is indeed suitable. The structure is evolving in the sense of temporal evolution – not the Darwinian understanding of evolution, although there are chemists who try to make systems evolve in the Darwinian sense, and such systems have to be out of equilibrium.

MF: Would you actually describe nonequilibrium, in the sense of instability, as being one of the crucial characteristics of active matter systems? Instabilities which actually prompt structure and its appearance?

J-FJ: Yes, because active systems mostly flow, and many active systems do not have a steady state where flow does not occur – and then one can structure, because of the flow; there is a strong interaction between activity and flow. But the steady state of a system, which is a state without flows, is generally unstable.

KK: To return once more to the problems of descriptions of nonequilibrium active systems, how concrete can the predictions for the behavior of such a dynamic system be? Is it mostly about statistical predictions? And what happens if in your analysis you want to go through different scales and find a description that includes these different scales?

J-FJ: I have no problem with predicting something in a statistical way – I mean where things fluctuate. There are systems at equilibrium which have large fluctuations as well. This is not true for macroscopic objects such as the surface of a table. If we look at small systems, a theory can be built up: the stochastic thermodynamics of a small system.[10]

And even if the system is fluctuating you can still predict what the average values are. You can also predict what the noise is around the average values. You should be able to measure that and to compare it to theories. It is something that you have to include in the game. I do not say it is easy, but you should be careful.

The question on the combination of different scales relates to something that we have not talked about yet: numerical simulations. I do not do these myself, but the people I work with do. For instance, I had a postdoc, Jens Elgeti, who started a

10 Stochastic thermodynamics uses stochastic variables to better understand nonequilibrium dynamics. Moreover, while in classical thermodynamics one of the key assumptions is that the system has a large number of particles involved, stochastic thermodynamics deals with smaller systems [Seifert 2012].

big program on the simulation of tissues. If you want to simulate a tissue, it is made of cells, so one has to describe the cell reasonably well too. In the case of a cell, you can put everything you know in the simulation, but if you saturate your computer with one cell, you will not be able to do anything at the larger scale. Hence, the way we treated cells – this might be a shock – was that the cell was represented by two points and, when it was growing, these two points repelled each other. When the distance between the two points is too big, the cell divides in two. Then we put two other points to obtain the two daughter cells, each represented by two points. However, when we show it we do not use points but we use small spheres, because then it looks like cells, and people are pleased. But in the calculations, cells are only points. In this case, it means that the small scale is not well treated, because the cell has a nucleus, DNA, proteins, and so on, and there is nothing like this in the simulation. What I strongly believe though is that, if you look at the properties of the tissues at scales larger than the size of the cell and at times larger than the cell division time, it should be fine. When I say 'fine,' I mean it is mostly suitable. There are, however, properties of the tissues that depend on explicit properties of the individual cells, and if I want to know those, I need to make a simulation at the scale of the cell – something that remains very difficult.

MF: Can one claim that one of the aims of the research in this field is to obtain a general description irrespective of scale? We already touched on this topic implicitly when talking about bone as active matter or, to give another example, about whether a cell is 'just' two spheres or just two points.

J-FJ: Yes and no. What one means by 'irrespective of scale' is that the typical scale in different systems is different but perhaps the theory is the same. For example, if I look at buffalo herds or if I look at bacteria the general behavior is very similar: bacteria have to swim in the same direction while buffalo all run in the same direction. There are details that are different but there is this similarity in effect. This is what I mean when I say 'independent of the scale.' Now if I look at tissues, I have several length scales. I have the scale of the atoms in nanometers – and we have good theories for this, such as quantum mechanics, but I do not want to take this into account. Then, we have all of the questions around proteins that can be found in the cell, for which we also have good theories. I can also add the scale of the cell. And furthermore, I can discuss the scale of the tissue, the scale of the animal, and so on. I think maybe the most difficult problem is that of determining the scale at which to describe the system. In order to be able to answer this question, I know that I will have to ignore everything that is too small, and I will also have to ignore everything that is too big. So ultimately, I will try to construct a theory that describes the behavior exactly at the length scale relevant for the question that you are asking.

To give a more elaborate example, sometimes the scales are coupled; that is, a material is investigated at two different scales. Take plastic for instance. Plastics are made of polymers, and we have very good theories to describe the structure of polymers based on the interactions between atoms, but if you put the problem in a computer to carry out a calculation, it will be much too large to be handled. Here, a solution is offered by multiscale calculations. One calculates microscopically the properties of one polymer, and then one employs coarse-grained modeling, that is, modeling the polymer using subunits, then defining their known properties and the interactions between the subunits. Finally, one can then think about investigating the material itself.

MF: This point brings me to topological structures, which also appear irrespective of scale, that is, of whether we are dealing with bacteria or a herd of buffalo. Does this not indicate that topology should also be considered in the framework of the research on active matter systems?

J-FJ: Speaking generally, we have these experiments where we look at the orientation of cells where defects appear. Of course, when you do that – and especially when you are in contact with biologists – you worry about what the significance of these defects could be for biology. What is at the back of your mind is the fact that nature has had some five million years to 'think' about it and that it evolved (this time in the Darwinian sense), that if something is here then it should be useful. And it is. For example, if you take cell layers, sometimes cells are extruded or cells die. Recent experiments show that they are extruded always at the place where there is a defect. Consequently, there is a correlation between the extrusion and the fact that there is a defect. I told you about this spontaneous shear that appears in these layers if you make bands. A colleague of mine, Pascal Silberzan, looked for that and found articles on cancer which describe sheets of cells on top of fibers. This is related to the way that cells go out of or into a tumor. We have not yet gone very far down this road, but now we are aware of it. Depending on the thing we want to do in the end, one wants to understand the biological question. The point of view of the biologist is as interesting as my point of view. Hopefully everything will come together in the end; otherwise, we are in trouble.

KK: Continuing this line of thought, what in your opinion are the future tasks of active matter research?

J-FJ: If we look for a moment at the beginnings of this research, there were at least several theories that in the end achieved a sort of consensus. Suffice to say theory went well beyond what was being carried out experimentally. You have all of these biological systems where you do not control all of the parameters, when you describe a phenomenon with sometimes fifty percent uncertainty or even more. However, we should be able to find model systems where we could verify that if you run one parameter it goes exactly as anticipated by the theory. What I mean by that exactly is not the possibility of answering positively or negatively but making it into a quantitative description. A lot of progress has been made in this respect. And the experiments here are moving in this direction. But there is a lot of work to be done in this area. There are also all kinds of developments about what we can learn about nonequilibrium physics. Can we attempt to use general laws for example? This is one direction in which we can go. The one I have tried to pursue concerns whether we can make a physics of biological tissues.

More generally, I want to have a description that is simple enough but that respects all of the laws that I know. What I mean by that is that I do not want a system with 25 parameters where I would have to explain everything. I would like a description coming from physics at the local scale by means of equations which I hope will be simple. I can bypass some of the steps, because I know that some of the symmetries have to be respected.

MF: You noted that you want the desired description to preserve the laws that you know, but one can underline that, in the framework of the research on active matter, there are several laws that may be broken, such as the law of large numbers.

J-FJ: It depends what you call broken. It is broken in the sense that, if you take systems at equilibrium, what describes the fluctuations is the law of large numbers. But it means that there is a difference between nonequilibrium and equilibrium systems in this instance. Thus, if you take for example all of these theorems on the law of large numbers itself, it is still valid of course. The central limit theorem is valid, there is no doubt about that, but in order to be valid it has to have some constraints. I mean that there are hypotheses to theorize about that, and this situation means that such hypotheses are not satisfied with these systems out of equilibrium.

MF: Hence, this situation actually implies that there may be new laws?

J-FJ: Yes, it does imply that there are new laws. But they all respect the basic principles of physics. Energy is conserved. Not by the system itself, since energy flows in and out; but, if you put the whole universe in there, you do not make energy out of nowhere. Also, forces are balanced. Newton's laws of motion are valid, even at the microscopic level. Perhaps you have to be at a large enough scale not to worry about quantum effects but, as long as you are at this limit, forces should be balanced – Newton's laws should be valid.

Bibliography

Betz T, Lenz M, Joanny J-F, Sykes C. ATP-Dependent Mechanics of Red Blood Cells. Proceedings of the National Academy of Sciences 2009, 106 (36), 15320–15325, DOI: 10.1073/pnas.0904614106.

Chaté H, Ginelli F, Montagne R. Simple Model for Active Nematics: Quasi-Long-Range Order and Giant Fluctuations. Physical Review Letters 2006, 96, 180602, DOI: 10.1103/PhysRevLett.96.180602.

Deisseroth K. Optogenetics. Nature Methods 2011, 8, 26–29. DOI: 10.1038/nmeth.f.324.

Duclos G, Erlenkämper C, Joanny J-F, Silberzan P. Topological Defects in Confined Populations of Spindle-Shaped Cells. Nature Physics 2017, 13, 58–62, DOI: 10.1038/nphys3876.
Finlayson BA, Scriven LE. Convective Instability from Active Stress. Proceedings of the Royal Society A 1969, 310 (1501), 183–219, DOI: 10.1098/rspa.1969.0071.
Joanny J-F, Ramaswamy S. A Drop of Active Matter. Journal of Fluid Mechanics 2012, 705, 46–57, DOI: 10.1017/jfm.2012.131.
Krueger D, Izquierdo E, Viswanathan R, Hartmann J, Pallares Cartes C, De Renzis S. Principles and Applications of Optogenetics in Developmental Biology. Development: For Advances in Developmental Biology and Stem Cells 2019, 146, 1–13. DOI: 10.1242/dev.175067.
Kruse K, Joanny J-F, Jülicher F, Prost J, Sekimoto K. Generic Theory of Active Polar Gels: A Paradigm for Cytoskeletal Dynamics. The European Physical Journal E 2005, 16, 5–16, DOI: 10.1140/epje/e2005-00002-5.
Kubo R. The Fluctuation-Dissipation Theorem. Reports on Progress in Physics 1966, 29 (1), 255–284, DOI: 10.1088/0034-4885/29/1/306.
Marchetti MC, Joanny J-F, Ramaswamy S, Liverpool TB, Prost J, Rao M, Aditi Simha R. Hydrodynamics of Soft Active Matter. Reviews of Modern Physics 2013, 85, 1143, DOI: 10.1103/RevModPhys.85.1143.
Schrödinger E. What Is Life? The Physical Aspect of the Living Cell. With Mind and Matter and Autobiographical Sketches. Cambridge, Cambridge University Press, 1992 [1944].
Seifert U. Stochastic Thermodynamics: Fluctuation Theorems and Molecular Machines. Reports on Progress in Physics 2012, 75 (12), 126001, DOI: 10.1088/0034-4885/75/12/126001.
Turlier H, Fedosov D, Audoly B, et al. Equilibrium Physics Breakdown Reveals the Active Nature of Red Blood Cell Flickering. Nature Physics 2016, 12, 513–519, DOI: 10.1038/nphys3621.
Willie B, Duda GN, Weinkamer R. Bone Structural Adaptation and Wolff's Law. In: Fratzl P, Dunlop JWC, Weinkamer R, eds. Materials Design Inspired by Nature: Function Through Inner Architecture. Cambridge, RSC, 2013, 17–45.

Jean-François Joanny has started his career as a CNRS researcher at Collège de France and then in Lyon. He was appointed as a professor at University Louis Pasteur in Strasbourg in 1989 and then at University Pierre et Marie Curie (UPMC) in Paris in 2003.

He is a theoretical physicist and has worked on various aspects of soft condensed matter physics and then of physics for biology. Between 2014 and 2018, he was the director general of Ecole Supérieure de Physique et Chimie Industrielles. He has been a professor at Collège de France since September 2018.

When Joanny moved to UPMC in Paris, he became director of the physics unit of Institut Curie. Since that time, he has worked on cell biophysics and fundamental cellular processes, the mechanics and growth of tissues and cancer physics. He describes all these biological phenomena with the concept of active matter.

Jean-François Joanny has received several awards including the bronze and silver medals of the CNRS or the Ampère prize of the Academy of Sciences. He was a junior and senior member of Institut Universitaire de France.

Michael Friedman, Karin Krauthausen, Barbara Mazzolai

Interview with Barbara Mazzolai: Plants, Plantoids, and Active Materials

KK: The research on active matter includes studies in smart materials, hence materials that sense and react to the environment. In your research, you examine the 'smartness' of plants and you use these studies for bioinspired robotics. Exemplary is your work on the plantoid: a robot that one might say replicates the behavior and mechanisms of plants. It is part of a new generation of plant-inspired robots – a plantoid does not have a humanoid form but, like a plant, it has not only a trunk, branches, leaves but also roots. These roots grow due to the addition of synthetic material printed in 3D from the inside, and move and orientate themselves in their surroundings thanks to sophisticated sensors positioned in the tips of the roots. Therefore, these robots are able to move by growing and explore the environment with their sensors. But before elaborating on your research on plantoids, maybe you can tell us how you consider robots and their connection to living beings and to plants in particular.

BM: Robots, and more specifically bioinspired robots, can be interestingly linked to living beings and used to better understand how living beings work. One of my research challenges is to use biorobotics as a tool to investigate and explain the strategies and functionalities of natural organisms. This understanding is fundamental to get closer to a full knowledge of a natural system and for its safeguard. With respect to plants, this closeness is particularly absent because, in school, plants are only considered for oxygen and food production, but are not really understood from the point of view of how they work. Robots could help in this respect. This must not be understood only in terms of the promising applications of plant-inspired robots – from exploration and environmental monitoring, to medical applications, to search and rescue in remote or extraterrestrial scenarios – but also with respect to the ways in which robots can bring us back to biology [Mazzolai, Beccai, and Mattoli 2014; Sadeghi et al. 2014; Lucarotti et al. 2015]. Having this in mind, I have investigated characteristics of plants such as growing capabilities, osmosis-based movements, strategies of carnivorous plants, movements based on circumnutations, just to mention a few. Robots, as physical and dynamic systems, could really help in the understanding of the physical interactions between an organism and its environment. In the future, robots could be a sort of tool for scientists who would use them to demonstrate the principle and working functionalities of living beings.

Acknowledgment: Michael Friedman and Karin Krauthausen acknowledge the support of the Cluster of Excellence 'Matters of Activity. Image Space Material' funded by the Deutsche Forschungsgemeinschaft (DFG, German Research Foundation) under Germany's Excellence Strategy – EXC 2025 – 390648296.

Open Access. © 2021 Michael Friedman et al., published by De Gruyter. (cc) BY-NC-ND This work is licensed under the Creative Commons Attribution-NonCommercial-NoDerivatives 4.0 International License.
https://doi.org/10.1515/9783110562064-007

Fig. 1: The plantoid: a bioinspired robot developed at the Italian Institute of Technology, Pisa, Italy, by The Plantoid Project led by Barbara Mazzolai. © Bioinspired Soft Robotics Lab, Italian Institute of Technology, Pisa, Italy.

KK: Following this, can one say that biorobotic engineering considers plants as smart material – in the sense that they are responsive to the environment – and therefore as active matter?

BM: Plants effectively adapt to the environment and respond to external feedbacks; and this is why are so relevant in biorobotics. Material for me is an essential part of this picture. We need materials which are more similar to natural materials to really interact with the environment by giving different properties and more abilities to the robots. Consequently, we have to look for non-traditional materials that are safer, even intrinsically so, when we need or want to develop a robot to interact with us or with another living system, or with the natural environment. This is something unaddressed in industrial robotics. The reason behind this is that the interaction with the human operator is usually limited. Industrial robots are rigid and strong, since this is the role one assigns them in a society based on production: robots produce the products in a factory. If one would like to see robots in another way, however, as something that shares our environment, our domestic environment – and I do not mean only for cleaning for example, but also for other means of assistance – you cannot make robots rigid. This is one of the trends which will be developed in the future, focusing for example on aspects of softness. This is also one of the important aspects of active material in order to achieve adaptive interaction, compliance, and safety. The material becomes really more and more fundamental in this type of robotics, which is called soft robotics. Perhaps this is the wrong term. In the past, we called it 'soft robotics' in order to distinguish it from traditional 'rigid' robotics.

One of our starting models among natural organisms was the octopus [Laschi et al. 2012; Mazzolai et al. 2019]. This animal is considered as a paradigm for soft robotics, since the octopus is completely malleable (Fig. 2), without any rigid structure, and has unique abilities of squeezing and morphing. At the same time, with its arms and suckers, the octopus is able to exert very high forces, to explore the environment, as well as to grasp and manipulate objects. It demonstrates that we do not need rigid structures in every case to carry out an action. Of course, we need to consider it in its environment: the octopuses' physical abilities and intelligence work very well in aquatic environment, but if we transfer them to our own terrestrial environment, they would certainly not be as smart. What we have to consider in this example is that environment is fundamental. A robot cannot be designed without considering the environment in which it has to act.

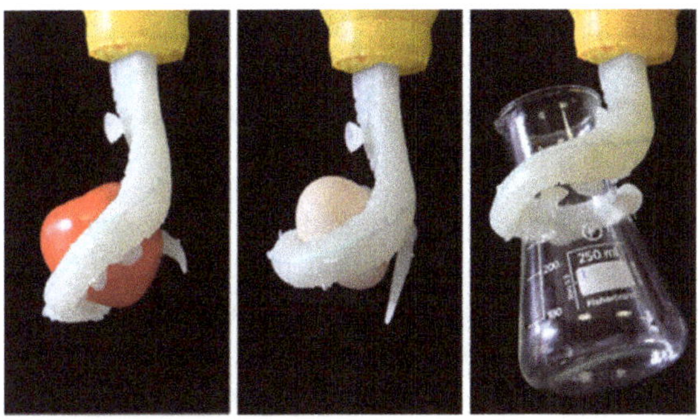

Fig. 2: An artificial octopus-like arm with suckers. See also: [Mazzolai et al. 2019]. © Bioinspired Soft Robotics Lab, Italian Institute of Technology, Pisa, Italy.

MF: So you would say that one of the characteristics of active materials is exactly this adaptation to the environment?

BM: Exactly. One may say that the mechanism is in the design because, in the design, environment is already included. This is something entirely characteristic of natural systems. If you consider plants as another paradigm of active materials, you can even see motion in dead organs. For example, the pine cone is dead but it continues to move. What is impressive from my point of view is that the life of plants is perpetuated by structures that are dead but continue to work. Energy use is anticipated in the design phase.

KK: Could one say that in the case of the pine cone, the energy is stored in its anisotropic material structure and when triggered by the environment in the right way this energy is consumed for a certain activity?

BM: Precisely. The pine cone is activated and releases the seeds when the environment provides the right conditions for this: in terms of humidity, in terms of temperature, and so on. The pine cone is thus 'designed' to release the seeds only when they will find the conditions to germinate. In this way, the pine cone spreads the seeds in the environment because the conditions are right. This is amazing because one can reduce the energy strictly to the design phase. Subsequently, pine cones take their energy from the environment.

More generally, of course, if you have to do very complex tasks, you need a control system and you need external energy. But you can optimize the movement, you can optimize the control using this material. This is something completely missing in current robotics because most of the robots that we use are either made of rigid material or now soft material but without particular properties. Currently, we use most of the silicones available; because they are soft, they can in some way express this softness. In many cases, however, they do not have the proper functionality. What I mean is that they cannot really interact with the environment.

KK: Would you say that the research on active materials and especially on bioinspired materials will change our understanding of nature – nature per se but also nature as a source and a counterpart for engineering? And how do the more classical biologists look at this kind of interdisciplinary research?

BM: There are certainly people who want to collaborate and who are attracted by this multidisciplinarity. But we also have to recognize that there is a lot of talk about multidisciplinarity while in actual fact it is hardly done at all. However, one has to remember that people in the past did work in a multidisciplinary fashion. Take Leonardo da Vinci for example, or even the entirety of the Renaissance. They carried out their work precisely in this way and created great things.

I know that much of the technology that we use is not bioinspired. I do not want to say that we can solve the problems of the world by means of bioinspiration. You have to select when it is needed. What I mean is that we should have a critical approach. In some cases, you quite simply do not need bioinspiration. A traditional approach can sometimes be more convenient. That said, I think there are many great bioinspired examples that we use every day. We simply do not recognize that they spring from such sources. Hence, this is the goal: to have something novel that can also be used in new contexts and that can open up new fields.

MF: I would like to return to the example we started with: the plantoid (Fig. 1), a robot that was developed in part as a result of a 'simple' question: how do roots behave? Can one argue that with this example there is a new conception of the plant? As you mentioned before, the plant was considered in the past as something that was not, so to speak, intelligent. What is this new conception that you are introducing here with respect to roots, and how is this then translated into robotics?

BM: I arrived at the study of plants through my experience in biophysics. When I moved from biology to biophysics I began by studying the impact of pollutants, particularly heavy metals, on human beings and the environment. I carried out experiments with natural samples, such as our blood, principally to detect and understand the impact of mercury, but also of other heavy metals. When I moved to engineering, the first idea I had was why not create a technology for detecting and understanding such things? Quite simply there were very poor technologies for monitoring in situ: for detecting polluted air, water, food, soil, and so on. Thus, I developed several robots. First, a sensor to detect mercury, then a sensory device, and only then did we move on to robots. We did so because robots are mobile. Hence, the idea was to have robots with sensory capabilities that can cover and detect a large area. One of the central problems, however, was that of the soil. Soil is so complex. How can we send a robot into the soil? This is the reason why I returned to biology. I determined that the best living beings that can carry out this task were in fact plants and not animals. Not only do plants penetrate efficiently into the soil like other animals such as the earthworm (e.g., the *Lumbricus terrestris*), the mole, and so on, but they also create a network. Plants really produce the best system to probe the soil because they develop a capillary system that penetrates it so thoroughly. We began by studying how plants accomplish this, and we used a robot to validate the biological system. We know from biology that plants grow at their extremities, but there is no evidence for why they do this. What we then did was to compare two identical robotic systems: one system able to move from the tip and the same system pushed from the top [Sadeghi, Mondini, and Mazzolai 2017]. In this way, we demonstrated that in the first condition we could save energy and reduce energy consumption during the soil penetration. We arrived at plants quite reasonably because in terms of application in the environment they are the best. With these insights, one can at least consider the possibility of using plants as a model for robotics.

KK: With plants as models for robotics, interesting features start to arise: first, that growth is a way of moving through the environment; second, the idea of a network, or maybe even a multitude; and third, the involvement of plant communication, or at least signaling ...

BM: ... and actuation, because there are also other actuators. It is a revolution simply because for the first time we can start to talk about growth as movement. This is a new paradigm for the concept of actuation. Usually, in bioinspired robots, the model is the animal; consequently, the model for the actuator is the muscle. Plants do not have muscles, but they are able to move and reach with their branches or roots points that are far from their point of germination. What we have demonstrated by contrast is that plants move by growing. They cannot move in a different way. If they have to move from here to there, they have to grow from here to there. Hence, growth is a new form of movement. And, while growing, plants adapt to their environment – which is also part of the novelty of this kind of robot. Thus, if we are really able to implement this mechanism from a technological point of view with active material – because the secret here is in the material – we can really have a robot that perceives the environment and makes decisions. This is the intelligence of a system that is bioinspired: being able to make decisions concerning movement and direction. There are two different ways to control this robot: either the operator can control it remotely and give it commands such as 'look for water,' 'look for survivors,' 'look for chemicals,' or it is the robot itself that decides on the direction on the basis of high-level control. Plants are the inspiration for this latter behavior. The question is how they can manage all of this information without a brain – but they do indeed do it.

Nor is the movement of plants totally random. One of the initial observations about plants was that they only move around, without a direction. But this is incorrect, since plants have sensors, which they, of course, use. There is no sensor in biology that is not employed, since this would simply be a way of losing energy. But of course, there is a part of the movement that can be random. This is also true for animals, for example, when they do not perceive smell, the pheromone, but move in a certain way until they do perceive it and subsequently modify their behavior. Moths are typical in this respect since, when they perceive the pheromone, they reduce their movement amplitude, their energy. It is the same with plants, which also move at random for as long as they do not perceive anything. For example, when a plant looks for water, it moves randomly for as long as it does not have a gradient or chemicals to follow. Once these are perceived or sensed, however, the plant of course goes in that direction.

KK: It seems to be a situated behavior, hence a behavior that is not, or not only, determined by a DNA program.

BM: Exactly. It is something that emerges by the interaction with the environment, just as in other natural systems [Gagliano et al. 2014].

MF: Can one say that the root system or plant system is always, in a sense, in a state of nonequilibrium since, on the one hand, it adapts to its environment but, on the other hand, it also takes materials from the environment? That is, it is always on the verge of coming into being as an emerging structure.

BM: Absolutely. Plants use chemicals and water for photosynthesis; there is a communication between the roots and the upper parts of the plant. It is a continuous interaction on the basis of internal needs, state of development, and external competition, because they have to compete. It is not possible to simply be and grow; there is also a struggle for resources, competition between plants as well as between plants and animals, or parasites, and so on.

There is also the aspect of redundancy, the fact that plants have built-in redundancy in their organs.

To state the obvious, plants cannot move like animals, they just stay in the place in which they germinate, and there they can only grow. And because of this sessile nature, they cannot escape in case of an attack. This is probably the reason why they do not have a brain, which would make them too much vulnerable. Differently, we could say that plants have distributed 'command centers,' as well as distributed sensitive organs, and a unique capability to regenerate and create new structures.

The problem is that we are currently lacking the appropriate terminology. Very often when you talk about behavior in plants people respond that behavior is for animals, or intelligence is for animals. If that is the case, then at least provide us with some terms to describe plants.

MF: Because all of these concepts are actually connected with the idea of a mechanism that controls everything in a centralized way – that is, the brain so to speak.

BM: Exactly. Hence, there is no terminology for presenting or describing this behavior in plants quite simply because all of the scientific terminology is for animals. And this is another issue. How can I describe this behavior if I cannot use the terms 'behavior' or 'intelligence,' as these terms are strictly preserved for animals?

KK: What kind of robot could be created if one thinks of this from the perspective of the plant? Should there be a new kind of communication or movement involved?

BM: Precisely. If one defines the robot as an artefact that can move, communicate, perceive, and control parts of its structure, then all of these features are also true for plants. To demonstrate that we can use plants in robotics, I just have to start to show that they are able to communicate, move, and sense. It must be added that all of these features are *intrinsic* to the plant itself. Thus, we can translate these features into artificial systems that are robots, since they perceive and control movement and actuation.

KK: This translation of the intrinsic features of the plant to robots could entail a completely new understanding of the machine. If you think of the nineteenth century, when one first began to develop, for example, the first self-governed machines, these machines were very simple systems. Admittedly, this is still what people have in mind. For example, we think of the car as a machine; already the idea of a self-driving car is somewhat overwhelming. What you are proposing, however, seems to go even further, in the direction of a 'growing' machine; hence, it involves a new understanding of the machine itself.

BM: What I would like to stress concerning this new idea of artificial machines is the importance of materials for their structure and functionalities. Of course, in robotics, we already talk about intelligence, and also about different types of intelligence, for learning, motion control and communication. It must be said that material is still in its infancy in robotics. This is particularly the case when considering soft materials.

We can introduce the concept of softness, but this is still not enough. Quite simply, silicone is inadequate, since it cannot interact with the environment. Thus, we really need to implement the design phase in the material. We should work at this level. This should not only be understood in terms of assembling the components. This is the classical approach where you assemble the sensor together with the actuator together with intelligence and the control, and then only at the end do you think about the energy. And you finally come to an idea of the energy consumption of the robot when it carries out some specific tasks. By feeding this information into the motor – so starting from the motor – you can have a rough idea of the energy consumption, but then the robot is not able to really operate in the real environment. Instead, we need to think about materials that are multifunctional – in terms of sensing, actuation, and energy use – and able to evolve or grow depending on environmental conditions and needed tasks. This is what drives toward a new understanding of machines design and development.

In this direction, a new trend called morphological computation or embodied intelligence is moving. This can be considered a new paradigm in robotics, which gives importance to the body, to the shape, to the material, to the interaction with the environment, and not only to the brain. But what really is the problem? Clearly, the issue is how we are to translate this principle into the rules of design. Currently, we do not know how we can design this kind of robot. We know more or less the principle but not how to design it as such. In my case, I have an AutoCAD model from which I can start, but because the robot needs to adapt I cannot design the final shape. I can just put the design in the material as the basic principle. Essentially, it has to develop by itself. The development should be immanent in the design.

MF: Can one then argue that there is no predefined algorithm that would predetermine all future actions? I would say that there are some algorithms being structured in the design phase of the robot. Perhaps the best way to put it is to say that they are adaptive in a way – which is also the challenge.

BM: Adaptation is the key and the grand challenge for robots operating outside in the real world. Adaptation is fundamental for animals, plants, and all living organisms, and so it is for artificial machines.

Now, for this, the challenge we find in the design phase shows that you need different people like engineers, computer scientists, biologists, material scientists, physicians, and so on from various disciplines working together. This is not easy. How can we collaborate effectively? How can we involve people that work in mathematics in this phase, or people developing the control or the material, together with biologists, together with engineers, in order to work together to structure the rules? This is quite simply a necessity: we need this cross-disciplinary interaction for the design of new adaptive robots.

Unfortunately, in our world, there are many disasters: earthquakes, floods, tsunami, and so on. When observing this fact, one can quickly come to the realization that there are currently no alternatives to humans or dogs when looking for survivors; there are no robots deployed in this fashion at all. Why is that so?

Were a robot to be deployed in a disaster environment, after a tsunami for example, it would have to have a different configuration [Murphy 2014], it would have to be able to adapt. For example, it would have to be amphibian, because it would need to move in water and on land. This model of robot would need to be able to adapt to conditions that cannot be predicted. After a disaster, the unstructured environment becomes the rule rather than the exception. In such a situation, external conditions cannot be known in advance. This is why a new model of adaptable robot is needed. A robot that can count on adaptive materials with integrated multifunctionalities, allowing it to have an evolving and growing body structure, capable to re-shape or re-form its parts, and with an efficient and sustainable use of energy. Energy is in fact also a key point, because if you collapse after an hour ...

MF: ... it is hopeless.

BM: But this is the situation. I blame it on the fact that we think in a very complex way. This is what makes our robots too complicated. We embed materials, sensors, and actuators, and then energy is the last point we address. But living beings do not work in this way.

MF: Hence, we are coming to the point where we realize that new robots, new machines should actually be much simpler and not just a very complex imitation of 'the way nature works.'

BM: Yes, I believe this is the real issue: to understand what is needed. We need to understand what is really useful and not just copy the shape of a human to have a humanoid. Why should our ideal robot be a humanoid? Why is the human form necessary in this environment? It is not the best. It cannot move in debris. Why do I want to develop a humanoid for rescue in an environment that has been destroyed? There are many living beings that are better than us. Why not then develop new creatures? In robotics, you can also go beyond nature. We should just take what is needed for some kind of application and not simply try to copy them. If I want to copy a plant, for example, I put a trunk in my plantoid because it is needed in order to relay a message, not just to demonstrate that it is a plantoid. Of course a trunk is clearly part of a plant and not of an animal, but the concept goes beyond the shape of the plant. The shape is just for pure representation. The message that I want to send requires growth, but the root does not need the trunk or the branches with the leaves. If I want to do something, to move in this environment, only the concept of growth is needed, whether what is grown be branches, roots, or something similar is not important.

MF: So, we could say that these new materials are inspired by nature, but also go beyond nature insofar as they are not simply copies of it.

BM: In the end, we use an approach very similar to a 3D printer to translate growth. But the difficulty lies in understanding the secret of growing as such. We developed several prototypes before trying a mechanism. Roots also inspire this design because they release dead cells and then move inside this interface they create. So we tried to imitate this because it was a really efficient system. We needed to be able to reduce the external friction because, when you dig in soil, after a few centimeters the system collapses from the internal friction. So, it was clear that this was not enough. In the end, we realized that growth was the key problem. But how can we imitate the addition of cells, since plants develop by means of cell division? By absorbing water from the external environment, plants also extend. Thus, they grow and then elongate.

For example, depending on the soil impedance, what occurs more: cell division or cell elongation? And for these two, what are the important factors for movement? We therefore attempted to implement growth in our robot [Mazzolai et al. 2020]. But before we had to understand that the 3D printer approach, the manufacturing approach, was in some ways a suitable solution, so we used it for the robotic root growing and bending. It took more than a year, several prototypes and experiments, and in the end, we used thermoplastic materials. The latter were not the best in terms of mechanical properties, but we could anyway use them to implement growing abilities in a machine, and demonstrate that if there was a uniform deposition in the soil, the robot could move (i.e., grow) vertically. Moreover, if we deposited the material in a differential way, bending occurred at the robotic root, as natural systems do where more cells are on one side with respect to the other one. So, imitating this feature, we implemented this differential growth also in our robot.

So, first we need to understand the key principles at the base of a functionality that we want to imitate from the natural system, and then how to translate them into an artificial system.

KK: In living systems, growth can mean different processes. There is the growth of trees, which adapt to internal needs or external conditions by adding new cells with various geometries and cell wall structures in relevant parts. There is the remodeling of bones in living organisms, which works quite differently because it includes degrowth. What kind of adaptation processes, and hence what kind of activity and agency, do you intend to translate, or maybe reinvent, in the technical sphere?

BM: So, once more the material and adaptation strategies come back to the center. Plants move in different environments, while they explore them. One of the movements they use with this aim is circumnutations, that is, elliptical or circular paths made by the apical regions of both roots and shoots. We study circumnutations in the aerial part and in the soil. For example, climbing plants use circular movements to look for support or to reduce the friction during movement itself. We are also implementing this behavior in our growing robot. Thus, the point is to understand how we can have a sensing capability in a system that is growing. First of all, you can add sensors in the tip. There are discrete elements: chemical sensors, sensors of temperature, and so on. Yet this part is fixed; it does not change. More difficult is to add sensors in the growing part, since it continuously changes. This is why the aim is to embed sensing properties in the structural material, used to create the robotic body. Consequently, we look for a 3D structure in which these materials have a specific functionality, like a sensing capability and so on.

The other challenge to consider is the interaction of the material with the environment in time. Thus, we can go beyond the concept of 3D printing, toward a 4D approach [Khorsandi et al. 2020]. The material can provide this 4D aspect, so the robot can change its 3D shape and behavior over time. In the current version of the robot, we miniaturized a 3D printer machine inside the robotic tip and we used thermoplastic material to create the robot's structure. In the next future, the goal is to develop materials that include sensing properties.

KK: I wanted to further emphasize this point concerning interaction and sensing by asking how such a robot would be controlled. Obviously, there has to be a new understanding of control: a local control and autonomous system – a kind of 'soft computer.'

BM: Exactly, we need to rethink how to embed the control, or better how to 'embody' it. It is something that must really be integrated into the material. The goal is not only to have sensing, which is fundamental, but also to associate this sensing to a form of behavior. Such mechanism is evident in plants, for example in the tropisms. This could be understood as tropism, as we have implemented it in the plantoid robot. Tropism means that the roots or branches move by turning toward or turning away from environmental stimuli. We have implemented this behavior in the roots of the plantoid, integrating sensors and using an adaptive material for growing. Another example is also in the case of behaviors that are independent of the direction of the stimulus, as is the case of the fast closure of the leaves of the *Dionaea muscipula*, the Venus flytrap. Its leaves have tactile hairs, and when a prey touches at least two hairs within a period of 20 s the plant closes, trapping the insect. The tactile feedback triggers the mechanism, and the movement is associated with the structure and properties of the leaves' material, to the system instability and to osmosis. There is therefore a motion associated with sensing – closing the leaves, movement by growth, and so on – and the 'soft' computer is embedded in the material.

In robotics, we still have to work on this to reach something similar. There have been a few examples, but these have occurred outside the field of robotics. The central issue is how can we merge these technologies in the robot in order to have skin/body, materials, and so on that really embed all of the needed functionalities?

MF: Can one say that the plant does not add more and more mechanisms in order to grow but that growth itself is already self-propelled in a way by the mechanisms intrinsic to the material itself? Is this the point of the design you were referring to?

BM: That is correct. I cannot add another motor because the material itself is the motor. And the sensing should also be in the material. Hence, I need a soft computer in order to implement the intelligence associated with the sensing capabilities. This is difficult, so we need to have some idea of how we can register tactile feedback in a body that is growing or in humidity and so on, because I have to perceive and register this stimulus and then implement the behavior. This could be as simple as giving one direction or another as well as the interactivity with the environment that is implied. But you are correct, the motor is not needed there, the motor is the material.

MF: In other words, to translate plant mechanics into soft robotics one does not need to centralize the control to have everything dictated or programmed in advance. The material does not even determine this, but rather reacts to the environment to make its own structure.

BM: Yes, I do not need to apply additional functionality, since it is already contained in the material. Of course, this is true for a form of functionality such as growth. But it could also be used in the case of sensing properties. The softness that we mentioned should also be present in terms of the interaction with other organisms, because I cannot work with a rigid robot.

KK: Reorienting robotics and mechanics on the basis of the research on plants provides a new paradigm for engineering. Does it also make a difference if you understand the living system as a multitude with distributed intelligence? What happens to robotics when you understand the inspiring material, the plant, no longer as a closed system but as an open one?

BM: This is precisely what one sees when one compares the root to animals that move in the same environment. There is another way to collaborate, to cooperate, not just as ants or fish or bees do, as the main examples of the swarm intelligence paradigm. There are groups of scientists working on the latter form of behavior to develop new algorithms for cooperation between robots based on this paradigm. In the case of roots, we also have this kind of cooperation for a specific task, but they are part of the same system. Thus, there are millions of agents that cooperate for the survival of the plant by having different aims. Insofar as I see it, it is much more efficient to develop a robot that can explore the environment. Why should we use just a probe?

KK: Could we think the root – its structure, properties, and in general its agency – through the environment? Could we understand the environment as part of the physiology of the plant?

BM: Plants and roots are also able to perceive vibrations, gradients. In the end, gradients are essential because plants follow them. It seems that plants also perceive vibrations. When you have an animal that moves in the soil, it does not have eyes since it has no need of them. The same is true of plants; they have no need of such things, since the roots move in soil. But all of these animals perceive vibrations, and plants perceive vibrations as well. This is new as a sensing capability. Thus, it is a further demonstration that these systems are adapted to the environment. One cannot say that there is here a less intelligent implementation and that for other living beings there is a more intelligent implementation: it depends on the environment in which one moves. And so for *that* environment, for that task plants are actually the best adapted. This is the best shape, the best thing that can be implemented for that environment. But very often, because they do not have eyes, because they do not have a centralized system as we have, we consider them ...

MF: ... inferior.

BM: Absolutely. And essentially because they are different from us. Plants have another way of using the environment. The solution is really effected by the environment. In fact, if you compare roots with an earthworm (e.g., *Lumbricus terrestris*), they show similar adaptive solution. *Lumbricus* also has this enlargement of the diameter; it anchors in the soil and then pushes from the tip. It has liquid inside its body and contracts muscles to move. Thus, the worm can elongate and expand according to the contraction of various muscles. Roots have a similar behavior: they enlarge their diameter, they anchor with hairs, and then they push the tip, simply because the environment affects the solution. Even if the body could be different, in the end most of the solutions that living beings implement are given by the environment and the morphology. If you compare an animal with legs to a root, of course it is different, but if you compare an animal without legs to a root you will find they are very similar. Not only is the body similar but the environment is also the same.

We must consider how the environment affects the morphology and the behavior. If our robot is to move in the soil, it needs to have a similar behavior to the worm and the root. The environment is actually the decisive factor. And this is true in the design phase as well.

MF: Can one therefore take movement as an indication of intelligence?[1] Can one in fact make this definition a bit broader and say that a design that already takes the environment into account is an indication of intelligence? That is to say, insofar as a design really does not necessitate a centralized form of control.

1 [Mazzolai 2016, p. 114]: "In animals, behavior usually refers to movements generated by muscles; plant intelligence on that basis does not exist. Movement is, however, the expression of intelligence it is not intelligence itself [...]. Nonetheless, plants respond to internal and external signals. Thus, a simple definition of plant intelligence could be adaptively variable growth and development during the lifetime of the individual. Exploiting adaptive abilities in plants could lead to the development of smart devices to monitor soil – not solely with the ability to sense but with the capability to follow stimuli/gradients and to take decisions to accomplish the needed tasks." See also: [Trewavas 2003; Trewavas 2004].

BM: I think this is very crucial because what is at issue here is the definition of intelligence that we use. This is an aspect that, with respect to terminology, is very narrow. What I mean to say is that our definition of intelligence is very narrow and it is tied to our own preconceptions of the human being.

We talk about smart or intelligent material. We say that something is intelligent because it may have sensing properties. What is meant, as we noted before, is that the control is in the material itself. Thus, one can say that plants are smart: they use the environment; they are intelligent, since they are aware of processes; they can adapt. More specifically, some scientists have recently demonstrated that plants really are capable of taking decision on the direction of growing or on other adaptation strategies, and do have a memory. Once more, it is not in the brain. And because there is no brain, the scientists cannot locate precisely where the memory is in the plant. They performed an experiment with a *Mimosa pudica*, the plant that closes its leaves once it is touched [Gagliano and Marder 2019]. After a while, the plant did not close its leaves. Some days later, they attempted the experiment once more and the plant still did not close its leaves. The plant in some way 'understood' that it is not in a dangerous situation. The question remains as to where the memory actually is. Since there is no brain, it is probably in the tissue of the plant – again in the material. Of course, the experiment does not demonstrate where the memory is located, but since it is certain that there is no brain another part should have this functionality, and probably this is in the material. This behavior associated with the interaction with the environment may be termed 'intelligence,' or at least it is another way of conceiving intelligence.

Obviously, the definition of intelligence is contingent. From another perspective, some people can claim that a plant is not intelligent, simply because by definition a plant (or material as such) is not intelligent. But what is intelligent? For me, the plant is intelligent, since another way of conceiving intelligence is in the way plants are adapted to the environment. Accordingly, I try to see the way in which this behavior is implemented by plants. What are the components that allow them to behave smartly in their environment?

Talking about plants as intelligent is probably the most difficult part in this kind of study. This is also due to a problem of observation, to the fact that we cannot see the movement, and we must remember that observation is the first step in the scientific method. Only now that we have technology that allows us to accelerate the movement can we start to talk about this as movement as such. Before this, it was quite simply impossible. Since scientists could not see the movement, plants did not move. Now that we have the technology, we are in the right position to change this preconception, to push its boundaries.

Bibliography

Gagliano M, Marder M. What a Plant Learns: The Curious Case of Mimosa Pudica. Botany One, 2019. (Accessed January 9, 2021, at https://www.botany.one/2019/08/what-a-plant-learns-the-curious-case-of-mimosa-pudica/.)

Gagliano M, Renton M, Depczynski M, Mancuso S. Experience Teaches Plants to Learn Faster and Forget Slower in Environments Where it Matters. Oecologia 2014, 175, 63–72, DOI: 10.1007/s00442-013-2873-7.

Khorsandi D, Fahimipour A, Abasian P, et al. 3D and 4D Printing in Dentistry and Maxillofacial Surgery: Printing Techniques, Materials, and Applications. Acta Biomaterialia 2020, DOI: 10.1016/j.actbio.2020.12.044.

Laschi C, Cianchetti M, Mazzolai B, Margheri L, Follador M, Dario P. Soft Robot Arm Inspired by the Octopus. Advanced Robotics 2012, 26 (7), 709–727, DOI:10.1163/156855312X626343.

Lucarotti C, Totaro M, Sadeghi A, Mazzolai B, Beccai L. Revealing Bending and Force in a Soft Body Through a Plant Root Inspired Approach. Scientific Reports 2015, 5, 8788, DOI: 10.1038/srep08788.

Mazzolai B. Biology for Robotics – A Multidisciplinary Approach to a Researcher's Life. IEEE Robotics & Automation Magazine 2016, 23 (1), 114–115, DOI: 10.1109/MRA.2015.2511687.

Mazzolai B, Beccai L, Mattoli V. Plants as Model in Biomimetics and Biorobotics: New Perspectives. Frontiers in Bioengineering and Biotechnology 2014, 2, 2, DOI: 10.3389/fbioe.2014.00002.

Mazzolai B, Mondini A, Del Dottore E, Sadeghi A. Self-Growing Adaptable Soft Robots. In: Koshima H, ed. Mechanically Responsive Materials for Soft Robotics. Weinheim, Wiley-VCH, 2020, 363–394, DOI: 10.1002/9783527822201.ch15.

Mazzolai B, Mondini A, Tramacere F, et al. Octopus-Inspired Soft Arm with Suction Cups for Enhanced Grasping Tasks in Confined Environments. Advanced Intelligent Systems 2019, 1, 1900041, DOI: 10.1002/aisy.201900041.

Murphy RR. Disaster Robotics. Cambridge, The MIT Press, 2014.

Sadeghi A, Tonazzini A, Popova A, Mazzolai B. A Novel Growing Device Inspired by Plant Root Soil Penetration Behaviors. PLOS ONE 2014, 9 (2), e90139, DOI: 10.1371/journal.pone.0090139.

Sadeghi A, Mondini A, Mazzolai B. Toward Self-Growing Soft Robots Inspired by Plant Roots and Based on Additive Manufacturing Technologies. Soft Robotics 2017, 4 (3), 211–223, 10.1089/soro.2016.0080.

Trewavas A. Aspects of Plant Intelligence. Annals of Botany 2003, 92 (1), 1–20, DOI: 10.1093/aob/mcg101.

Trewavas A. Aspects of Plant Intelligence: An Answer to Firn. Annals of Botany 2004, 93 (4), 353–357, DOI: 10.1093/aob/mch059.

Barbara Mazzolai is associate director for robotics and director of the bioinspired soft robotics laboratory at the Istituto Italiano di Tecnologia (IIT). From February 2011 to March 2021, she was the director of the IIT Center for Micro-BioRobotics (CMBR). She obtained in 2011 the PhD in microsystems engineering from the University of Rome Tor Vergata. In the years 1999–2009, she was research assistant and then assistant professor in biomedical engineering at the SSSA Center for Research in Microengineering (CRIM Lab, now the BioRobotics Institute). In 2013, she started research line on bioinspired soft robotics, for which she is the principal investigator.

Her research activities are in the areas of biologically inspired robotics and soft robotics. In this context, she is the pioneer of the fields of plant-inspired robots and growing robots. With the motto 'from Nature to Nature,' her research is a mutual combination of biology and engineering, pursuing the development of new technologies and robotic systems inspired by biological models – specifically plants and soft animals – and the advancement of scientific knowledge on living organisms and natural ecosystems.

Michael Friedman, Karin Krauthausen, Robert Shepherd

Interview with Robert Shepherd: On Soft Robots, Biomimetics, and Beyond

KK: Before starting with the general discussion on active matter, we would like to discuss more specifically the field you are researching, soft robotics, and how this field is situated in relation to active matter. How do you see yourself in this field? How do you consider the various materials or the various categories of matter – adaptive matter or programmable matter – which are being employed within this field of research?

RS: Adaptive matter and programmable matter are two different characterizations, and they are also employed differently in soft robotics. Here one has to recall that this differentiation is not exhaustive; there are several characterizations to add to this description of materials, namely, smart materials, adaptive (or adapted) matter, programmable matter, and so on.[1] For me, programmable matter and adaptive matter are different things. Programmable matter is prescribed to respond in a particular way. Adaptive matter should be able to change how it responds. To give an example for programmable matter, one can take origami [Holmes 2019]: you fold it so that, when you apply particular stress and then release it, it unfolds into the same shape. One can get better at making it respond differently, but traditionally it is mostly programmed to do a task based on one input. Adaptive structure can assess what is happening to it and change how it responds based on this assessment. This is how I differentiate adaptive and programmable matter. The term 'smart matter' is a bit of a catch-all. All of these materials are smart, one can say. What soft robotics does is that it takes advantage of all of these properties and puts them into a device for performing work that you would think a robot would perform. We want these robots to be autonomous, in the sense that they can make decisions based on their environmental conditions with the least amount of human input as possible. But, having said that, you can still use programmable matter to form the basic motion of the device. Take,

1 On the blurry distinction between 'matter' and 'materials' in this discourse, see the introduction to this volume.

Acknowledgment: Michael Friedman and Karin Krauthausen acknowledge the support of the Cluster of Excellence 'Matters of Activity. Image Space Material' funded by the Deutsche Forschungsgemeinschaft (DFG, German Research Foundation) under Germany's Excellence Strategy – EXC 2025 – 390648296.

Open Access. © 2021 Michael Friedman et al., published by De Gruyter. (cc) BY-NC-ND This work is licensed under the Creative Commons Attribution-NonCommercial-NoDerivatives 4.0 International License.
https://doi.org/10.1515/9783110562064-008

for example, our work on a walking quadruped (see Fig. 1),[2] which I would say is programmable. When you apply pressure to it, it obtains curvature. You could say that it is programmable matter, since based on that input it has this or that curvature. But if you apply a different amount of pressure in another place, and all of the sudden it obtains a different shape, perhaps then one may call it adaptive. It is a blurry line at this point. But it is all smart. There is a mechanical intelligence programmed in here – or rather even a mechanical computation, or a material computation. It is as if there is a material computer that inputs stress and outputs a mechanical force and strain profile depending on the situation. That is what makes it different than 'normal' robots, which for example are moving something; when another object gets in the way they just crush it – for example, when a banana is in the way, it just squishes it. The robots in soft robotics are very different than normal robots because their material is doing the computation.

MF: You have mentioned several subjects to which we will return later. I want to concentrate on the first topic you mentioned: soft robots. You describe soft robots as those composed of soft mechanisms and soft chassis.[3] As you said, hard machines, the 'regular,' 'normal' robots, can move objects from one place to another but cannot cope with changing conditions and environments. Soft robotics offers another point of view. How do you see this new conception of the machine stemming from soft robotics? In some of your papers you give several characteristics that are not usually associated with machines, like self-healing or being composed of autonomous materials [Mishra et al. 2020; Wallin et al. 2017; Tolley et al. 2014; Bekey 2005, pp. 1–25]. What are the differences between the old conception of the machine and the new one?

RS: One may say that the machine is something like a lever, it is stiff, and this mechanical advantage allows you to lift heavy things. These are traditionally the conceptions people associate with machines. But one should make a distinction between robots and machines. The reason I have done that is to work in the robotics community, as they have a particular vision of what a robot is. The main difference between robot and machine is the existence of sensors for robots. As I see it, the usage of 'robot' means sensors integrated with a computer to control the response of the machine. To me that

2 See [Shepherd et al. 2011, p. 20400]: The quadruped is a "soft robot, composed exclusively of soft materials (elastomeric polymers), which is inspired by animals (e.g., squid, starfish, worms) that do not have hard internal skeletons. Soft lithography was used to fabricate a pneumatically actuated robot capable of sophisticated locomotion (e.g., fluid movement of limbs and multiple gaits). This robot [...] uses no sensors, only five actuators, and a simple pneumatic valving system that operates at low pressures [...]. A combination of crawling and undulation gaits allowed this robot to navigate a difficult obstacle."

3 Note by Robert Shepherd: The difference between soft robots and 'normal' ones may be that they (i.e., soft robots) routinely combine the chassis and the mechanisms into one (embodiment).

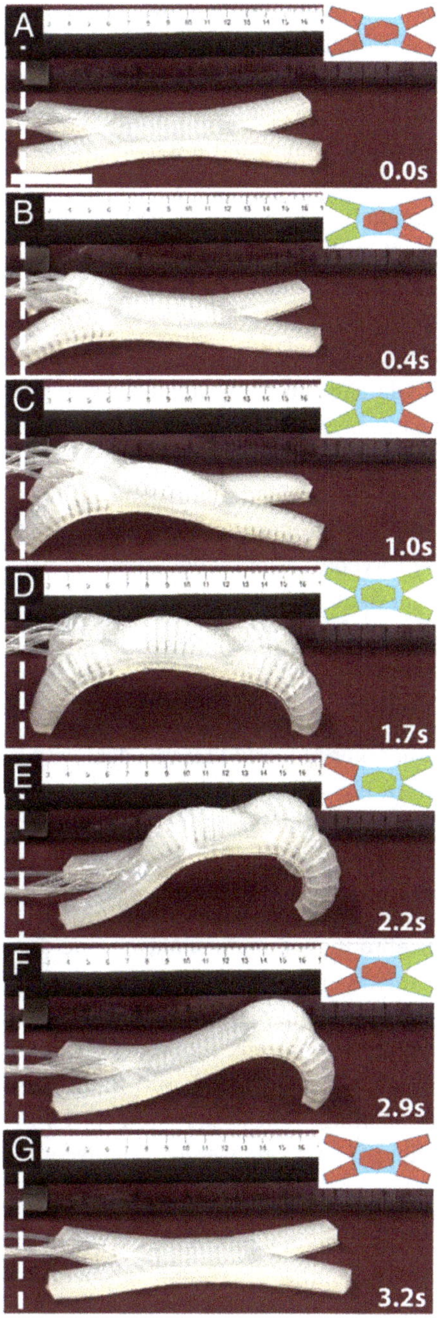

Fig. 1: Cycle of pressurization and depressurization of a quadrupedal soft robot, which results in undulation. From: [Shepherd et al. 2011, 20401, Fig. 2]. © The Authors of [Shepherd et al. 2011].

is a robot. For the sake of simplicity, one could say that a 'machine' has no sensors. You apply input and it operates according to this input. Whereas for robots there are sensors that can respond differently based on the situation – in order to enable better control. In a nutshell, machine without sensors; robot sensors with integrated computer. The question hence arises: how are these soft machines and soft robots different from existing machinery and existing robots?

To give an example: One can consider human beings as skeletons with skin, organs, sensors, and a brain. Our brain is not just in the head, it is distributed. For a long time I was thinking we need to make a soft robot that is totally soft, like an octopus, as this is very adaptive. If one adds a skeleton, it is not adaptive anymore. But in fact there are a lot of advantages to having deformable tissue on a hard skeleton. We have the ability to sense very well. If I damage myself, I can detect it. If I get a blunt trauma, my skin can actually be fine but my bone on the inside will crack. Being able to feel that damage is important.

If you have these endoskeletal soft robots moving around, interacting with people, one of the things one would wish or require from these robots is an ability to self-heal, which is a challenge for robotics. Everybody wants robots that can walk around and operate for days. But they may get damaged. And when they get damaged, they are going to behave differently. Their control systems are going to adapt to that. But this ability to sense inside of a robot's 'flesh' does not exist right now. We will need this and self-healing for better control.

MF: You have just talked about self-healing, another property that you deal with in your research is shape memory activation (see Fig. 2).[4] These are properties that might seem to be inspired by nature. What is actually the relationship between nature, biology, and your approach to robotics? Does one need to imitate all of nature's architecture or is it enough to be inspired only by several 'good' properties?

RS: We obviously do not need robots with reproductive organs. Maybe one does not need all these extra abilities either. But the crucial point is that at best, rather than mimicking nature, we are inspired by it. We cannot make muscle. Some people are growing it into robots, but we certainly cannot mix some chemicals together and then have something that acts like muscle. Having said that, I also believe that almost nobody is able to do something that is non-bioinspired because this inspiration is around us all the time. Everything we do is by necessity nature-inspired.

KK: I would like to focus on the relation between structural hierarchies in biological materials and in synthetic materials. One may say that biological systems operate as

4 See, for example: [Van Meerbeek et al. 2016].

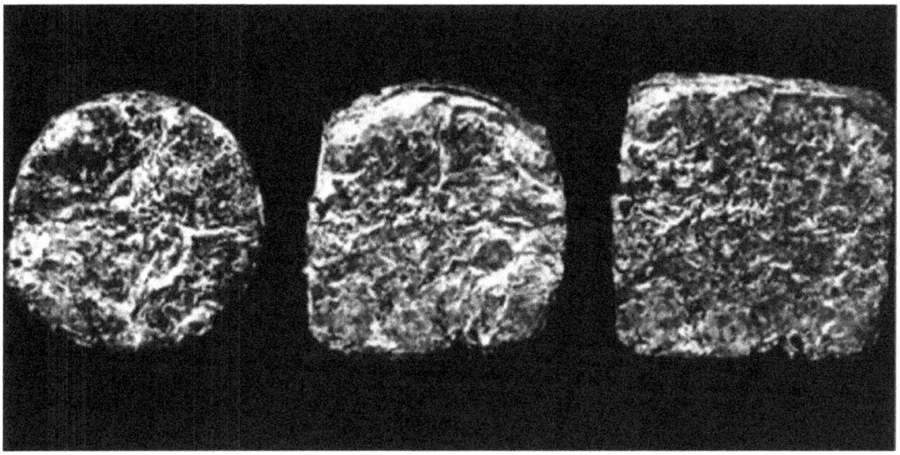

Fig. 2: Shape memory actuation: a compressed cylinder of metal-elastomer composite expanding into a cuboid after melting the metal foam. Adapted from: [Van Meerbeek et al. 2016, p. 2804, fig. 3]. © 2016 Wiley-VCH Verlag GmbH & Co. KGaA, Weinheim.

multiscale chemomechanical feedback processes on various levels and through various hierarchies. They can adapt and develop because there is no predetermined program and not everything is controlled. Hence, an emergent structure may appear, as there is a kind of instability in the system itself. Does the new approach you are delineating with soft robotics also point toward this complexity of structure?

RS: Complexity and emergence of complexity certainly arise in nonequilibrium structures. The problem is that the number of devices and elements in our designs right now is just too small to expect an emergence of anything totally unexpected. Right now, we are not dealing with emergence or complexity this way. Eventually – hopefully – we will be able to do that. I think we can use additive manufacturing to get out a lot of the bottom-up assembly that happens in nature, but it would not have the resolution or complexity found there.

I believe a way to reformulate your question would be to ask what is the future of smart materials? To give an example, in our bodies we are constantly using fluid transport to operate. So we deliver sugar and we burn that sugar and our muscles contract. That is impractical in a car and in most things today because the surface-area-to-volume ratios are too low. For us to operate we have the very high surface-area-to-volume ratios of our blood. Our arteries, veins, and capillaries are distributing energy and removing energy (or removing waste) very efficiently. But if you were to ask an engineer to build something like us they would say that it is not possible or that it would not work well. But we human beings can have these abilities, because we self-assemble into these 'impossible' structures. So a possible solution to this problem would be to construct materials or robots with soft batteries, or in other words to think about embodied energy (see the lecture: [Shepherd 2019]). Explicitly, embodied energy

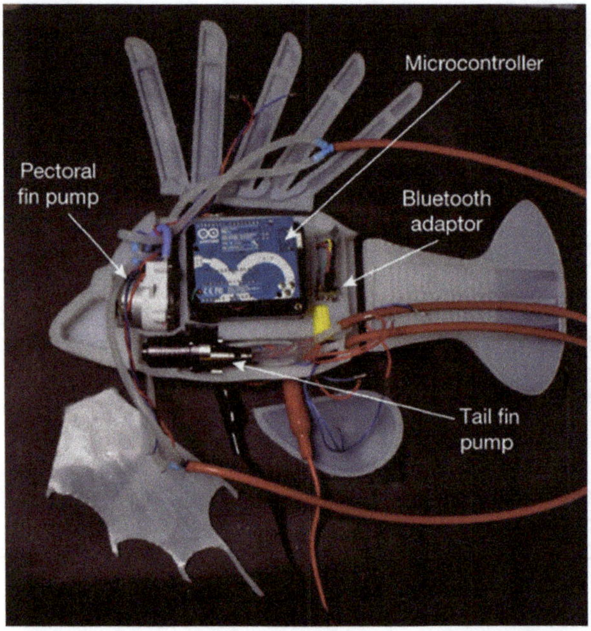

Fig. 3: One half of a disassembled 'fish' robot showing how the pumps and control hardware are housed internally. Material from: [Aubin et al., 2019, p. 55, fig. 5b], Springer Nature.

is a composite where the energy is part of the composite. We are working on a project on robot blood where the fluids are used to transmit force (see Fig. 3).[5] You inflate this leg and it bends – this is done with air but you could also do it with liquid. If the liquid can also store energy it becomes multifunctional. You could also imagine doing this with a solid-state energy storage system in a rubber housing. A lot of work in the future is going to be on storing and packing as much energy as you can into every cubic millimeter of volume.

KK: Can one consider it as structurally stored energy?

RS: Yes, that is one example. But the structure is more than just a structure; it is also the actuator or the sensor, or ideally structure–actuator–sensor altogether. Which is also one of the benefits of a soft robot: it is the structure but it is also the actuator.

5 [Aubin et al. 2019, p. 51]: "Modern robots lack the multifunctional interconnected systems found in living organisms and are consequently unable to reproduce their efficiency and autonomy. Energy-storage systems are among the most crucial limitations to robot autonomy, but their size, weight, material and design constraints can be re-examined in the context of multifunctional, bioinspired applications. Here we present a synthetic energy-dense circulatory system embedded in an untethered, aquatic soft robot. [...] this synthetic vascular system combines the functions of hydraulic force transmission, actuation and energy storage into a single integrated design."

The idealized case is that the material is already structured, and the whole assembly of functions is already inscribed in it. We have to make compromises currently, but ultimately that is what we would like. I think 3D printing allows us to get there too.

MF: I want to discuss 3D printing in more detail. Three-dimensional printing may be seen as biomimicry in a certain way. This brings me back to an earlier question: to which level do we want to mimic nature? And, if we return to the question on emerging structures, can one imagine a scenario where one only prints the basic structure, when afterward something, a more complicated structure, will emerge?

RS: I think the future is in the application of 3D printing to robotics. The past has been about programmable matter. The present is a combination of the two. People have been growing tissue on 3D printed scaffolding for a long time. Not just through 3D printing but also through molding things. There is a difference between additive manufacturing and 3D printing. Three-dimensional printing is always additive manufacturing but additive manufacturing is not always 3D printing. Any time you are adding a material to something, that is, additive manufacturing. But 3D printing offers the most freedom in design. But one has to recall the idea of growing muscle on a scaffold and then using the muscle as an actuator in a robot: this is new.

Another concept in 3D printing is when you try to embed dynamics in the structures so it will move – to embed a fourth dimension, some people would say, as it responds differently based on inputs. I certainly believe that future research will be more about 3D printing structures where there are also actuators, which are also sensors.

To your point, however, using a 3D printed structure to nucleate more complex or nanostructured architectures is an exciting area of research.

MF: What you are proposing here is that the 3D printed material would also be an actuator, in a sense that it can be considered as four-dimensional (4D) printing, that is, as printing a material which is adaptive, specific to the environment, which does not have a fixed, 'programmed' behavior from the beginning.

RS: Yes, this is what I would say 4D printing is now. Honestly, though, I do not really like the term. I think it implies something more grandiose than what is mostly done. That is not saying it is not exciting, but does not it sound like you are warping through space? Anyway, yes, if you incorporate the ability for the structure to also sense and change its actuating response based on what it is sensing, you are not only printing a robot, you are actually printing a robotic material.

KK: If I may highlight another aspect of this adaptivity, another aspect of 3D printing concerns the geometry of what is printed. Three-dimensional printing is ultimately printing geometry. What are the geometrical properties that are taken into account when printing? Should one print the whole geometric structure? Especially when it comes to biological materials, if one prints an organ. At what level of detail do we actually need to print?

RS: When one talks about the 3D printing of tissues, one can definitely consult the work of Jennifer Lewis (Lewis Lab), who works in that direction quite a bit. An area that this community is working on is organoids, such as groups of cells that are representative of an organ, but they are not the whole organ. Similar to a bunch of liver cells, when one makes tests and sees how the cells respond. Concerning printing geometrically, concerning what we do in synthetic structures, I would say you can print the most complex shapes with stereolithography very quickly. The issue is that it is only one material. So how do we program the geometry and the response for what we want without having access to multiple materials? The answer is to just print different porosities, something akin to metamaterials.

MF: As you indicate, you are printing only with one material. But obviously that is not how nature works. Does one also think in the direction of heterogeneous architecture? Different materials are combined with each other, with each material having a different function.

RS: This is something that does happen now, but not at the level of sophistication needed. We are working on a project right now where we are embedding carbon nanotubes into elastic structures to give us a particular stress response. But it is clear that this is not even near to the level of biology.

There is an opportunity here, though, to merge your other notion with this one. Printing a hard structure and nucleating the polymerization and growth of a soft one off of it for endoskeletal–soft tissue constructs (or, perhaps more difficult, the inverse).

MF: The question therefore may be whether and how we have to follow biological structures or be inspired by nature. In one of your papers you note that nature has a variety of ways to demonstrate these stretchable structures, whereas we only have the ability to inflate a balloon [Pikul et al. 2017] (see also Fig. 4). What can we actually learn from nature in respect to these stretchable materials?

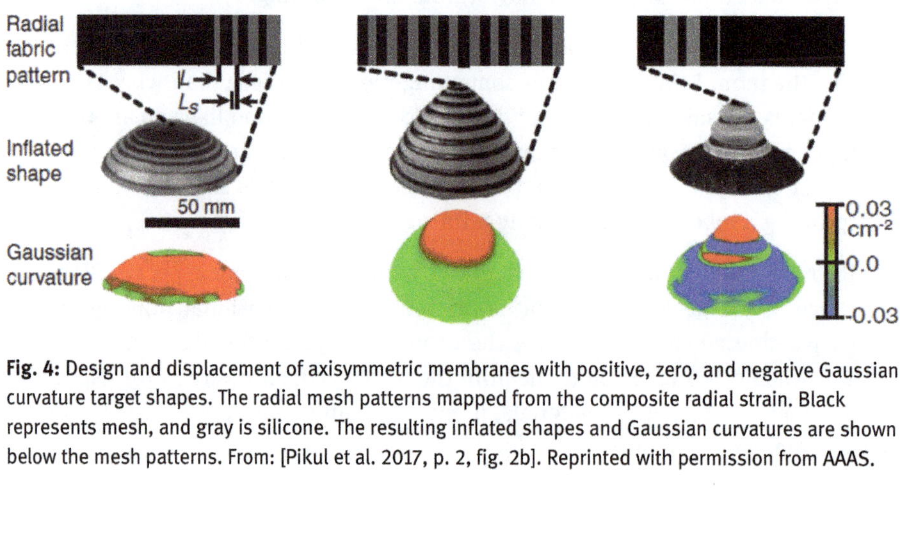

Fig. 4: Design and displacement of axisymmetric membranes with positive, zero, and negative Gaussian curvature target shapes. The radial mesh patterns mapped from the composite radial strain. Black represents mesh, and gray is silicone. The resulting inflated shapes and Gaussian curvatures are shown below the mesh patterns. From: [Pikul et al. 2017, p. 2, fig. 2b]. Reprinted with permission from AAAS.

RS: Nature is very good at not making the best of anything but does adapt to lots of situations. Nature is always dealing with compromises. One way to get a good compromise is in mechanical compliance. Leaves want to absorb as much light as possible. They should be very broad, very thin, because all the light is absorbed in the first few layers. So why even be thick? In addition, when it is very windy, if they were rigid and fixed to a branch, the whole tree would fall down. How nature deals with that is by making the leaf actually fold up. The aerodynamic profile is greatly reduced. That is a very simple example of how compliance is used in nature as an engineering compromise. That is what bioinspiration is: seeing what nature is using compliance for.

MF: That points to one of the essential topics of the research on active materials. We program or manufacture materials that blur the line between the natural and the artificial, between biological and human materials. How does the community of scientists redefine essential concepts for these two domains, concepts such as 'life' or 'intelligence,' if for example the concept of self-healing is now blurred?

RS: As you point out, the problem here is defining what words may mean. Scientists have been 'hijacking' language. But, to speak more concretely, do we have a robot that can self-heal like a gecko? No. If you pull the tail off of a gecko, it can grow another tail (morphogenesis). Mostly what we are talking about when we talk about self-healing is – when there is a fracture the two surfaces recombine. We (re)define the word to capture that. But in nature, self-healing usually means regrowing. There are very few, if any, examples of that in soft robotics.

MF: Can one say that the current research points at least toward a change of the definition of 'intelligence'? That intelligence is not necessarily connected to human beings.

RS: So maybe a Turing test for mechanics? Certainly there is a material computation which is intelligent: intelligently programming your material to respond to stress input.

KK: In the second half of the twentieth century, there was an emphasis on genetic biology and the expectation that it would be able to decipher and later even to (re)write the code of life [Kay 2000]. Obviously, these great expectations were not fully met. However, the 3D printing of living cells reminds me of this idea of reading and/or writing life. Hence, on the one hand, we are still thinking with our old concepts, but on the other hand there maybe also a new element that comes from robotics, namely that intelligence is not something related only to human beings or to animals, that intelligence is something that comes from diversity, from a play between forces.

RS: You have a point there – though I think that on an individual level we would not be able to do anything other than mimic biology. I also think that CRISPR is perhaps an example that shows we have kept up with our potential in working purely within biology. But perhaps as a group something could emerge that is non-biomimetic. AI has a good chance of producing something non-biomimetic.

MF: To finish I would like to discuss again in more detail the example of the biomimetic robot you mentioned at the beginning of the interview. When we are talking about bodies of machines – for example, the soft locomotive robots or the quadrupedal soft robots you have developed (see Fig. 1) – this does not look like any animal or at least like any human being – and one may very well say that it walks in a way that is unpredictable. Can we manufacture materials that can act in unpredictable ways? Was the quadruped's way of crawling and walking predictable beforehand?

RS: I predicted that it could move. But after I made it I had to figure out the best way to make it move. I developed the gait after we had the body.

MF: Were you able to predict how it would move?

RS: In a very general sense. We knew that with the four legs of the quadruped, we would have to use some of them for stability but others for moving forward. We knew that if we made the angle between the legs and the body 90° it would not move forward, it would move sideways. If we made them just straight ahead, then we felt turning would be harder to do. It could have been 45°. But the exact angle between 0° and 90° is just a guess.

MF: By designing some legs to be stiffer, for example, does not one already preprogram a state of equilibrium? If all the legs would have the same functions, the same stiffness, and properties, one could say that it is in an equilibrium state. And not only that but that it would hardly move in that state.

RS: The question is: what does it mean to be out of equilibrium? I myself am out of equilibrium. If I take the legs away from the quadruped, it is going to be on the ground. Now it is clearly to be found more in a lower energy state but it can get even lower, since I could even burn it and it would release energy. So, when one says that something is out of equilibrium, the question is: relative to what? 'Out of equilibrium' may mean a highly dissipative structure; that is, one can put energy into it and that energy is going to be used to overcome some kind of energy barrier to ultimately lower the energy of the structure. Then there is a cascading effect of taking energy in and redistributing it to make the whole system reach a more energetically favorable state.

When we apply this to the quadruped, it becomes more out of equilibrium when I pressurize it. By pressurizing it, I increase its energy levels, so when I relieve the pressure it wants to dissipate that energy. My point is that there are degrees of how out of equilibrium a system is. For example, we as humans are so out of equilibrium that it is almost impossible to believe that we exist. Our proteins are constantly moving around but our body temperature stays the same. Every part of us at every level is just constantly in motion. It is amazing that we live as long as we do, or that we even exist.

Bibliography

Aubin CA, Choudhury S, Jerch R, Archer LA, Pikul JH, Shepherd RF. Electrolytic Vascular Systems for Energy-Dense Robots. Nature 2019, 571, 51–57, DOI: 10.1038/s41586-019-1313-1.

Bekey GA. Autonomous Robots: From Biological Inspiration to Implementation and Control. Cambridge MA, MIT Press, 2005.

Holmes D. Elasticity and Stability of Shape-Shifting Structures. Current Opinion in Colloid & Interface Science 2019, 40, 118–137, DOI: 10.1016/j.cocis.2019.02.008.

Kay L. Who Wrote the Book of Life? A History of the Genetic Code. Stanford: Stanford University Press, 2000.

Lewis Lab. The President and Fellows of Harvard College, 2021. (Accessed January 19, 2021, at https://lewisgroup.seas.harvard.edu/.)

Mishra AK, Wallin TJ, Pan W, et al. Autonomic Perspiration in 3D-Printed Hydrogel Actuators. Science Robotics 2020, 5 (38), eaaz3918, DOI: 10.1126/scirobotics.aaz3918.

Pikul JH, Li S, Bai H, Hanlon RH, Cohen I, Shepherd RF. Stretchable Surfaces with Programmable 3D Texture Morphing for Synthetic Camouflaging Skins. Science 2017, 358 (6360), 210–214, DOI: 10.1126/science.aan5627.

Shepherd RF. Embodied Energy for Enduring Robots. ACM Symposium on Computational Fabrication. Lecture, June 17, 2019. (Accessed January 19, 2021, at https://www.youtube.com/watch?v=eb8DH9yswql.)

Shepherd RF, Ilievski F, Choi W, et al. Multigait Soft Robot. PNAS 2011, 108 (51), 20400–20403, DOI: 10.1073/pnas.1116564108.

Tolley MT, Shepherd RF, Mosadegh B, et al. A Resilient, Untethered Soft Robot. Soft Robotics 2014, 1 (3), 213–223, DOI: 10.1089/soro.2014.0008.

Van Meerbeek IM, Mac Murray BC, Kim JW, et al. Morphing Metal and Elastomer Bicontinuous Foams for Reversible Stiffness, Shape Memory, and Self-Healing Soft Machines. Advanced Materials 2016, 28 (14), 2801–2806, DOI: 10.1002/adma.201505991.

Wallin TJ, Pikul JH, Bodkhe S, et al. Click Chemistry Stereolithography for Soft Robots that Self-Heal. J Mater Chem B 2017, 5 (31), 6249–6255, DOI: 10.1039/C7TB01605K.

Robert F. Shepherd received his BS (2002) and PhD (2010) in materials science at the University of Illinois, where his research focused on developing polymeric and colloidal suspensions as 'inks' for 3D printers. He also fabricated microfluidic devices to synthesize single micron- to millimeter-scale parts (e.g., glass and silicon microgears). Since 2013, he is an assistant professor and since 2018 an associate professor at the Cornell University. Since 2018, he is the co-founder of Organic Robotics, Corporation.

Robert F. Shepherd is interested in developing disruptive manufacturing technologies (e.g., 3D printing, replica molding, and microfluidics) and functional materials to enable new devices and user experiences. He is particularly interested in simultaneously increasing the speed, dimensionality, resolution, and materials capability of free-form fabrication techniques, and developing soft actuators that mimic biological functions, but not necessarily their mechanisms: These interests are synergistic and will, perhaps, enable more efficient and life-like machines. His main research themes are energy-dense robots, engineered living materials, haptics and prosthetics, underwater robots, agricultural robots, and additive manufacturing.

Michael Friedman, Ramin Golestanian, Karin Krauthausen

Interview with Ramin Golestanian: Living Matter, or What Is Life?

MF: Active matter is basically a relatively new area of research. I would like to begin with a question about the history of this research, since you have made an important contribution to this field yourself in your own research, and you have also reflected on it as the editor of a special issue on active matter [Golestanian and Ramaswamy 2013]. Against the background of your own discipline, physics, what questions and research objects do you associate with the beginning of active matter research? And what were the questions and objects that you started out with yourself in this area?

RG: The starting point, as I know it, was the application of statistical physics to the field of animal behavior [Vicsek et al. 1995; Toner and Tu 1995]. Animal behavior is a very active field in which scientists study the collective behavior of animals when responding to environmental changes, shortages of food, threats of predators, as well as other phenomena such as migration. While it is not surprising that animals, which have brains and the ability to process environmental signals, are capable of such behavior, it is remarkable, if you think about it, that even bacteria can exhibit such collective behavior [Wong et al. 2021]. Thus, very simple organisms or life forms can 'decide' or 'decide together' what to do as a whole in response to an external cue. This affects their mode of existence, and as a consequence, you can also observe changes on the large-scale level of behavior. Statistical physicists are very well equipped to deal with this kind of phenomena because the modeling of minimal many-body systems is pivotal in statistical physics. It provides us with descriptions on the microscopic scale of a system, and from there we can go through the different levels of coarse-graining and predict a large-scale behavior that is related to the interaction of microscopic elements. Historically statistical physics can be traced back to the work of James Clerk Maxwell, Ludwig Boltzmann, and others in the second half of the nineteenth century and at the turn of the twentieth century. They were already trying to understand how one could bridge the different scales and hence make one's way from the statistical description of individual atoms and molecules at the microscale to thermodynamic behaviors on the macroscale. When studying a gas, we measure various macroscopic properties (such as pressure, volume, and temperature), which we can relate to the microscopic elements that we know make up the system. For example, the temperature in this room is related to the average kinetic energy of individual molecules – this is a fascinating insight. Because statistical physics is such a powerful tool it has made its way into many different

Acknowledgment: Michael Friedman and Karin Krauthausen acknowledge the support of the Cluster of Excellence 'Matters of Activity. Image Space Material' funded by the Deutsche Forschungsgemeinschaft (DFG, German Research Foundation) under Germany's Excellence Strategy – EXC 2025 – 390648296.

Open Access. © 2021 Michael Friedman et al., published by De Gruyter. (cc) BY-NC-ND This work is licensed under the Creative Commons Attribution-NonCommercial-NoDerivatives 4.0 International License.
https://doi.org/10.1515/9783110562064-009

fields over the years: into chemistry and chemical physics in the study of colloids, polymers, polymer solutions, and various forms of complex liquids and fluids [Jones 2002]; into biology with the study of DNA, proteins, and of the membranes of various biological components;[1] into the study of epidemic spreading [Bittihn et al. 2021]; and into economics (in a field called econophysics) to understand fluctuations in the market as well as economic changes at a global scale.[2]

Because statistical physics has ventured into so many different fields, it is natural that it has also found its way into the field of animal behavior. The first model that was proposed by a number of physicists was based on the flocking of birds [Vicsek et al. 1995; Toner and Tu 1995].[3] These physicists basically studied flocking by assuming some sort of local interaction and developed a simple and appealing understanding of the seemingly complicated behavior of birds. In statistical physics, we tend to focus on simple models and try to develop a deep understanding of the phenomena that they entail. Thus, for some years, the focus was on pushing the models on animal flocking behavior further. The next turning point was the idea of synthetically or artificially making microscopic self-motile systems, hence self-propelled elements. Motility is one of the very visible aspects of life, which is why physicists have studied biological molecular motors in order to design synthetic molecular motors.

This was something I was very much excited about (and involved with) from the very beginning. We started with what I call 'swimming' because it is a free movement in a liquid environment. By looking at these self-propelled systems in water or solution, we were able to propose several models.[4] These microswimmers essentially had to be nonequilibrium systems. Because of the small sizes of the elements in this model system, one has to worry about the laws of statistical physics and fluid dynamics, which in that environment is predominantly viscous. An important aspect is providing a source of energy. We know that biological molecular motors convert chemical energy directly into mechanical work – there are proteins that undergo conformational changes when they are catalyzed and break up the very energetic molecule called adenosine triphosphate into lower-energy components. It is the combination of enzymatic activity and mechanical work that gives rise to the very interesting behavior of these molecular motors. My fascination with molecular motors entered a new level during a workshop on this topic at the Isaac Newton Institute in Cambridge in March 2004. I remember that I bought a copy of the most authoritative book on the topic at the time by Jonathan Howard from a local bookstore in

1 For the biophysics of the cell, see: [Boal 2002].

2 The term 'econophysics' was coined by the statistical physicist H. Eugene Stanley from Boston University, who also wrote the first textbook on the new interdisciplinary field of econophysics [Mantegna and Stanley 1999].

3 For a report on recent approaches, see: [Cavagna, Giardina, and Grigera 2018].

4 On the first model systems, cf. [Najafi and Golestanian 2004; Golestanian and Ajdari 2008; Golestanian 2010]; for a review, see: [Elgeti, Winkler, and Gompper 2015].

Cambridge and read the introduction that included a very interesting statement: "[...] motor proteins are unusual machines that do what no man-made machines do – they convert chemical energy to mechanical energy directly, rather than via an intermediate such as heat or electrical energy" [Howard 2001, p. 1]. This immediately sparked an interest in me: can we take on this challenge and design a model system – the microswimmer – that could convert chemical energy directly to mechanical work [Golestanian, Liverpool, and Ajdari 2005]? There have been experimental realizations of our prototypical microswimmers which have proved that it is possible to build small-scale systems that behave like living systems [Leoni et al. 2009; Tierno et al. 2008; Grosjean et al. 2016]. Naturally, the nonequilibrium activity of the system physically changes the environment, and a mechanistic knowledge of this change will be important for understanding nonequilibrium interactions between different components. For example, if the system consumes a certain chemical fuel for its activity, it will act as a sink for this chemical, and that would have the long-range effect of depleting the chemical concerned. The local concentration of the fuel would be 'sensed' by the other chemicals involved, and this could provide a mechanism for the molecular model systems to communicate and then exhibit emergent traits [Soto and Golestanian 2014].

MF: Some of the model systems operate with sensors and signals, which they use to organize themselves, or which they use to develop collective behavior [Bäuerle et al. 2018] – was that the idea?

RG: It is very fascinating to study the collective effects in these systems, which are similar to what we could attribute to collective effects in microorganisms such as bacteria or in higher organisms such as birds. Microswimmers operate under the condition of consuming a certain chemical energy, which provides them with a flux for maintaining their nonequilibrium activity. They could then naturally communicate with others by means of the laws of the physical medium – for example, momentum conservation in the fluid will basically lead to hydrodynamic interactions,[5] and the chemicals being depleted will create fluxes that will naturally be felt by everyone [Saha, Golestanian, and Ramaswamy 2014]. Coming back to your question, at the basis of my research is the question of what identifies living systems as *living* systems. What distinguishes them from synthetic systems, even if we can model a few attributes of life – mainly motility, operating in nonequilibrium conditions, sensing, and signaling – or construct these on the nano- or microscale? This can be compared with some developments in science in the nineteenth century. The botanist Robert Brown took an interest in minute moving particles suspended in water that he could observe under a microscope. With his careful experiments Brown

5 See the review on studies in synchronization driven by hydrodynamic interactions in: [Golestanian, Yeomans, and Uchida 2011].

could show that an inorganic form of matter such as dust particles would perform this kind of jittery motion in a fluid medium [Brown 1827]. Thus, being 'alive' is not a necessary requirement to exhibit some of the behaviors associated with living systems. Interestingly, understanding the Brownian motion was the starting point for molecular theories of matter, since it is the movement of the molecules that form the fluid which enables the motion of the inorganic particles.[6] Even at thermal equilibrium, when there is basically no consumption of energy, matter is constantly moving due to fluctuations.

The study of active Brownian particles is part of the research on active matter [Gompper et al. 2020]. Our models provide a nonequilibrium analogue to Brownian motion, which was initially observed at equilibrium conditions. We use simple models to show that synthetic systems can behave similarly to living systems. Naturally, this does not mean that there is nothing special about living systems. In our approach, we would like to make living systems from the bottom up in order to find out more about the specificity of what constitutes life and where the boundaries of synthetic systems are – this is the goal of my new Department of Living Matter Physics at the Max Planck Institute for Dynamics and Self-Organization in Göttingen[7] (see Figs. 1 and 2 for examples of recent research at the Department of Living Matter Physics).

KK: Motility, gradient sensing, signaling, and replication are traditionally seen as properties of living systems. If science is able to reproduce some of these life-related agencies in a synthetic system the result – at least at the moment – will still be a technical thing. Nevertheless, the concept of 'living matter' and the physical understanding of matter and material that comes with it is influencing our idea of life and subtly blurring the boundary between living and dead matter. I assume not all biologists approve of the physical description of life?

RG: Biologists have a very specific view on living systems, which regards all the details as important. Physicists tend to use abstraction and simplification, and subtract all the elements that are nonessential. It is natural for a biologist to look for differences because in biology the diversity of species is the fundamental law of life. The fact that species vary in space and time and that a small difference can often lead to a new species inspired Charles Darwin to his principle of natural selection, that is, the significance of variation for the survival of the species. Darwin noticed that the beak

6 These molecular movements could only be modeled with the help of statistical physics. At the beginning of the twentieth century, Albert Einstein was able to quantitatively predict Brownian motion, and Jean Perrin was able to experimentally prove these theoretical deductions [Einstein 1905].

7 At the time of the interview, Ramin Golestanian was on the point of leaving his post as professor of theoretical condensed matter physics at the University of Oxford to direct the Department of Living Matter Physics at the MPIDS in Göttingen. He subsequently built up the department in Göttingen and organized the research on living matter into the areas mechanical activity, information flow, and dense active matter (https://www.ds.mpg.de/lmp, accessed February 4, 2021).

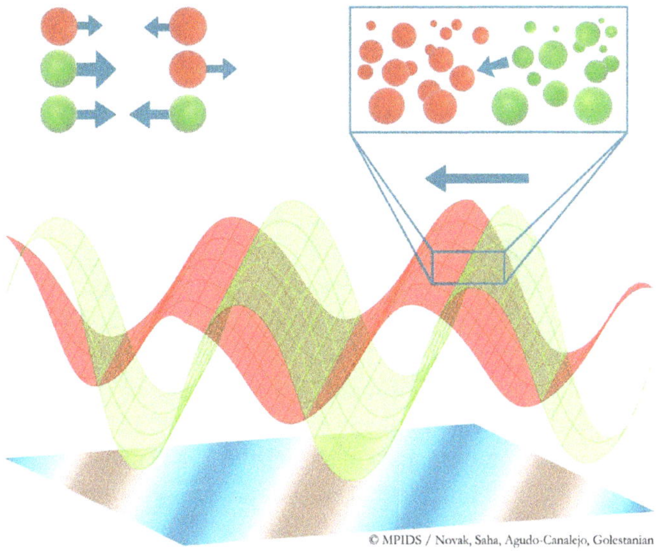

Fig. 1: Particles of two types (red and green) interact with each other. Whereas particles of the same type necessarily experience reciprocal attraction or repulsion, particles of different types can interact nonreciprocally. Here the green particles chase after the red particles. At large scales, high-density bands of green particles chase after bands of red particles, leading to the emergence of global order and net motion in the system. From: [Golestanian 2020]. Image and description: © MPIDS / V. Novak, Saha, J. Agudo-Canalejo, R. Golestanian.

Fig. 2: Like pieces of a puzzle, the proteins (yellow) can fit together to form a complex. Only then are they functional and able to bind to the target molecules (red). The proteins diffuse faster as singlets and form complexes when they need to perform their function. Image and description: © MPIDS / V. Novak, J. Agudo-Canalejo & R. Golestanian.

of the Galapagos finches differed from island to island, and the ornithologist John Gould confirmed that these variations did not belong to one species but indicated that different species lived on adjacent islands.[8]

Biologists build their knowledge on observed differences and details. By contrast, physicists look for similarities in order to find some of the fundamental rules constraining all forms of life. These two approaches are complementary and can help each other immensely, provided there is adequate communication between the two communities. We can understand this better from a simple argument: if our quest to understand living systems can be regarded as a search in a very high-dimensional space of possibilities, then we can narrow down the possibilities – and make the task easier – if physics can tell us which constraints to include. For example, any creature living in the physical world should satisfy the physical conservation laws. Thus, when we are confronted with many different possibilities, we need to look for the subset of the configuration space that satisfies these physical constraints, and it is only physics that will provide that information.

KK: In the twentieth century, molecular biology wanted to decode DNA in order to understand the 'program of life.' This research and its attendant ideas have changed what we understand by 'life.' To put it simply, the genetic code stood for an abstract structure that informs and predetermines all concrete manifestations of life.[9] Despite the necessary reduction that accompanies modeling in statistical physics, can one say that the environmental conditions of a living system, the constant exchange between this nonequilibrium system and its environment, and finally also the history of this system play a greater role in active matter research than in genetic biology?

RG: The genetic code needs to interact with the physical world in order to realize its information, and that interaction involves all kinds of things. Firstly, it involves the physical constraints that I talked about, namely the conservation laws. Secondly, there are differences in the environments that influence the system – this you can call the history. For example, a bacterium that lives on the surface of our tooth will develop differently than the same bacterium swimming happily in isolation in its planktonic state. In both cases, the bacterium is using its genetic information, but it is expressing different genes. And sometimes, the same genes change because they are influenced by the environment. What biologists would call the possible impact of the environment on the genetic code, I would call physics, because the creature

8 Charles Darwin noticed the differences, but it was only some years after his return to England and John Gould's examination of the collected specimens that Darwin would understand the significance of variation for the survival of species. In the end, the collected Galapagos finches indeed led Darwin to his principle of divergence and to his famous *On the Origin of Species by Means of Natural Selection* (1859) [Sulloway 1982].

9 On the history and wide impact of this idea in the twentieth century (from the 1940s to the 1960s), see: [Kay 2000]; for a more recent gene-centric view of life, see: [Dawkins 1982].

that is made of this information is interacting with its physical world, and that interaction in turn will inevitably have consequences that are different in different circumstances. One also has to take into account the intrinsic stochasticity: if one repeats the same action over and over, one will encounter slightly different initial conditions and slightly different thermal fluctuations or local viscosity. The influence of these changes can escalate because the physical world includes turbulences, chaos, and the tendency to be sensitive to the environmental conditions and how they can unfold over time. It is not difficult to imagine that the genetic code of a living being can unfold in different ways and lead to different results. I would say that physics is able to explain this, whereas the traditional view of biology is not. However, the two fields will become complementary and will have to work hand in hand in order to predict what is basically the state of living matter. Active matter is a starting point for this very challenging task.

MF: In Gautam I. Menon's paper on active matter, the term 'code' does not appear [Menon 2010]. Therefore, I would like to ask, firstly, is that term even relevant? And, secondly, is what active matter actually proposes not in fact changing the definition of life? Instead of reducing life to some sort of genetic code, we might say that life is exactly this list of traits: motility, replication, gradient sensing, and signaling. Hence, can one argue that the choice to distinguish between inanimate and animate becomes dubious?

RG: The research on active matter reminds us in the first instance that we should be a bit more refined in our attempt to define life, because we now know that it will be possible to design synthetic systems that will do what living beings also do. The aim is to shed light on the question of how much of a certain behavior of a living being is simply physics. This could enable a clearer understanding of the specific features that are needed in order to have life, as well as of the features that, simply speaking, can be fabricated.

One can see similarities between active matter research and the history of flight. Humans have always wanted to fly; we have always watched birds with admiration, and there are many accounts of people who have attached artificial wings and jumped from towers – some even managed to fly. However, when we eventually succeeded in building flying machines, it was not by simply mimicking what a bird does, which involves very sophisticated flapping movements of the wings. In the end, all of that was irrelevant. What we really needed was to understand the laws of fluid dynamics, the Navier–Stokes equations (which were found in the nineteenth century), then the development of the science of buoyancy, lift force, and the implementation of this scientific knowledge through engineering. What really did it for us was coming to a fundamental understanding of fluid dynamics, and hence of the science that was behind the movements of the bird. And the field of active matter is trying to extend this scientific knowledge to the microscopic scale, from the scale of individual molecules to the level of organisms and colonies.

In addition to the question of the specific features that constitute life, there is another major question in biology that is yet to be answered and where active matter research could be really helpful, namely the question of the multiscale, hierarchical nature of organisms. We know that changing a small molecular feature can lead to systemic changes at the level of the whole organism, say from a healthy state to a metastatic cancerous state for example. This can be viewed as a question in statistical physics because it relates to a many-body system at the microscale with elements that communicate. We have to understand the phase diagram and the boundaries that correspond to phase transitions between different possible emergent states. Such a problem is not very different from superconductivity or magnetism or superfluidity. Take for example iron. Iron conducts electricity already at normal temperature, but when you cool it down to near absolute zero it becomes a superconductor, which is a state in which the electric current flows without resistance. How do you explain the changing properties of such a material? The interaction of elements – atoms and electrons – leads to phase transitions and to emergent properties. You can see similar effects in living systems, which can be understood using similar conceptual approaches.

KK: What part does nonequilibrium thermodynamics play in active matter research?

RG: The language of thermodynamics is only used for the macroscopic description of a system. Statistical physics is not a separate entity; it provides the formalism that bridges the different scales and takes us from individual molecules and atoms that are fluctuating to the macroscale and predicts their collective behavior in terms of some simple quantities that can be measured. Pressure, for example, is the average of all the kinetic kicks of the molecules bouncing off the wall. With statistical physics you learn how to average those random forces and calculate the pressure that the gas is exerting on the wall of its container. Thermodynamics describes the macroscale – it tells us that when we compress the system, the pressure rises, and when we expand the system the pressure falls. Statistical physics gives us the microscopic picture and explains why a certain behavior is observed at the macroscale, because with statistical physics we can directly trace the specific activity of the molecules that leads to the resulting macroscale behavior of interest.

Now at a fundamental level we can apply statistical physics for the description of systems under nonequilibrium conditions. We only need to define the specific conditions – for example, by defining a temperature gradient or other forms of nonequilibrium drive. While it may be the case that a complete description of all nonequilibrium systems in statistical physics is still not available, I believe the available knowledge has not been used to its full potential. For example turbulence is a part of nonequilibrium statistical physics about which a significant amount is known. If we want to understand the turbulent airflows around airplanes, or predict the weather, we have to apply nonequilibrium statistical physics, since we are predicting a stochastic process that involves the motion of a fluid, in this case the atmosphere. To understand living systems, we also need to apply those tools, and that constitutes a program of research that has not reached its full potential.

In his 1944 (lectures and) book *What Is Life?*, Erwin Schrödinger asked how to explain the living cell with physics – and he admitted that the physical tools needed for this task were not yet sufficient [Schrödinger 1944]. This was before physicists had fully developed what we now call condensed matter physics, which is the application of statistical physics to understanding dense phases of matter. In these phases, the molecules are so close to one another that the interactions between them dominate their behavior. These interactions create complex emergent properties such as superconductivity, magnetism, and much else. Condensed matter physics predominantly flourished in the 1940s, 1950s, and 1960s, and it has continued to prosper until today [Kohn 1999]. It ventured into soft condensed matter and polymers already in the 1960s and 1970s in the work of Pierre-Gilles de Gennes and many others.[10] Now we have a very strong toolbox, because we know that life is made of soft condensed matter, of polymers, lipids, proteins, and so on, and operates far away from equilibrium. Thus, I am very optimistic about what our field can now contribute to Schrödinger's question: what is life?

MF: You mentioned two turning points in the research on active matter. The first is about applying the tools of statistical physics to flocks of birds for example, or to schools of fish, in order to describe their behavior. This is what Tamás Vicsek and others are concerned with. The second turning point would be the self-propelled materials that we ourselves can produce. Can one say that the third turning point is instability considered from a mathematical point of view? What I mean by that is a consideration of the topological structures, defects, and singularities which emerge while dealing with active matter. As you note in one of your papers, in some of the systems of active matter, "because of the continuous input of energy, defects in [...] [these] systems can be formed in pairs and subsequently move apart giving rise to a steady state where topological defects are continually being created and destroyed" [Thampi, Golestanian, and Yeomans 2014, p. 1].

RG: This is a very interesting aspect of active matter. Structures such as singularities and topological defects have in fact been quite an important part of the studies of condensed matter physics in the twentieth century. Take for example the studies on phase transitions by David Thouless, Duncan Haldane, and Michael Kosterlitz. They received the Nobel Prize in 2016 for their description of topological phases of matter, hence when matter can assume strange states such as those found in superfluids or thin magnetic fields. They developed the idea that a very specific two-dimensional model of magnetism shows phase transition via the creation of pairs of defects [Kosterlitz and Thouless 1972; Kosterlitz and Thouless 1973; Haldane 1983a; Haldane

10 Pierre-Gilles de Gennes received the Nobel Prize in Physics in 1991 [De Gennes 1991].

1983b; Thouless et al. 1982]. These transitions happen because of thermal fluctuations – at some point the thermal energy provided to the system will be such that it can afford to create a pair of defects that would otherwise come together and be annihilated. If you maintain the system at that temperature, these defects will be constantly created and annihilated in an average equilibrium description. This was a very significant turning point, because the whole field of topological insulators was built on it.

In living systems, understood as active matter systems, it is not the thermal fluctuations that create these vortices and defects but the elements that feed energy at the small scale. A solution of bacteria will look like a turbulent system, and this is not because energy is injected from the larger scale and cascades down to the small scale, which is the traditional description of turbulence, but because the bacteria individually pump in energy and this energy cascades upward to the larger scale to create the collective effect. This is an important point, because you can inject energy via the element itself. Thus, there are subtle differences, but the description of living matter essentially follows a formulation that was already developed for liquid crystals, for elasticity, turbulence, and superfluidity. It is very appealing to be able to build a mathematical formulation to describe the key properties of matter effectively in terms of defects and topological singularities [Thampi, Golestanian, and Yeomans 2015].[11]

I would say that the important turning point for active matter research was to see that nonequilibrium activity in the form of individual agents bringing in energy could basically excite modes in the system which would otherwise be excited from the outside by injecting energy or having a heat bath to provide the energy. That was the new insight, and it continues to be fruitful in the field of active matter.

KK: Going beyond the topological description of instability, what significance has the aspect of instability for the research on active matter?

RG: For me instabilities have always been very fascinating, because the physical account of the microscopic world, for example the interior of a living cell, describes a very thick, very viscous, very slow environment. There is this quantity called the Reynolds number that tells you in relation to your scale how viscous the environment is felt to be. If the objects are small, they move around in the fluid and they feel the environment as if they were larger objects moving in a much more viscous environment. For us the experience of moving in honey would essentially be the equivalent of how a bacterium feels when it moves in water: it is constantly battling with the environment and the viscous, thick, and sticky nature of it, which slows it down all the time. If we need to move quickly in this kind of environment, how

11 For a more recent modeling of living system properties by means of topology, see: [Tang, Agudo-Canalejo, and Golestanian 2021].

would we do it? On our length scale, we are used to Isaac Newton's first law of motion and force, which is primarily about a system with a certain momentum that will keep going at the same velocity. But a small influence can modify its course, and in a stickier environment, we would have to inject energy to keep the system going. To change the course of motion, we would need to inject even more energy. This is not a very efficient way to instigate change in a system. Living systems have found efficient solutions to this problem that are mostly built around instabilities that are bound to happen but are kept under control, that is, within a small margin of parameters that prevent the instabilities from becoming pervasive. In such a situation, we would only need a small trigger to push a controlled instability across a threshold, at which point it can get completely out of control. Hence, this mechanism can instigate a large change by bringing a small perturbation into the system. Living systems exploit this mechanism widely, since they operate near these instability thresholds. I have studied the properties of living systems many times and used these properties as an inspiration for synthetic systems because they are an incredibly efficient way to enable switches between different traits and behaviors. Instabilities are going to be extremely important in active matter research, since we see them everywhere in the living environment.

MF: I would like to focus on the changes which active matter research is introducing into the natural sciences. It seems that there is an incredible emphasis on modeling with numerical simulations.

RG: The field of active matter is essentially a subfield of condensed matter physics. It arose in the 1990s out of a theoretical impetus, and for some time the theory has been a little more advanced than the experiments. Over the last few years, however, a lot more work has been done on the latter, and these are catching up [Sanchez et al. 2012; Bricard et al. 2013]. The introduction of experiments on synthetic systems has been an important step because it has become possible to compare the results and the theories. Ultimately any field of condensed matter physics – electronic properties, semiconductors, magnetism, insulators, topological insulators – will predict something theoretically and try to realize it experimentally by taking experimental measurements. Only when both parts coincide can the scientific object be said to be clear and understood.

MF: How would you describe your evolution over the last 10 to 15 years in the field of active matter? I suppose it is hard to compare your views 15 years ago with your views today, but how do you consider the changes in the field itself? And how has the field changed with respect to both the academic institutions and the funding institutions?

RG: The biggest change for me personally has been the transition from the scattered activity of seemingly independent fields with a very low profile to this new situation. Fifteen years ago, when I was working in the field of active matter, I was interested in very simple and very limited questions. I was not thinking about life and how it could be explained with respect to active matter. This kind of ambition comes only after one dares to take one step and then, depending on the outcome, perhaps another step. When there is a consistency in what you see, you develop more courage. The first time I talked about living matter was in 2013 when I was asked to give a public lecture to the alumni of the University of Oxford.[12] I described the physical view on living systems, my understanding of the historical developments, and how we might be able to explain more in the future. In the last 4 or 5 years, I have followed the perspective that I proposed in this lecture. Unfortunately, the resources I had in the UK were limited, and several of my proposals for funding were not successful. The situation has changed – now I can see scraps of those initial ideas coming out in different places. On the whole, I would say that things went in the right direction.

From an institutional perspective, I would say that active matter research has always been an interdisciplinary field. It grows out of physics but also aims to address other disciplines, and as such, it has some intrinsic difficulties. Any interdisciplinary field will have an identity problem – are you a physicist, or a biologist? And there is also the question of approval or acceptance – are the science departments hiring people in this field? I think physics as a community views active matter research in a particular way, since physicists can see that this field has been growing and now plays a major role in the American Physical Society. There are now many conferences and sessions on active matter, and clearly there is a large influence.

Biologists certainly view the research on active matter or living matter with skepticism, since they do not think it is biology. And they are right – the aim of this field is not to explain or replace biology but to provide new information. Active matter research could answer some of their questions which are related to physical principles. It could spare them energy and resources. Eventually, in 10 years, this will feed into biology, but at the moment it may not be viewed that way.

Bibliography

Bäuerle T, Fischer A, Speck T, Bechinger C. Self-Organization of Active Particles by Quorum Sensing Rules. Nature Communications 2018, 9, 3232, DOI: 10.1038/s41467-018-05675-7.

Bittihn P, Hupe L, Isensee J, Golestanian R. Local Measures Enable COVID-19 Containment with Fewer Restrictions due to Cooperative Effects. EClinicalMedicine 2021, 32, 100718, DOI: 10.1016/j.eclinm.2020.100718.

Boal DH. Mechanics of the Cell. Cambridge, New York, Cambridge University Press, 2002.

Bricard A, Caussin JB, Desreumaux N, Dauchot O, Bartolo D. Emergence of Macroscopic Directed Motion in Propulsions of Motile Colloids. Nature 2013, 503, 95–98, DOI: 10.1038/nature12673.

12 The podcast is still available [Golestanian 2013].

Brown R. A Brief Account of Microscopical Observations Made in the Months of June, July and August 1827, on the Particles Contained in the Pollen of Plants; And on the General Existence of Active Molecules in Organic and Inorganic Bodies. Philosophical Magazine 1827, 4 (21), 161–173, DOI: 10.1080/14786442808674769.

Cavagna A, Giardina I, Grigera TS. The Physics of Flocking: Correlation as a Compass from Experiments to Theory. Physics Report 2018, 728, 1–62, DOI: 10.1016/j.physrep.2017.11.003.

Dawkins R. The Extended Phenotype. The Gene as the Unit of Selection, Oxford, Oxford University Press, 1982.

Einstein A. Über die von der molekularkinetischen Theorie der Wärme geforderte Bewegung von in ruhenden Flüssigkeiten suspendierten Teilchen. Annalen der Physik 1905, 17, 549–560, DOI: 10.1002/andp.200590005.

Elgeti J, Winkler RG, Gompper G. Physics of Microswimmers – Single Particle Motion and Collective Behavior: A Review. Reports on Progress in Physics 2015, 78 (5), 056601, DOI: 10.1088/0034-4885/78/5/056601.

Gennes P-G de. Soft Matter. Nobel Lecture, December 9, 1991. Stockholm: Nobel Prize Outreach AB, 2021. (Accessed 3 March, 2021, at https://www.nobelprize.org/uploads/2018/06/gennes-lecture.pdf.)

Golestanian R. Synthetic Mechanochemical Molecular Swimmer. Physical Review Letters 2010, 105 (1), 018103, DOI: 10.1103/PhysRevLett.105.018103.

Golestanian R. How Order Emerges when Matter Comes Alive. 2020. Munich: Max-Planck-Gesellschaft, 2021. (Accessed: June 2, 2021, at https://www.ds.mpg.de/3666576/201028_living_matter_order.)

Golestanian R. Living Matter: A Theoretical Physics Perspective. Online Lecture at Oxford University in September 2013. Munich: Max-Planck-Gesellschaft, 2021. (Accessed: February 4, 2021, at https://www.ds.mpg.de/lmp/lectures/oxford-2013.)

Golestanian R, Ajdari A. Mechanical Response of a Small Swimmer Driven by Conformational Transitions. Physical Review Letters 2008, 100 (3), 03810, DOI: 10.1103/PhysRevLett.100.038101.

Golestanian R, Liverpool TB, Ajdari A. Propulsion of a Molecular Machine by Asymmetric Distribution of Reaction Products. Physical Review Letters 2005, 94, 220801, DOI: 10.1103/PhysRevLett.94.220801.

Golestanian R, Ramaswamy S. Active Matter. The European Physical Journal E 2013, 36 (6), 67, DOI: 10.1140/epje/i2013-13067-2.

Golestanian R, Yeomans JM, Uchida N. Hydrodynamic Synchronization at Low Reynolds Number. Soft Matter 2011, 7, 3074–3082, DOI: 10.1039/c0sm01121e.

Gompper G, Winkler RG, Speck T et al. The 2020 Motile Active Matter Roadmap. Journal of Physics: Condensed Matter 2020, 32, 193001, DOI: 10.1088/1361-648X/ab6348.

Grosjean G, Hubert M, Lagubeau G, Vandewalle N. Realization of the Najafi-Golestanian Microswimmer. Physical Review E 2016, 94 (2), 021101(R), DOI: 10.1103/PhysRevE.94.021101.

Haldane FDM. Continuum Dynamics of the 1-D Heisenberg Antiferromagnet: Identification with the O(3) Non-Linear Sigma Model. Physics Letters A 1983a, 93 (9), 464–468, DOI:10.1016/0375-9601(83)90631-X.

Haldane FDM. Nonlinear Field Theory of Large-Spin Heisenberg Antiferromagnets: Semiclassically Quantized Solitons of the One-Dimensional Easy-Axis Néel State. Physical Review Letters 1983b, 50 (15), 1153, DOI: 10.1103/PhysRevLett.50.1153.

Howard J. Mechanics of Motor Proteins and the Cytoskeleton. Sunderland, Massachusetts, Sinauer Associates, 2001.

Jones RAL. Soft Condensed Matter. New York, Oxford University Press, 2002.

Kay L. Who Wrote the Book of Life? A History of the Genetic Code. Stanford, Stanford University Press, 2000.

Kohn W. An Essay on Condensed Matter Physics in the Twentieth Century. Reviews of Modern Physics 1999, 71 (2), 59–77, DOI: 10.1103/RevModPhys.71.S59.

Kosterlitz JM, Thouless CJ. Long Range Order and Metastability in Two Dimensional Solids and Superfluids. (Application of Dislocation Theory). Journal of Physics C: Solid State Physics 1972, 5 (11), L124, DOI: 10.1088/0022-3719/5/11/002.

Kosterlitz JM, Thouless DJ. Ordering, Metastability and Phase Transitions in Two-Dimensional Systems. Journal of Physics C: Solid State Physics 1973, 6 (7), 1181, DOI: 10.1088/0022-3719/6/7/010.

Leoni M, Kotar J, Bassetti B, Cicuta P, Lagomarsino MC. A Basic Swimmer at Low Reynolds Number. Soft Matter 2009, 5 (2), 472–476, DOI: 10.1039/B812393D.

Mantegna RN, Stanley HE. An Introduction to Econophysics: Correlations and Complexity in Finance. Cambridge, Cambridge University Press, 1999.

Menon GI. Active Matter. In: Krishnan JM, Deshpande AP, Kumar PB, eds. Rheology of Complex Fluids, New York, Springer, 2010, 193–218.

Najafi A, Golestanian R. Simple Swimmer at Low Reynolds Number: Three Linked Spheres. Physical Review E 2004, 69 (6), 062901, DOI: 10.1103/PhysRevE.69.062901.

Saha S, Golestanian R, Ramaswamy S. Clusters, Asters, and Collective Oscillations in Chemotactic Colloids. Physical Review E 2014, 89 (6), 062316, DOI: 10.1103/PhysRevE.89.062316.

Sanchez T, Chen DTN, DeCamp SJ, Heymann M, Dogic Z. Spontaneous Motion in Hierarchically Assembled Active Matter. Nature 2012, 491 (7424), 431–434, DOI: 10.1038/nature11591.

Schrödinger E. What Is Life? The Physical Aspect of the Living Cell. Based on Lectures Delivered under the Auspices of the Institute at Trinity College, Dublin, in February 1943. Cambridge, Cambridge University Press, 1944.

Soto R, Golestanian R. Self-Assembly of Catalytically Active Colloidal Molecules: Tailoring Activity Through Surface Chemistry. Physical Review Letters 2014, 112, 068301, DOI:10.1103/PhysRevLett.112.068301.

Sulloway FJ. Darwin and his Finches: The Evolution of a Legend. Journal of the History of Biology 1982, 15 (1), 1–53, DOI: 10.1007/BF00132004.

Tang E, Agudo-Canalejo J, Golestanian R. Topology Protects Chiral Edge Currents in Stochastic Systems. Condensed Matter: Statistical Mechanics 2021, arXiv: 2010.02845, in press at Physical Review X, July or August 2021.

Thampi SP, Golestanian R, Yeomans JM. Instabilities and Topological Defects in Active Nematics. Europhysics Letters 2014, 105 (1), 18001, DOI: 10.1209/0295-5075/105/18001

Thampi SP, Golestanian R, Yeomans JM. Driven Active and Passive Nematics. Molecular Physics 2015, 113 (17-18), 2656–2665, DOI: 10.1080/00268976.2015.1031840.

Thouless DJ, Kohmoto M, Nightingale MP, Nijs MD. Quantized Hall Conductance in a Two-Dimensional Periodic Potential. Physical Review Letters 1982, 49 (6), 405. DOI: 10.1103/PhysRevLett.49.405.

Tierno P, Golestanian R, Pagonabarraga I, Sagués F. Controlled Swimming in Confined Fluids of Magnetically Actuated Colloidal Rotors. Physical Review Letters 2008, 101, 218304, DOI: 10.1103/PhysRevLett.101.218304.

Toner J, Tu Y. Long-Range Order in a Two-Dimensional Dynamical XY Model: How Birds Fly Together. Physical Review Letters 1995, 75 (23), 4326, DOI: 10.1103/PhysRevLett.75.4326.

Vicsek T, Czirók A, Ben-Jacob E, Cohen I, Shochet O. Novel Type of Phase Transition in a System of Self-Driven Particles. Physical Review Letters 1995, 75 (6), 1226–1229, DOI: 10.1103/PhysRevLett.75.1226.

Wong GCL, Antani JD, Lele P et al. Roadmap on Emerging Concepts in the Physical Biology of Bacterial Biofilms: From Surface Sensing to Community Formation. Physical Biology 2021, online ahead of print, DOI: 10.1088/1478-3975/abdc0e.

Ramin Golestanian obtained his PhD in 1998 from the Institute for Advanced Studies in Basic Sciences (IASBS) in Zanjan, Iran, under the remote supervision of Mehran Kardar from the Massachusetts Institute of Technology, USA. He was subsequently an independent postdoctoral research fellow at the Kavli Institute for Theoretical Physics at the University of California, Santa Barbara (UCSB). He has held academic positions at the IASBS, Sheffield University, and Oxford University, rising through the ranks to become a full professor in 2007. Since 2018, he has been director at the Max Planck Institute for Dynamics and Self-Organization (MPIDS), heading the Department of Living Matter Physics in Göttingen, and Honorary Professor at the University of Göttingen. His department is engaged in a wide range of theoretical research aimed at a multiscale understanding of the dynamics of living systems from a physical perspective. The research topics cover, broadly speaking, *chemical* and *mechanical* nonequilibrium activity in living matter across the scales. Golestanian is distinguished for his work on active matter and, in particular, for his role in developing microscopic swimmers and active colloids. He is an elected fellow of APS and IoP, an elected member of the Göttingen Academy of Sciences and Humanities, and a recipient of the Holweck Medal of the Société Française de Physique and the Institute of Physics, the EPJE Pierre-Gilles de Gennes Lecture Prize, a Martin Gutzwiller Fellowship of the MPI-PKS, a Nakamura Lecturer Award of UCSB, and the 50th-Anniversary Most Distinguished Alumni Award of Sharif University of Technology. Ramin Golestanian is currently a Chairman of the IUPAP C6 Commission for Biological Physics and Divisional Associate Editor of the *Physical Review Letters*, the journal of APS.

Nikolaus Correll, Michael Friedman, Karin Krauthausen

Interview with Nikolaus Correll: Robotic Materials

MF: We would like to begin with how you perceive yourself in the field of active matter and active materials. There is a wide range of adjectives that are attached to these active materials, like 'smart,' 'robotic,' 'autonomous,' and so on. How would you describe the development of your research in this field? And how would you characterize robotic materials – a category that you and Richard Voyles introduced into materials science already in 2014?[1]

NC: I see myself as a proponent of computation in the field of smart materials. But the word 'smart' is overloaded; there is also a lot of discussion about active materials. I created the term 'robotic materials' because I come from robotics and my research is on materials for robots. Robots are placed in the real world and must therefore respond to an often uncertain environment. Robotic materials can take an active part in this challenge since they integrate acting and reacting into the material itself [Mengüç et al. 2017]. I think a robotic material is the ultimate smart material because it actually behaves like a robot – for example, it can move, it can change its shape, and its appearance [Hughes, Heckman, and Correll 2019]. It is about a new class of metamaterials that tightly integrate sensing, actuation, computation, communication, and power routing in a periodic fashion.[2]

To return to the first part of the question, I am a proponent of computation and I have a training in electrical engineering, where you learn about signals, systems,

1 For the first mention of the concept of robotic materials as a new class of materials, see: [Correll and Voyles 2014]. Since then the research field of robotic materials has become more and more important. Cf. the documentation of the workshop "Robotic Materials" (Washington, DC, April 23–24, 2018) in [Correll et al. 2018, pp. 1–16]. Apart from at Correll's laboratory at the University of Colorado Boulder (since 2009, online: http://correll.cs.colorado.edu/ (Accessed May 5, 2021) research is also being carried out at the new Robotic Matter Lab in Evanston, Illinois, led by Ryan Truby at Northwestern University (starting fall 2021) and the new department Robotic Materials led by Christoph Keplinger at the Max Planck Institute for Intelligent Systems in Stuttgart (also starting in 2021).

2 The term 'metamaterials' is referring to materials that "overcome the limitations of classical composite materials" [Correll and Voyles 2014]. Since first presenting the concept of robotic materials in 2014, Correll has written on various occasions about this class of materials (together with Richard Voyles and others but also on his own). Robotic materials are defined as: "[...] an emerging class of metamaterials that tightly integrate sensing, actuation, computation, and communication. Such 'robotic materials' extend the class of composite materials by providing them with autonomous functionality. In particular, they enable off-loading signal processing and control into the material" [Correll 2019, abstract].

Acknowledgment: Michael Friedman and Karin Krauthausen acknowledge the support of the Cluster of Excellence 'Matters of Activity. Image Space Material' funded by the Deutsche Forschungsgemeinschaft (DFG, German Research Foundation) under Germany's Excellence Strategy – EXC 2025 – 390648296.

Open Access. © 2021 Nikolaus Correll et al., published by De Gruyter. (cc) BY-NC-ND This work is licensed under the Creative Commons Attribution-NonCommercial-NoDerivatives 4.0 International License.
https://doi.org/10.1515/9783110562064-010

and controllers, as well as mathematical analysis or flow diagrams. Very few people in the materials sciences would speak about programming in the way an engineer would. When they speak about programming material, they usually mean something like heating a material up and getting a phase transition and freezing it in this state. Then, when you heat it up again, the material restores itself. There is a huge disconnect when I say that I program the material because what I have in mind is to describe the response algorithmically. This is actually a continuum. What I like to use as an example is a simple feedback controller, that is, I measure something and then decide whether it has to go faster or slower. You can do this in many different ways, either purely mechanically or by using something that swells and then shuts off, acting like a valve. When there is too much of a liquid there, it will shut off the flow of liquid. That is a feedback controller. Or you could have a centrifugal governor, which is a simple flywheel mechanism like the one used to regulate the rotational speed of a steam engine. The flywheel pulls up a weight, which in turn regulates the steam. By changing the weights, you can regulate the speed at which the engine is throttled. However, I could also say I can measure it and run it through an electronic circuit, such as an operational amplifier, that implements this computation, but it would not be very variable at all. And, of course, I can really program it using computer language using a microcontroller that is connected to sensors and actuators.

One can take the human body as an inspiration, because it is carefully tuning where the signal processing happens. Does the signal processing happen in the brain or is it autonomous, like in the colon? Is it integrated in the material or is a nervous system involved?

MF: Hence, in your research, you are asking where the computation is actually taking place. I think that this a crucial aspect of robotic materials: the program does not have to be centrally controlled. Do you see this as some sort of design shift or just a shift in how materials are actually being perceived?

NC: The people who make active materials – for example, something that exhibits phase transitions like changing color – would probably model an active material with differential equations. I can describe this same transition computationally. But irrespective of how you model it, either using equations or computation, the active material implies computation – and nevertheless it is not a computer. For me, the centrifugal governor is also about computing, even a clock is about computing. If you build a clock, you will think about gear ratios, so you actually hard wire the computation. When does it become a computer? I think the transition is very fluid. If you think about the Turing machine or, more explicitly, the universal Turing machine, this is a mathematical model for any kind of computation,[3] a simple device,

3 A 1936 paper on computing by Alan M. Turing proposed an answer to the so-called *Entscheidungsproblem* (decision problem) raised in 1928 by the mathematicians David Hilbert and Wilhelm

but with it Turing was able to demonstrate the properties of computation in general. Whatever can be computed can already be computed with a Turing machine, and what is not computable with a Turing machine is not computable at all. The original Turing machine is about a central processing unit that controls all data manipulation. The universal Turing machine is even more abstract, as it is the simulation of an arbitrary Turing machine and allows a second-order observation on the machine and the input.

Hence the question is: why or when is a computation a Turing machine? If you take the NAND gates in digital electronics, you build a computer up with NAND gates in a simulator; then you have a universal Turing machine and can build the rest of the computer, that is, you can program *Tetris* to run on it (see Fig. 1).[4]

A NAND gate is a NOT–AND gate, hence a universal binary logic gate that has the property of functional completeness, which means that any Boolean function can be implemented by using a combination of NAND gates.[5] A NAND gate can be made from two, sometimes four, transistors. Therefore, a NAND gate, and hence a universal Turing machine, can be constructed from anything that works like a transistor, for

Ackermann. Hilbert and Ackermann doubted that there could be a formal procedure (an algorithm) for deciding on the universal validity of a mathematical axiom – nevertheless they still asked for one. On the *Entscheidungsproblem*, cf. [Hilbert and Ackermann 1950 (1928), pp. 112–124]. See also [ibid., p. 112]: “From [the previous] considerations [...], there emerges the fundamental importance of determining whether or not a given formula of the predicate calculus is universally valid.” Turing was able to prove that such a formal procedure is not possible. With his model, he formalizes (among other things) mathematical procedures, that is, the calculation of numbers. And although Turing speaks of a concrete “computer and his tape” and sees the processing as a “physical system,” this computer is an abstract machine in the sense that for any computational work of a computing machine one can construct another machine that does the work of the first computing machine by observing the first machine and reading the work instructions [Turing 1937, p. 250]. See also the famous definition of computer in [ibid., p. 232]: “If an *a*[utomatic]-machine prints two kinds of symbols, of which the first kind (called figures) consists entirely of 0 and 1 (the others being called symbols of the second kind), then the machine will be called a computing machine.”

4 A NAND gate is a “primitive logic gate” (a logic governor) based on Boolean algebra that can be used to build other logic gates in order to get a standard set of gates that can be used to construct a computer’s processing and storage chips [Nisan and Schocken 2005, p. 7]. See also [ibid.]: “Every digital device – be it a personal computer, a cellular telephone, or a network router – is based on a set of chips designed to store and process information. Although these chips come in different shapes and forms, they are all made from the same building blocks: Elementary logic gates. The gates can be physically implemented in many different materials and fabrication technologies, but their logical behavior is consistent across all computers.” For Simon Schocken and Noam Nisan’s courses “From NAND to Tetris: Building a Modern Computer from First Principles,” which became quite famous and formed the basis of their publication, see: https://www.nand2tetris.org/ (accessed April 10, 2021).

5 Functional completeness means that any Boolean function (hence any other logic function and even an entire processor) can be implemented by using a combination of NAND gates [Mano, Kime, and Martin 2008, pp. 81–85].

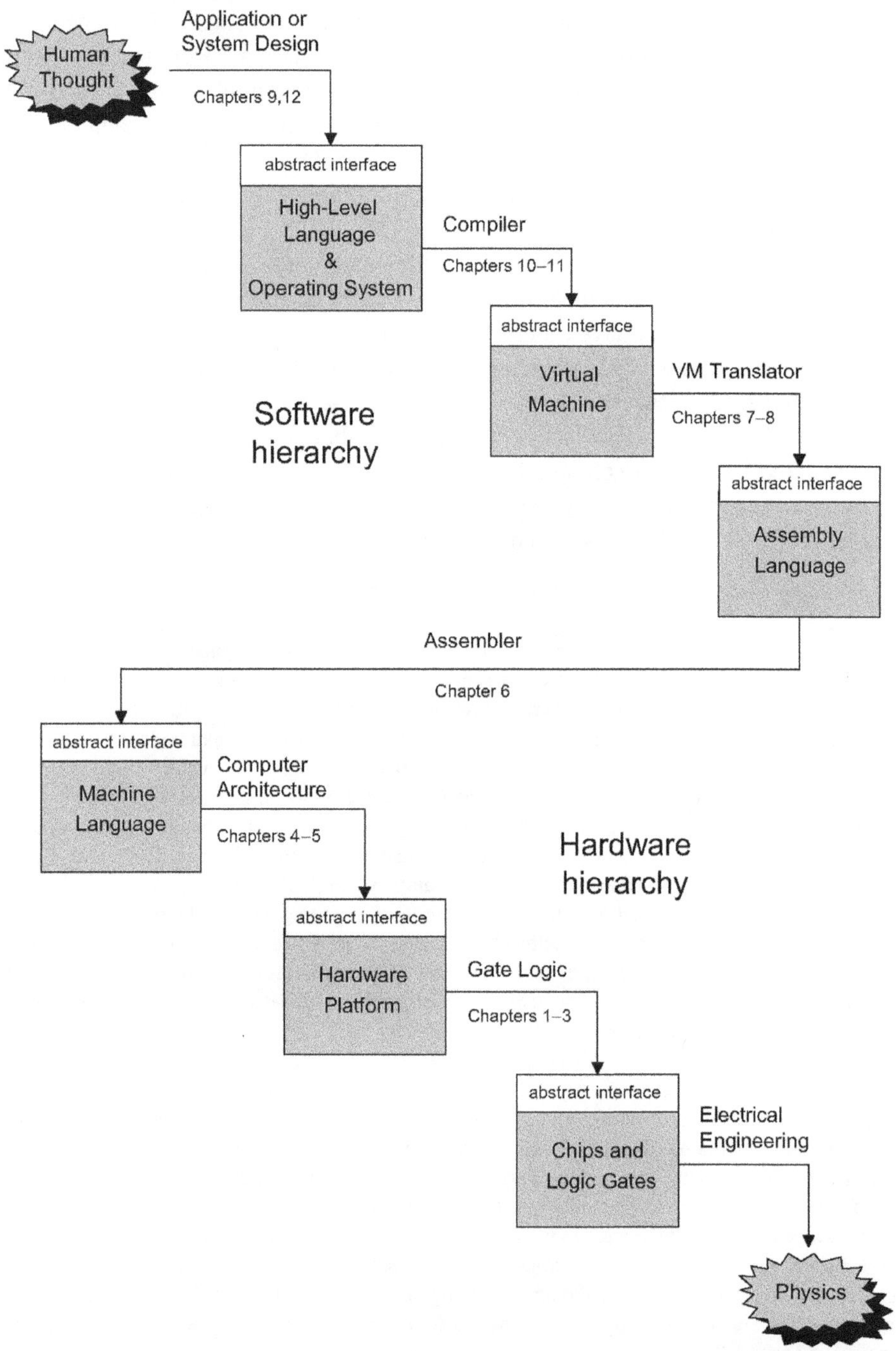

Fig. 1: A computer architecture (from physics and hardware to software). From: [Nisan, Schocken 2005, p. xvi, Fig. I.1]. © 2005 Massachusetts Institute of Technology, by permission of The MIT Press.

example, something that swells like a switch and by doing so controls the emission and flow of signals. You could also have something like a chemical process that does this kind of switching and works like a transistor. This kind of computation is complementary to morphological computing [Pfeifer and Gómez 2009], where a body's structure and morphology can contribute to control and cognition in natural or artificial systems, and the two can work together nicely.[6] An example for this task is the artificial 'sensing skin' that you have seen in my laboratory (see Fig. 2).

Fig. 2: A soft, amorphous texture-sensitive skin mounted on the back of a Baxter robot. From: [Hughes and Correll 2015, fig. 1]. © 2015 IOP Publishing Ltd.

This computational metamaterial is for texture recognition. The idea comes from the human skin, which includes many different sensors for measuring static and dynamic information with highly varying bandwidth, like pressure, shear, temperature, and textures. To replicate such a sensing system is a challenge for engineering. We constructed a robotic material that samples and processes high-bandwidth information locally and addresses a central processing unit only when an event occurs. Our design considers the skin as an amorphous material capable of processing stimuli within the material itself. Omnidirectional microphones serve as vibration sensors, and we collocate microcontrollers with these sensors (see Fig. 3) [Hughes and Correll 2015]. If you mount this texture-sensitive skin on the back of a Baxter robot, it will augment the robot's environmental awareness when navigating through an environment and when interacting with humans in order to assist them.

6 See the special issue on morphological computing in *Artificial Life* 19 (1) (2013). For a review and examples, see: [Müller and Hoffmann 2017].

The sensitive skin does a discrete Fourier transform of the frequencies it feels. In other words, it puts the different frequencies into different bins. Then I can look up the height of each of these bins and, as every material would create distinct frequencies, say this is wood or this is metal.

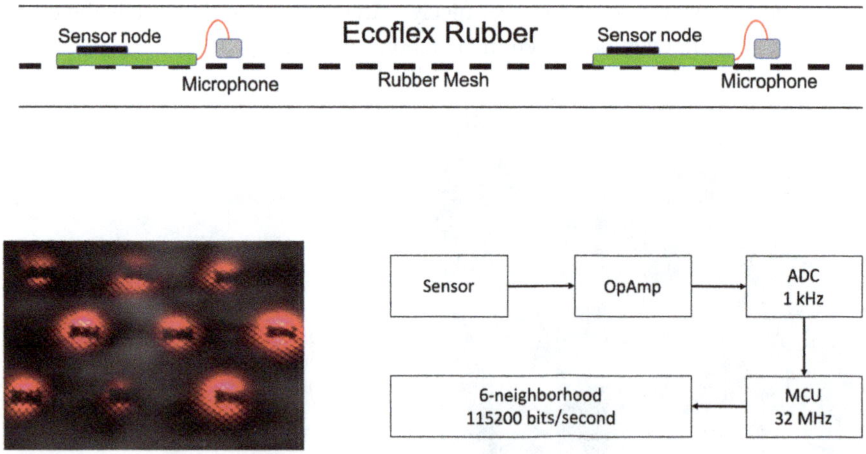

Fig. 3: System overview over a computational metamaterial that can identify and localize textures rubbed against it. Printed circuit boards and microphones are embedded in flexible rubber and suspended on a neoprene mesh. Figure is provided by Nikolaus Correll (a similar but slightly different figure was originally published in [Correll and Voyles 2014, fig. 1]). © Nikolaus Correll.

The sensor node network does exactly what the cochlea in the human ear does. The cochlea looks like a trumpet, that is, a channel that starts out wide and gets narrower and narrower, and the sound comes in and gets reflected inside. Since the function of the frequency that comes in depends on this geometry, you have different points in this grafted channel where you have resonance. The ear puts nerve ends into this channel and in this way is able to grab the frequencies. If you have a cochlea implant then it does exactly that, but in a computer, it takes the sound, it bins the frequencies, and then it excites the nerve ends.

For me, the interesting point of the sensing skin is the computation. I actually have a little computer that implements what is known as the fast Fourier transform (FFT). We programmed this – it is an algorithm that is very well understood and available in any textbook, such as *Numerical Recipes in C* [Press et al. 1992] – or I go and actually mechanically put in a grafted channel that is made for the frequencies that I am interested in, which is in the 250,000 Hz range, and physically put sensors at the right places to get my Fourier transform. Then I put little vibration sensors at these different points and I pull it into my computer. And then my computer does not have to do all these thousands of lines of code for the FFT. Unfortunately, the frequency range of vibrations is around 10 times larger than that of the audible spectrum, making the wavelengths ten times shorter, and therefore an appropriate mechanism to measure vibrations 10 times smaller than the cochlea, which is difficult to make. Indeed the Cochlea in the human ear has a length of 9 mm when stretched out, whereas the Pacinian corpuscle, which is responsible for detecting vibrations in the human skin has a size of only 1 mm.

So, how and where to implement computation is a complete continuum spectrum with a wide variety of available tools.

MF: What you describe seems to be an interplay between discrete and continuum computations, because the skin is only computing the sound on several occasions, but it is not taking everything into account. This raises the following question: what is the correct modeling for the local sensing of the skin? If I rub my skin here, it does not matter what happens at the same time in my foot for example.

NC: You mean what is the correct distribution of where the computation should be? I always argue that, if you have high-bandwidth information and you have many sources thereof (like the textures that you want to detect), the Pacinian corpuscle, which is one of the four major types of mechanoreceptor cell in the human skin and especially sensitive to vibrations, receives the signals.[7] That is the resonance frequency of the Pacinian corpuscle. It is still unclear what exactly the Pacinian corpuscles are doing. What we do know is that they take the continuous signal and turn it into spikes of neurons. I argue that there must be an information reduction, because it would be a complete waste to wire this kind of bandwidth to the brain. What the information reduction actually is we do not know, but it does take place in a certain way. And the Pacinian corpuscle does not send information when it does not have any. A computer camera always sends information even when there is nothing. The sensing body works differently: if nothing happens, it does not send anything to the brain. It is like when you get used to your clothes and do not feel them anymore. This is because the cells (they are called 'slow adapting') adapt to the stimulus and the fast adapting cells (like the Pacinian corpuscles) give you the changes. Thus, the body has all of these things, but there is very little computation. The skin is autonomous like the retina or the colon. They all preprocess things.

MF: And what you are attempting to build in your laboratory are materials that decide by themselves, materials that are decentralized and in this sense autonomous?

NC: That is exactly what I want. As an engineer, I create hypotheses by building. At the moment, we are building a camouflage skin for robots. The inspiration for this distributed particle system comes from the octopus, because the cephalopod skin has extraordinary camouflage capabilities. Octopuses change the color of their skin

7 On the mechanoreceptors of the human skin and their signaling, see: [Kandel, Schwartz, and Jessell 2000, pp. 430–441].

by using small structures known as chromatophores. These devices can expose pigments of a certain color by activating an annular muscle structure.

When we read the biological and physiological literature, we were quite confused: where is the camouflage pattern generated? It is not like a display – hence, the octopus is not controlling every pixel with its brain. Rather, the pattern is locally generated and self-organized.[8] It is a Turing pattern, a differential equation, since it is a function of some parameters [Turing 1952]; different pattern schemes can be triggered by providing only a few parameters. That is what makes the cephalopod skin autonomous. We simulated the chromatophores with a miniature robotic platform, the 'droplets,' which look like ping-pong balls but can change their color using a built-in light-emitting diode. They are, of course, also equipped with a color sensor in order to register colors and patterns in the environment. We used a distributed computation for the information exchange between the droplets. For collectively performing a certain pattern type, the droplet swarm then uses a discrete version of Turing's reaction-diffusion models (see Fig. 4).

The inspiration is not only going from the animal world to engineering. I think by looking at it from an information-theoretic perspective, you can in turn generate hypotheses for the physiologists. There is a nice paper by Barbara Webb in which she wonders how constructing robots can help us to understand insects and other small animals [Webb 2020]. You have a hypothesis on how their sensory motor system works, you build a robot that way, it works, you can show that in the mechanical world this information is sufficient to realize a certain function, so maybe the animal does it in a similar way. The robotic engineering generates a new hypothesis that physiologists and biologists can then test. I think such an exchange is valuable.

KK: With their capacity for coordinated sensing, information processing, and response activity, living systems are an important inspiration for the research on robotic materials. As you said, this stunning agency of living systems is not always derived from a central processing unit such as the brain; it can also happen locally, for example in the organs and hence in the 'material' itself. How would you describe the relation between 'natural' active materials like the skin of an octopus and 'artificial' active materials like the robotic materials that you are working on? And as a follow-up question: with these new kinds of robotic materials, what is or will be the status of robots?

NC: A robot is very complicated to make and we do not have any autonomous robots, except vacuum cleaners, that are doing anything reasonable. Biological systems are not made from the homogenous materials that are currently used to construct robots, such as links, gears, and joints. If you want to build something, you choose your

8 The camouflage of the octopus is not (or not only) driven by the animal's visual system, as this is color blind, nor is it driven solely by the brain; rather it relies on local sensing and control [Correll 2019; Li, Klingner, and Correll 2018].

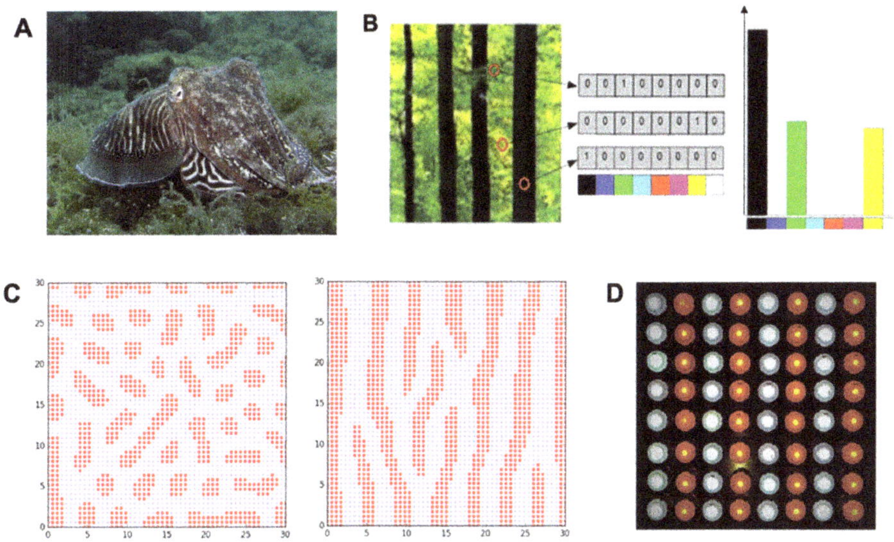

Fig. 4: Distributed particle system for artificial camouflage. (A) Chromatophores in the cephalopod skin show typical self-organized patterns. CC0 Public Domain (https://creativecommons.org/publicdomain/zero/1.0/). (B) Artificial camouflage can be realized by selecting both a pattern (here vertical stripes) and a color scheme by finding dominant colors in the environment. Reprinted by permission from Springer Nature: [Li, Klingner, and Correll 2018, p. 1639, fig. 2]. (C) Different patterns can be realized by a fully distributed, self-organized process, here showing speckle and striped patterns. Reprinted by permission from Springer Nature: [Ibid., p. 1643, fig. 8]. (D) Implementation on a swarm of miniature robots that can only communicate locally. Reprinted by permission from Springer Nature: [Ibid., p. 1657, fig. 12].

materials based on functional requirements. You go to Home Depot and buy materials because you have a function that needs to be satisfied. If you want to construct a table you probably buy wood – you will not use stone because the way you can work with wood is different, the weight of wood is different, it is lighter yet stiff enough for your application, and it maybe even has aesthetic benefits. However, if you want to have a really complex functionality then it helps to use materials where you can off-load functionality – if you want to build a robot that can manage the inflow of information from tactile events for example, as with sensing skin. If I did this in a conventional way, then all these signals would be wired to a central processor, as to the brain. The brain would get a lot of data all the time from millions of sensors and it would have to decide all the time which data to ignore. However, if you transfer this decision to the material – for example, the human skin or the artificial sensing skin – then it provides you or the robot with that functionality, and you do not have to do anything. And, when something happens, you get a high-level piece of information which says: I got touched on my back and I should probably look. I can then compose the robotic system out of these functional components in a much simpler way. Such modularization makes things simpler. And that is how I think about robotic materials: they are multifunctional composites with autonomous functionality.[9]

If you take any materials science textbook on advanced composite materials, it will usually end with the bone. The authors of these books celebrate the bone as a very smart material and they will probably say that you should go and design your bridge like a bone structure. However, the bone that you have in your body is much more functional than any structure that one can build. The bone regenerates, and if you load it more it will grow stronger. And it also stores energy in terms of fat that you can draw on – people eat bone marrow from animals because it is high-energy and it generates white blood cells. The bone is actually a machine with sensing capabilities, and it even does information processing in some hard-coded way, that s, with lots of cells measuring the strain and then emitting a chemical substance that creates or attracts calcium. We need to have these kinds of materials for robotics in order for it to really advance.

KK: In this respect, do robotic materials blur the distinction between device and material, and maybe even between machine and material? You mentioned the first possibility in one of your articles on robotic materials.[10]

NC: Robotic materials are blurring the dynamics of the system and the control algorithm. In a way, this is already happening in the centrifugal governor – and I could build a centrifugal governor that is so small that you cannot see it, but I use the same principle. That is where the blurring happens.

All these biological systems are machines. You have atoms that make molecules and these are already machines, molecular machines – or materials. These molecules make cells, and then the cells make materials such as a bone. And now I have to correct myself: the bone is not a machine; it is a material and a device. There are many devices in the body that make it a complex machine, and all of these devices are only materials. The computer is also only a material. If I gave it to you and I

9 On the necessary conditions for an engineered structure to be called a "robotic material" see [Correll 2019, p. 297]: "[A robotic material should] 1. Consist of many homogeneous elements that are arranged in either an amorphous or a grid-like fashion. 2. Integrate computation with sensing or actuation in multiple elements. 3. Function independently of its size, that is, the number of constituent elements, for example, even when cut in half or rearranged in an arbitrary shape (scalable). 4. Not lose its capabilities even if individual units fail (robust)."

10 See [Correll and Voyles 2014, footnote 1]: "Advances in polymer science, miniaturization of computing, and manufacturing techniques have enabled a new class of robotic devices that embed sensing, computation, actuation and communication at high densities. This embedding can be so tight that the distinction between device and material blurs. This new class of materials has the ability to change its physical properties such as stiffness, density, weight, shape or appearance in response to external stimuli, and is able to adapt and learn. Here, the boundary between the dynamics of the physical system and control algorithms blurs, which allows trade-offs between morphological and silicon-based computation."

gave you all the tools to analyze it, you would never figure it out. It is like giving you a bunch of atoms and then asking you what is going on? Why is it not exploding? Why cannot you just shake the atoms out?

MF: Would you claim that the structural point of view on bones that one often finds in materials science reduces the complexity of what is actually going on?

NC: Yes, that is what the materials community is stuck in: they see the stunning properties of the octopus arm and then think about making a liquid that could alter its stiffness or its color in a similar way, but they do not appreciate that the octopus has a nervous system. If you want to build something like an octopus arm then you have to put in a nervous system. Given the continuous nature of computing, you can do this in many ways, for example with polymer electronics and a neuromorphic architecture. Or you could implement the mathematical function – I can type it into the computer and compile it into a Turing machine, and then I have a million transistors that do that. That is actually amazing.

MF: This brings me to another question: what is the relation of your research on robotic materials to nature? Are there materials that nature either cannot combine together or that nature simply does not have? And then what kinds of materials do we get? You would not call these 'bioinspired' would you?

NC: You get supranatural materials. For instance, wireless communication is supranatural. If I build a gripper that has all the tactile information that you have as a human, but which can also see its environment and even the 3D geometry of any object, then with this robotic material I actually get supranatural grasping and manipulation abilities.

MF: I will explain why I asked the question. You mentioned Barbara Webb, a professor of biorobotics at the University of Edinburgh. She argues that bioinspiration works in two ways: robotics can learn from biology a lot about smart materials and biology can learn from robotics about how to understand animal physiology and behavior.[11] The question that came to my mind was: why should this be an argument for robotics? Why cannot we just build robots that function better than earlier devices?

NC: The question is: does building robots contribute to anything or are we just wasting public money? Webb argues for people like me who build cockroach-like robots. She says that this is actually an efficient way to generate hypotheses and advance

11 See [Webb 2020, p. 244]: "It is an engineer's dream to build a robot as competent as an insect at locomotion, directed action, navigation and survival in complex conditions. But as well as studying insects to improve robotics, in parallel, robot implementations have played a useful role in evaluating mechanistic explanations of insect behavior, testing hypotheses by embedding them in real world machines."

our understanding of natural systems. I ask myself the same question: what can we contribute to biology? We build these swarm droplets that can form camouflage patterns, we show that they can recognize their environment and then decide what camouflage pattern to perform in a fully distributed way. This is not what the octopus does when exhibiting camouflage, because the octopus does it in the brain. It has a semantic understanding of its surroundings, which means that it knows that it is in a cave or near food or in a kitchen, and then it decides what to do, and this is high-level reasoning. Our swarm works differently. However, I still think we can contribute to biology – for instance when the famous octopus researcher Benny Hochner wonders how the octopus can control millions of tactile and chemical sensory cells (activators) distributed all over the body, arms, and suckers [Hochner 2012].[12] In neurobiology, one tries to understand how the central nervous system can work as a controller and coordinate sensory feedback and motor commands (see Fig. 5). This is not so far away from robotics.

So how do I know what to do with hundreds of activators? The robotics people have written down all the equations which are very complicated and very difficult to compute. Then how can the octopus compute the huge amount of sensory information? The biologists do not know. I was thinking about a similar problem when building a multisegment shape-changing material that consists of variable-stiffness elements. Each was equipped with a networked computing element, and I wanted a simple way of computing it. I asked myself: what happens if I only change one of the cells in this robotic material? In my laboratory, we then tried to get closer to that and found a distributed algorithm for calculating the inverse kinematic solution for the resulting *N*-body system [McEvoy and Correll 2018]. We also developed a communication model based on a central element collecting information from its neighborhood (see Fig. 6).

The way the algorithm works is to compute how a single piece of the arm would need to change so that the tip would get closer to its goal. We are doing this for every piece from the beginning to the end of the arm, and then start over, until the arm reaches its desired end point. That is, the computation literally travels along the arm in wave-like patterns. And then I read that the octopus's nervous system activation pattern actually has something like this [Gutfreund et al. 1996; Gutfreund et al. 1998; Mc Evoy and Correll 2018]. Does it work exactly like that? We do not know, but we have provided an explanation that is actually feasible given what else we know about the octopus nervous system. In general, we roboticists feel that neurobiology is extremely important in helping us to see where robotic material should go.[13] Because

12 Hochner is referring to the concept of 'embodied organization,' which is important for roboticists. According to Hochner the term "implies the dynamic interplay of information and physical processes between four components comprising the embodied creature: the controller, the mechanical system, the sensory system and the task-environment [...]" [Hochner 2012, p. R887]. On 'embodiment' see also: [Pfeifer, Lungarella, and Iida 2007].

13 For a review, see: [Webb 2017].

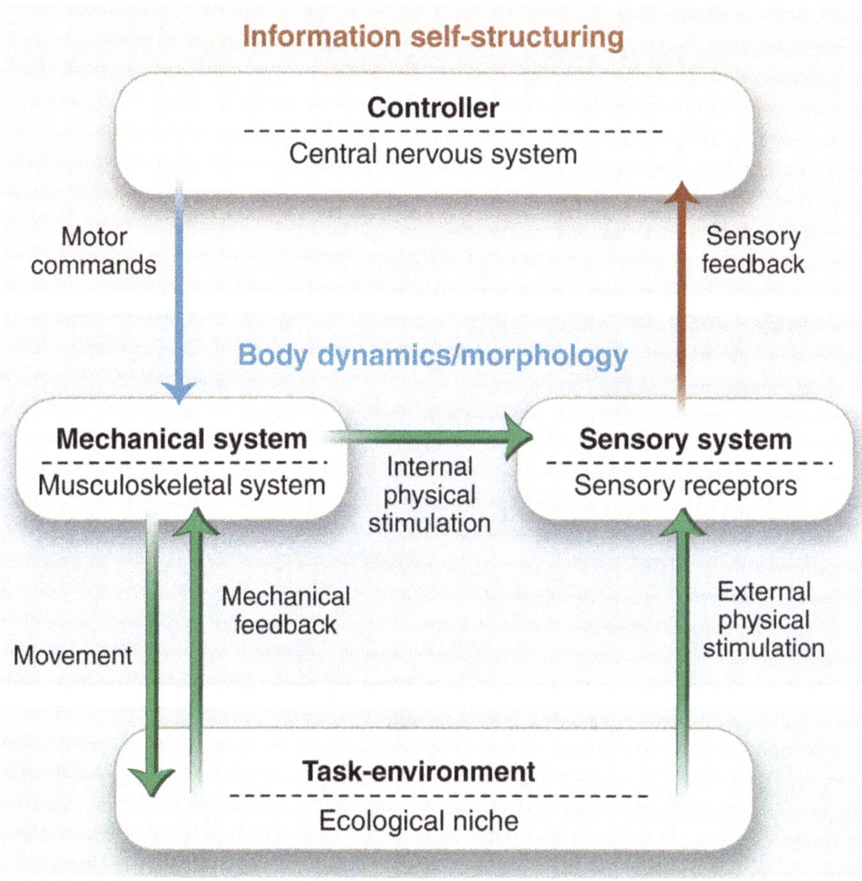

Fig. 5: Scheme of embodied organization in a living system. From: [Pfeifer, Lungarella, and Iida 2007, p. 1089, fig. 1]. Reprinted with permission from AAAS.

neurobiologists study the nervous system and how the animal is built. If I did not know that two-thirds of the neurons are in the octopus's arms, I do not think I would be so excited about pushing to make materials that way.

MF: Your research includes a lot of mathematical modeling, and you model in very different ways, for example with partial differential equations or with probabilistic methods. How do you actually combine this mathematical modeling with robots?

NC: Yes, the modeling and implementation is very difficult. Take, for instance, the two-bridge experiment with Argentine ants that Jean-Louis Deneubourg, a professor at the Université libre de Bruxelles, did. In order to understand when and how the transition from unorganized exploring to organized pathfinding happens in the ant colony, the researchers installed a diamond-shaped bridge; hence, a bridge that allows two different paths, between the ant's nest and a source of food. Initially

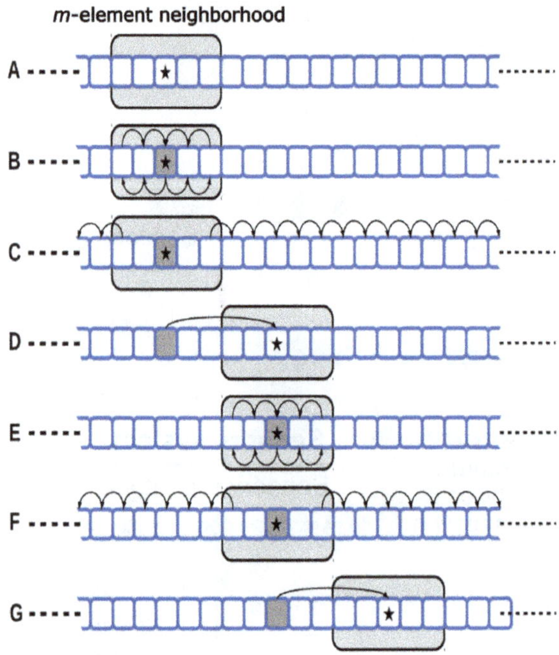

Fig. 6: Communication model of a robotic material beam with shape-changing properties. From: [McEvoy and Correll 2018, fig. 5]. © 2018, Mary Ann Liebert, Inc.

choosing between both paths at random, the ants quickly converge onto the shortest path.[14] At its heart, this can be modeled by a nonlinear system with an attractor for the mathematical equivalent of the shortest path [Deneubourg et al. 1990].[15] Obviously, nonspatial differential equations can describe very well the number of ants on each bridge.

In a similar experiment, we let a swarm of cockroaches choose between two shelters in a circular arena [Halloy et al. 2007]. Here, the same equations as with the bridge experiment are used to model the number of cockroaches underneath each shelter. But that assumes that I have a uniform distribution and that the animals' behavior is so simple that they end up in a Brownian motion and are creating this uniformity. Now the animals' behavior is actually not that simple because the cockroaches like to wall-follow, running in circles around the arena instead of congregating under the shelters. Hence, I have to think whether to put in a third equation that models the number of insects that wall-follow. Or I start modeling it in a spatial way so that I have a partial differential equation that describes the distribution over

14 The researchers modeled the ant behavior with a Monte Carlo simulation.

15 For the thermodynamic view on instabilities in reaction-diffusion systems (far-from-equilibrium systems) and the role of attractors, see: [Prigogine and Stengers 1984, pp. 146–153].

space. What is needed is spatial computing or amorphous computing.[16] In our research, we actually found a couple of problems which do require that.

In my thesis at the École polytechnique fédérale de Lausanne in 2007, I had a jet turbine inspection case study with a swarm of robots [Correll and Martinoli 2009]. The swarm should inspect the turbine blades by going around them (Fig. 7). You can model this and uniformly distribute the robots. Basically, you draw random blades and put the robots in and see how long they take for this task. That is the simplest model. We also built more complex models, but when we tested them in experiments they did not match. Hence, you have to take into account the actual geometry of this environment. The robots were programmed to hit the blades at some point and then always do a right turn and go around, then go around one more time, and once they detected the tip they should leave. With this, we wanted to make sure that the swarm covers the entire turbine. Now if the robots always leave in the direction of the tip you get a very obvious distribution that is highly skewed to this result [Prorok, Correll, and Martinoli 2011]. Whereas when you release them in a uniform way, the chance that this blade is covered is much lower.

The problem is that you are locked into certain mathematics and you can only make it more versatile by simulating the system in a physical environment. You have to simulate every individual robot and how it moves. We also had robot simulators, and we would see that the swarm behaves the same, because the robots would also leave. The simulator is like a 3D computer game. We can use the computer game as a predictor. And the hope with this 'multilevel modeling,' as my advisor Alcherio Martinoli called it, is always that you have many layers of abstraction, and in the end, you get this one equation that is the takeout.

When you do research on ant behavior, it is similar: you have to write down the differential equation, but you also have to think about the geometry of the environment. At the end of the day, you have to model every single aspect including the ant's emotional state and if it is hungry or not hungry. It is a Pandora's box of problems. I got a little bit away from this because the models are so limiting. However, I

---

16 On spatial computing as a manipulation of referents to real objects and spaces, see: [Greenword 2003]. For the first manifesto on amorphous computing, see [Abelson et al. 1996, n.p.]: "A colony of cells cooperate to form a multicellular organism under the direction of a genetic program shared by the members of the colony. A swarm of bees cooperates to construct a hive. Humans group together to build towns, cities, and nations. These examples raise fundamental questions for the organization of computing systems:

- How do we obtain coherent behavior from the cooperation of large numbers of unreliable parts that are interconnected in unknown, irregular, and time-varying ways?
- What are the methods for instructing myriads of programmable entities to cooperate to achieve particular goals?"

The authors of the manifesto refer to [Prigogine and Stenger 1984] and [Turing 1952]. For an introduction with examples, mathematical tools for computing, and programming styles, see: [Stark 2013].

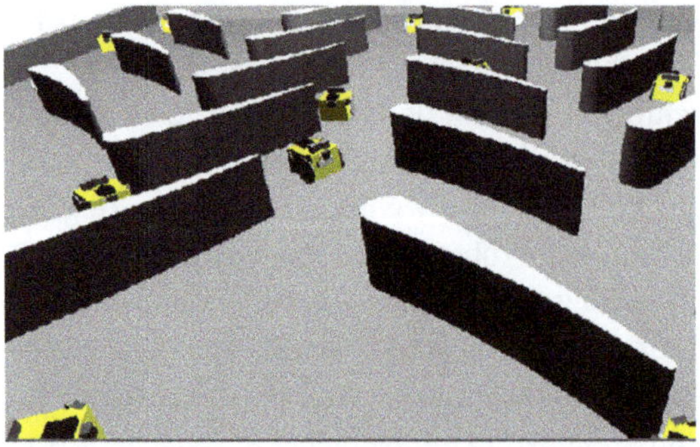

Fig. 7: Simulation of swarm robots on an inspection tour between the blades of a turbine. From: [Correll and Martinoli 2009, p. 110, fig. 10]. © 2009, IEEE.

often use the Gillespie algorithm, a variant of the Monte Carlo method [Gillespie 1976]. With that, you basically just simulate and see what happens in your complex system. But I have to admit that I prefer actually to build stuff [Correll et al. 2022].

KK: That brings me to our last question: how do you see the relation between fundamental research and engineering (or applied science), especially in the field of active matter?

NC: There has always been a tension between science and engineering, which actually exists less now because the engineers try to be hypothesis driven too. The design loop is to identify a problem and then think about solutions and then carry out tests. So the engineers actually do the same thing as the scientists, but they are goal oriented. I think it is called teleological research science.

My research on robotic materials is sometimes misunderstood, as one might think we are only putting things together without a more fundamental approach. I can say that I am doing fundamental algorithms – not the mathematics of nonlinearity and dynamics but fundamental algorithms. I am trying to prove that I can build anything mechatronic by putting together smart particles like the droplets, akin to cells in biology. However, it is not possible to build everything with such an approach. You can just build the subset that biology does. You cannot build an animal that is the size of the Eiffel Tower. It does not exist in nature because of some physical limitations.

I think today's materials are so interdisciplinary that they require integration and the integration is by default engineering (vs. science). In my view, what we need to have is an academic center for the science of the integration of materials – preferably materials that integrate sensing, actuation, and communication. It would change the world completely to make active materials the way I want them to be. One has to pull it off.

Bibliography

Abelson H, Knight TF, Sussman GJ, friends. Amorphous Computing. 1996. (Accessed April 15, 2021, at http://groups.csail.mit.edu/mac/projects/amorphous/white-paper/amorph-new/amorph-new.html.)

Correll N. Robotic Materials for Robot Autonomy. In: Walsh SM, Strano MS, eds. Robotic Systems and Autonomous Platforms: Advances in Materials and Manufacturing. Duxford, Woodhead Publishing, Elsevier, 2019, 295–307, DOI: 10.1016/B978-0-08-102260-3.00012-3.

Correll N, Baughman R, Voyles R, Yao L, Inman D. Robotic Materials. 2018, arXiv: 1903.10480.

Correll N, Hayes B, Heckman C, Roncone A. Introduction to Autonomous Robots: Mechanisms, Sensors, Actuators, and Algorithms. Cambridge MA, MIT Press, 2022 (forthcoming).

Correll N, Martinoli A. Multirobot Inspection of Industrial Machinery. IEEE Robotics & Automation Magazine 2009, 16 (1), 103–112, DOI:10.1109/MRA.2008.931633.

Correll N, Voyles R. Robotic Materials. From Smart Polymers to Computational Metamaterials. Proceedings of the Robot Makers Workshop. Robotics: Science Systems Conference 2014. (https://www.seas.upenn.edu/~nicbezzo/RoMa2014/, not accessible anymore.) Cambridge MA, Woodhead Publishing, Elsevier, 2019, 295–307, DOI: 10.1016/B978-0-08-102260-3.00012-3.

Deneubourg J-L, Aron S, Goss S, Pasteels J. The Self-Organizing Exploratory Pattern of the Argentine Ant. Journal of Insect Behavior 1990, 3 (2), 159–168, DOI: 10.1007/BF01417909.

Gillespie DT. A General Method for Numerically Simulating the Stochastic Time Evolution of Coupled Chemical Reactions. Journal of Computational Physics 1976, 22 (4), 403–434, DOI:10.1016/0021-9991(76)90041-3.

Greenwold S. Spatial Computing. Master's thesis, Massachusetts Institute of Technology, 2003. (Accessed April 15, 2021 at https://acg.media.mit.edu/people/simong/thesis/SpatialComputing.pdf.)

Gutfreund Y, Flash T, Fiorito G, Hochner B. Patterns of Arm Muscle Activation Involved in Octopus Reaching Movements. Journal of Neuroscience 1998, 18 (15), 5976–5987. DOI: 10.1523/JNEUROSCI.18-15-05976.

Gutfreund Y, Flash T, Yarom Y et al. Organization of Octopus Arm Movements: A Model System for Studying the Control of Flexible Arms. Journal of Neuroscience 1996, 16 (22), 7297–7307. DOI: 10.1523/JNEUROSCI.16-22-07297.

Halloy J, Sempo G, Caprari G et al. Social Integration of Robots into Groups of Cockroaches to Control Self-Organized Choices. Science 2007, 318 (5853), 1155–1158, DOI: 10.1126/science.1144259.

Hilbert D, Ackermann W. Principles of Mathematical Logic. New York, Chelsea, 1950 [1928].

Hochner B. An Embodied View of Octopus Neurobiology. Current Biology 2012, 22 (20), R887–R892, DOI: 10.1016/j.cub.2012.09.001.

Hughes D, Correll N. Texture Recognition and Localization in Amorphous Robotic Skin. Bioinspiration & Biomimetics 2015, 10, 055002, DOI: 10.1088/1748-3190/10/5/055002.

Hughes D, Heckman C, Correll N. Materials that Make Robots Smart. The International Journal of Robotics Research 2019, 38 (12-13), 1338–1351, DOI: 10.1177/0278364919856099.

Kandel ER, Schwartz JH, Jessell TM. Principles of Neural Science. 4th ed. New York, McGraw-Hill, 2000.

Li Y, Klingner J, Correll N. Distributed Camouflage for Swarm Robotics and Smart Materials. Autonomous Robots 2018, 42, 1635–1650, DOI: 10.1007/s10514-018-9717-6.

Mano MM, Kime CR, Martin T. Logic and Computer Design Fundamentals. 4th ed. Upper Saddle River, NJ, Pearson Prentice Hall, 2008.

McEvoy MA, Correll N. Shape-Changing Materials Using Variable Stiffness and Distributed Control. Soft Robotics 2018, 5 (6), 737–747, DOI: 10.1089/soro.2017.0147.
Mengüç Y, Correll N, Kramer R, Paik J. Will Robots Be Bodies with Brains or Brains with Bodies? Science Robotics 2017, 2 (12), eaar4527, DOI: 10.1126/scirobotics.aar4527.
Müller VC, Hoffmann M. What is Morphological Computation? On How the Body Contributes to Cognition and Control. Artificial Life 2017, 23, 1–24, DOI: 10.1162/ARTL_a_00219.
Nisan N, Schocken S. The Elements of Computing Systems: Building a Modern Computer from First Principles. Nand to Tetris Companion. Cambridge MA, London, MIT Press, 2005.
Pfeifer R, Gómez G. 2009. Morphological Computation: Connecting Brain, Body, and Environment. In: Sendhoff B, Körner E, Sporns O, Ritter H, Doya K, eds. Creating Brain-Like Intelligence: From Basic Principles to Complex Intelligent Systems. Berlin, Heidelberg, Springer, 2009, 66–83.
Pfeifer R, Lungarella M, Iida F. Self-Organization, Embodiment, and Biologically Inspired Robotics. Science, 2007, 318 (5853), 1088–1093, DOI: 10.1126/science.1145803.
Press W H, Teukolsky SA, Vetterling WT, Flannery BP. Numerical Recipes in C: The Art of Scientific Computing. 2nd ed. Cambridge, Cambridge University Press, 1992.
Prigogine I, Stengers I. Order out of Chaos. Man's New Dialogue with Nature. Toronto, New York, London, Sydney, Bantam Books, 1984.
Prorok A, Correll N, Martinoli A. Multi-Level Spatial Modeling for Stochastic Distributed Robotic Systems. International Journal of Robotics Research 2011, 30 (5), 574–589, DOI: 10.1177/0278364911399521.
Stark WR. Amorphous Computing: Examples, Mathematics and Theory. Natural Computing: An International Journal 2013, 12 (3), 377–392, DOI: 10.1007/s11047-013-9370-0.
Turing AM. On Computable Numbers, with an Application to the Entscheidungsproblem. Proceedings of the London Mathematical Society 1937, s2-42 (1), 230–265, DOI: 10.1112/plms/s2-42.1.230.
Turing AM. The Chemical Basis of Morphogenesis. Philosophical Transactions of the Royal Society of London. Series B, Biological Sciences 1952, 237 (641), 37–72.
Webb B. From Insects to Robots. Arthropod Structure & Development 2017, 46 (5), 687–688, DOI: 10.1016/j.asd.2017.08.002.
Webb B. Robots with Insect Brains: A Literal Approach to Mechanistic Explanation Provides Insight in Neuroscience. Science 2020, 368 (6488), 244–245, DOI: 10.1126/science.aaz6869.

Nikolaus Correll is an associate professor in the Computer Science Department at the University of Colorado Boulder and the director of the Interdisciplinary Research Theme "Multi-Functional Materials" at the College of Engineering and Applied Science. Nikolaus obtained his PhD in computer science from the École polytechnique fédérale de Lausanne (EPFL) in 2007, working with Alcherio Martinoli, and a diploma in electrical engineering from the Eidgenössische Technische Hochschule Zürich in 2003, with stints at the Technical University of Munich, Lunds Tekniska Högskola, and Caltech. From 2007 to 2009, he was a postdoc at the Massachusetts Institute of Technology, working with Daniela Rus. His research interests are materials that integrate sensing, actuation, computation, and communication, for which he coined the term 'robotic materials,' with particular emphasis on applications in robotics, including smart skins that can camouflage and sense touch, tactile sensing, shape changing materials, robotic muscles, as well as grasping and manipulation. Nikolaus's research has been awarded with an NSF CAREER Award, a NASA Early Career Faculty Fellowship, the Provost Achievement Award, and multiple Best Paper Awards at international conferences.

Essays: Part 1: Cultural, Scientific, and Design Perspectives

Mohammad Fardin Gholami, Lorenzo Guiducci, Susanne Jany, Khashayar Razghandi

Rethinking Active Matter: Current Developments in Active Materials

1 What Is Activity? An Introduction

The question of what activity is seems to stem from the realm of biology. Activity, a *biologist* would answer, is a characteristic of life; living beings are able to move, to grow, to reproduce. In the last half a century or so, *physicists* have developed an interest in these phenomena, in forms of movement, growth, self-assembly, or self-organization, which can be described in physical terms. At the core of the physicist's approach are forms of activity that are not exclusively bound to the living. Under the notion of 'active matter,' a new field of research has appeared in physics that studies forms of activity in the material world.

One of the current definitions of active matter is that it is a system of internally driven components which together show a collective behavior. The interplay between the local interaction rules and the conversion of energy at the scale of individual components gives rise to new, highly dynamic behaviors far from thermodynamic equilibrium.

Bacterial suspensions: a famous example of active matter
One of the frequently addressed examples of such active matter systems is the collective behavior of certain bacterial colonies. The bacterium *Escherichia coli* has long been an object of study in biology, but it is only recently that their colonies have been investigated under the label 'active matter' for their capacity to move and behave collectively. *E. coli* swim by moving their flagella in a corkscrew-like motion, and in this way can be considered as self-propelled particles – which are described as the basic units of active matter systems. Under conditions of high density, the bacteria tend to align their direction of movement and spontaneously form turbulent-like dynamics and complex patterns. This swarming behavior is not directed by any individual or central intelligence or control; rather, the global motion emerges from the vast number of local interactions and alignments that can be described in physical terms. These bacterial swarms can be observed on a microscopic scale (focusing on the interactions between individual bacteria), as well as on a macroscopic scale (focusing on the emergent collective behavior). In the macroscopic perspective, the bacterium is examined not for its biological features, but for its dynamic properties in relation to other bacteria. The collection of bacteria as self-driven swimmers and their surrounding fluid medium are conceived as an active colloid, that is, as a fluid material that changes its properties due to the behavior and direction of its constitutive particles [Menon 2010].

Acknowledgment: All authors contributed equally to this work. The authors acknowledge the support of the Cluster of Excellence 'Matters of Activity. Image Space Material' funded by the Deutsche Forschungsgemeinschaft (DFG, German Research Foundation) under Germany's Excellence Strategy – EXC 2025 – 390648296.

Open Access. © 2021 Mohammad Fardin Gholami et al., published by De Gruyter. (cc) BY-NC-ND This work is licensed under the Creative Commons Attribution-NonCommercial-NoDerivatives 4.0 International License.
https://doi.org/10.1515/9783110562064-011

A relatively heterogeneous research field that has been forming during the past two decades within biology and soft matter physics has been summoned under the umbrella term 'active matter,' comprising a variety of systems such as flocks of birds, schools of fish, migrating cells, the self-organization of microtubules and microfilaments in cytoskeleton, or synthetic microswimmers.

One may suggest that the beginning of this interest started when computer graphic expert Craig W. Reynolds proposed a theoretical model for the flying behavior of birds in flock formation [Reynolds 1987, p. 25]. By demonstrating that the emergence of flocking patterns derives from simple rules – namely, (1) collision avoidance, (2) velocity matching, and (3) vicinity to flock – he showed not only that a central control was not necessary but also that the swarming behavior of living entities could be described in simple mechanical terms. Later, Vicsek et al. proposed a mathematical model for the collective behavior of self-propelling particles with constant velocity [Vicsek et al. 1995]. At high enough density (or low enough noise), the model reproduces a phase transition from a disordered gas-like behavior to an ordered swarm. Given the broad validity of its defining characteristics (discreteness, constant velocity particles), the Vicsek model could be attributed to a wide range of phenomena (from the obvious example of schools of fish to bacterial suspensions) that exhibit cooperative motion (rotation, flocking, etc.) in order to survive under unfavorable conditions [ibid., p. 1226]. In this way, the micro-perspective on the interactions between particles in a system was combined with the macro-perspective that concentrates on "processes such as aggregation, viscous flow, or biological pattern formation" [ibid.] as a central characteristic of what was later defined as 'active matter.' Inspired by Vicsek, John Toner proposed a model of flocking based on standard hydrodynamics equations and the integrating elements of a discrete system (interacting ferromagnetic particles) [Toner and Tu 1995]. As a result, a more general continuum description of swarming systems was achieved, which contained Vicsek's model as a special case.

Whereas the development of such models of self-propelled particles was motivated by the need to understand the rich behavior of living systems present in nature, their wide applicability (from bacteria to animals) encouraged the design of artificial analogues. The first decade of the twenty-first century highlights the beginning of the experimental phase within active matter research [Popkin 2016, p. 17]. Following early attempts using biological motile components of cells – from microtubules and motor proteins [Nédélec et al. 1997] to synthetic microswimmers based on bimetallic nanorods [Howse et al. 2007] or Janus particles [Paxton et al. 2004] – many experimental platforms have been developed to test and benchmark active matter theories.

Against the background of these developments, active matter became a success story for multiple reasons. It provided a way to describe and explain the behavior of material systems that do not follow classical thermodynamic rules of externally driven systems. The so-called transduction of energy in these systems is defined rather generally and does not specify the energy source – whether this be, for instance, a metabolic biochemical process (as in cells) or a chemical one (as in synthetic microswimmers) – which enables the term to be applied equally to living and nonliving systems. Finally the definition does not imply a specific scale, which has led to its versatility in and application to a vast variety of systems (from the macroscopic flock of birds to crowded bacterial suspensions).

Parallel to this, in current materials science research, many more phenomena coming from different fields and occurring at various length scales are being investigated that deal with the notion of activity. This body of research uses terms such as 'shape morphing,' 'passive actuation,' 'self-regulation,' and 'autonomous devices,' or attributes to materials characteristics such as 'responsive,' 'adaptive,' and 'interactive' [Bertoldi et al. 2017; Burgert and Fratzl 2009; Wani, Zeng, and Priimagi 2017].

In view of this scope, the definition of active matter mentioned at the beginning [Menon 2010] appears relatively narrow, since it fails to cover such natural phenomena, while also overlooking current developments in physics, materials science, and biology. Through this work, we argue that such systems deserve to be considered and conceptualized under an umbrella category that covers a specific realm of material activity in order to form a new category of active matter. By challenging and going beyond current conceptions of activity and thinking of it in the broader sense of the changes in state of a system and its relation to the structure and function at hand, this extension of the term could enrich our understanding of the concept and draw a more comprehensive picture of material activity.

To provide the context for this proposal, in the next part we explore three case studies from contemporary scientific developments that, in our understanding, are concerned with active materials in nature, engineering, and design: (1) actuation systems in plants, (2) reconfigurable mechanical metamaterials, and (3) responsive 2D material systems. Subsequently, we discuss these examples against the background of the following questions: With what form of activity are we concerned? In what way do we need to extend the current definition of active matter in order to include these and similar objects of research?

2 Case Studies: Active Materials from Nature, Engineering, and Design

2.1 Actuation Systems in Plants

Response, in conventional thinking, is considered to be an organism's reaction, originated and initiated from the internal processing of the environmental information obtained through sensory systems, and the expenditure of metabolic energy to process the relevant information and regulate the next viable move (e.g., sensing, processing, and response in the sensorimotor system of mammals).

Passive response, on the other hand, is about embedding the necessary information at the material level of the system in order to minimize or externalize the processing of the information (sensing), the energy expenditure, and/or the regulation and control of the response (actuation).

In this regard, plants occupy a unique niche, since they have evolved various mechanisms and material solutions to utilize water for stress generation and for movements that are essential to a variety of their needs, such as growth, spatial orientation, acquiring nutrition, or seed dispersal. In these plant hydro-actuated movements, an elaborate makeup of the material structure at various molecular, cellular, tissue, and organ scales mediates and translates the water–plant material interactions into a regulated response to environmental stimuli.

The snapping closure of the leaves of the famous carnivorous Venus flytrap is one of the fastest movements among plants (approx. 100 ms; Fig. 1a). The moment a big enough insect enters and touches the sensitive hairs on the inner side of the leaves, a biochemical response triggered by some of the motor cells induces an action potential that results in water flow and the inflation of the cells on the outer side of the leaves. The consequent differential volume change of the leaf tissue leads to the rapid morphing from a convex to a concave shape and the closing of the trap. It has long been argued that water inflow and inflation is too slow to account for the fast movement required for the entrapment of the prey, and that the snapping behavior can only be explained through the mechanical and geometrical constraints of the structure, resulting in a prestressed metastable system. Initially, the cells' turgor pressure holds the leaves in a mechanically metastable outward-curved state. Upon the stimulation of the hair sensors, the biochemically induced water flow from the inner to the outer side of the doubly curved leaves induces a change of the curvature parallel and perpendicular to the midrib of the leaves. A close interplay and coupling of these two curvature changes lead to the destabilization of the metastable geometry and force the system to follow an elastic relaxation by abrupt curvature conversion and the closure of the trap (Fig. 1a) [Razghandi, Turcaud, and Burgert 2014; Guiducci, Dunlop, and Fratzl 2016; Burgert and Fratzl 2009; Forterre et al. 2005]. Here, triggered by the prey, physiologically induced water inflow and inflation of the cells trigger the response, yet it is the elaborated material structure at different length scales that turns this volumetric change of the cells into the snapping closure of the trap at the macroscale.

Another well-known example of plant actuation is the hydro-responsive movement of pine cone scales. The cone scales are closed in the wet state, and bend and open up as they dry to reveal their seeds (Fig. 1b). This reversible hydro-responsive movement is a famous example of how simple passive swelling and shrinkage can be translated into a movement at the macroscale through an elaborate hierarchical design from the cell wall structure to the tissue level. Each scale is made up of two layers of dead cells, each having a different response to water absorption and desorption: the top layer only slightly swells in width, while the bottom layer swells more along the scale axis. To compensate for the different responses of the two sides, the scale bends

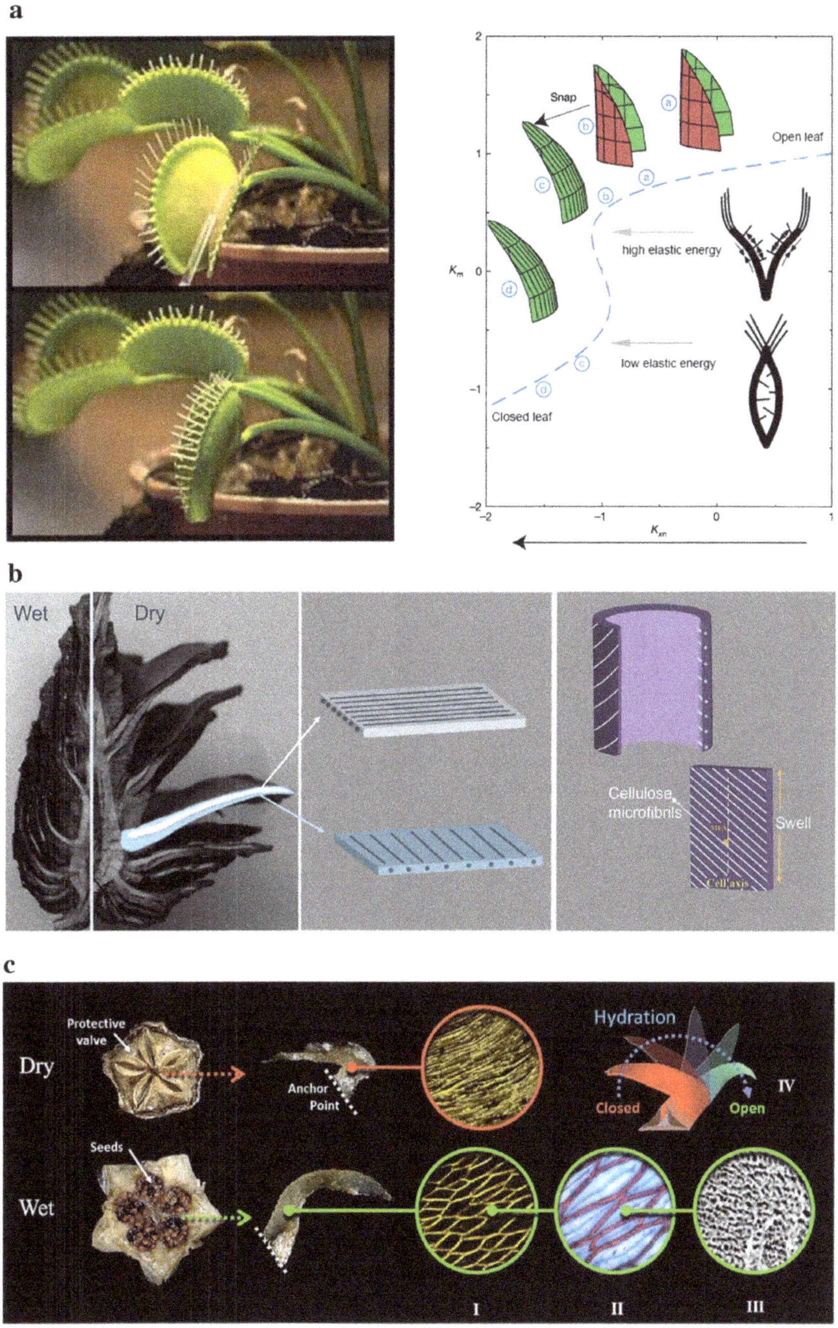

Fig. 1: (a) Snapping closure of Venus flytrap leaves (left). When the hair sensors inside the leaves are triggered by the presence of an insect, biochemically induced water flow and volume change of the cells at two sides of the leaves lead to change in the curvature of the doubly curved leaves.

correspondingly upon wetting and drying cycles. The thick dead walls of the cells are a composite structure consisting of stiff cellulose microfibrils arranged inside a swellable matrix of macromolecules such as hemicellulose and lignin. During growth, plants can control the specific arrangement of the 'hard' fibrils inside the cell wall, thus ensuring a level of control in the swelling and shrinking response of the wall to changes in environmental humidity conditions even after the cell's death. In this case, the cellulose microfibrils in the bottom layer are arranged perpendicular to the cell's axis, resulting in the swelling of the dead cells in the longitudinal direction, while a more parallel orientation of the fibers to the cells axis in the upper layer restricts the swelling due to the mechanical hindering of the non-swelling cellulosic fibers (Fig. 1b) [Razghandi, Turcaud, and Burgert 2014; Burgert and Fratzl 2009; Dawson, Vincent, and Rocca 1997]. Here, an elaborate design of the cell wall structure enables a predefined swelling behavior that is utilized to build up the 'right' architecture and enable the functional response at the higher scale, bringing an element of activity into the nonliving material.

The hydro-actuated unfolding of ice plant seed capsules is yet another example of how a dead tissue can respond to environmental stimuli to realize a functional movement. In the presence of water, the seed-containing valves of the ice plant seed capsule undergo an origami-like unfolding to reveal and release the seeds. The sophisticated structure and mechanism underlying this hydro-actuated movement can

Fig. 1 (continued)
An interplay between curvature change perpendicular and parallel to the midrib of the leaves results in a metastable state in which the leaves bypass gradual deformation and go through a rapid elastic relaxation path, flipping from a concave to a convex shape (right). Adapted by permission from Springer Nature: [Forterre et al. 2005, p. 422, fig. 1a, p. 423, fig. 3]. (b) Reversible hydro-responsive opening and closing of pine cone scales (left). Each scale is made up of two layers. While the top layer with cell with cellulose microfibrils running parallel to the cell wall axis undergoes minor dimensional changes upon water absorption and desorption, the bottom layer with cellulose microfibril orientation perpendicular to the cell wall axis undergoes a unidirectional swelling and shrinkage along the cell and tissue axis. The differential response of the two layers results in the outward and inward bending of the scales upon drying and wetting cycles (middle). The schematic representation on the right shows how the arrangement of cellulose fibrils inside the cell wall matrix can influence the swelling behavior of the cell wall. Re-sketched with permission of The Royal Society (U.K.), from: [Burgert and Fratzl 2009, p. 1547, fig. 3a,b]; permission conveyed through Copyright Clearance Center, Inc.; from: [Razghandi, Turcaud, and Burgert 2014, p. 186, fig. 8.9]. © 2014, John Wiley and Sons; and from: [Goswami et al. 2008, p. 536, fig. 5]. © 2008 The Authors of [Goswami et al. 2008]. Journal compilation © 2008 Blackwell Publishing Ltd. (c) Reversible hydro-actuated unfolding of ice plant seed capsules. The seed capsules with five seed-containing valves in the dry (closed) and wet (open) state (I). The hygroscopic muscle responsible for opening the valves consists of a hydro-responsive cellular tissue attached to an inert backing making up a bilayer structure that flexes upon wetting and drying cycles (II, IV). The hydro-responsive honeycomb tissue made up of an array of elongated hexagonal-shaped cells with a highly swellable cellulosic inner layer that undergoes a fourfold unidirectional expansion and shrinkage upon water absorption and desorption (III) From: [Guiducci et al. 2016, fig. 1]. License: CC by 4.0 (https://creativecommons.org/licenses/by/4.0/).

be seen in Fig. 1c. At the cell level, the lumen of the cells of the tissue responsible for the actuation is filled with a highly swellable cellulosic inner layer that, when exposed to sufficient hydration, passively swells and induces an inner pressure on the cell walls. The specific elongated hexagonal shape of the cell walls translates the pressure generated from the isotropic swelling of the inner layer into an anisotropic inflation of the cells perpendicular to their longer axis. An array of these cells organized in a honeycomb pattern generates a hygroscopic tissue that undergoes a unidirectional expansion or shrinkage in response to wetting and drying cycles respectively. An inert backing tissue attached to one side of this responsive 'muscle' restricts its expansion and shrinkage and transforms the linear hydro-responsive deformation of the honeycomb into the bending of the whole 'bilayer' structure and the unfolding of the seed-containing valves [Harrington et al. 2011]. Here, we have yet another hydro-responsive mechanism based on the differential response of the nonliving material structure to the environmental trigger at different length scales.

The mechanism behind such plant hydro-actuated movement has inspired exploration of the design of autonomous systems in which the desired response to a specific stimulus is embedded at the material level. The abstraction of such principles into simple models makes it possible to explore alternative architectures with different response solutions. In the aforementioned cases, for example, the behavior can be abstracted as follows: a differential response of different parts of the structure within a system can lead to a compromise between the differently reacting constituents and the consequent novel global behavior of the system as a whole.

In a simple bilayer structure consisting of two layers with different swelling behavior, the compromise between an expanding and an inert layer can be tailored into a variety of different responses of the system as presented in Fig. 2. At a lower scale, changing the inner structure (e.g., fiber orientation) gives a degree of control over the response of each of the layers (Fig. 2a left), while, at higher levels, playing with the geometry of the flat bilayer structure (Fig. 2a right) or different arrangements of the two layers in 2D cross-sectional space (Fig. 2c) provides the means to control the response of the system as a whole. Moreover, the bending bilayer concept can be implemented at smaller length scales of the hierarchical structure, as in the case of the autonomously deforming bilayered honeycomb structure shown in Fig. 2b. Each cell of this honeycomb structure is made up of two walls, each built based on the bending bilayer concept (inert layer: paper; active layer: spruce veneer) responding to changes in environmental humidity. As the bilayer walls bend upon wetting and drying cycles, the cells and the overall honeycomb structure undergo a hydro-actuated unidirectional expansion/contraction.

Such abstractions help to detach from the boundary conditions of the biological model system and explore other aspects of the response. For instance, with the right choice of material, the underlying concept of such an autonomously deforming system can be adapted to respond to other stimuli, such as change in temperature, pH, or magnetic field.

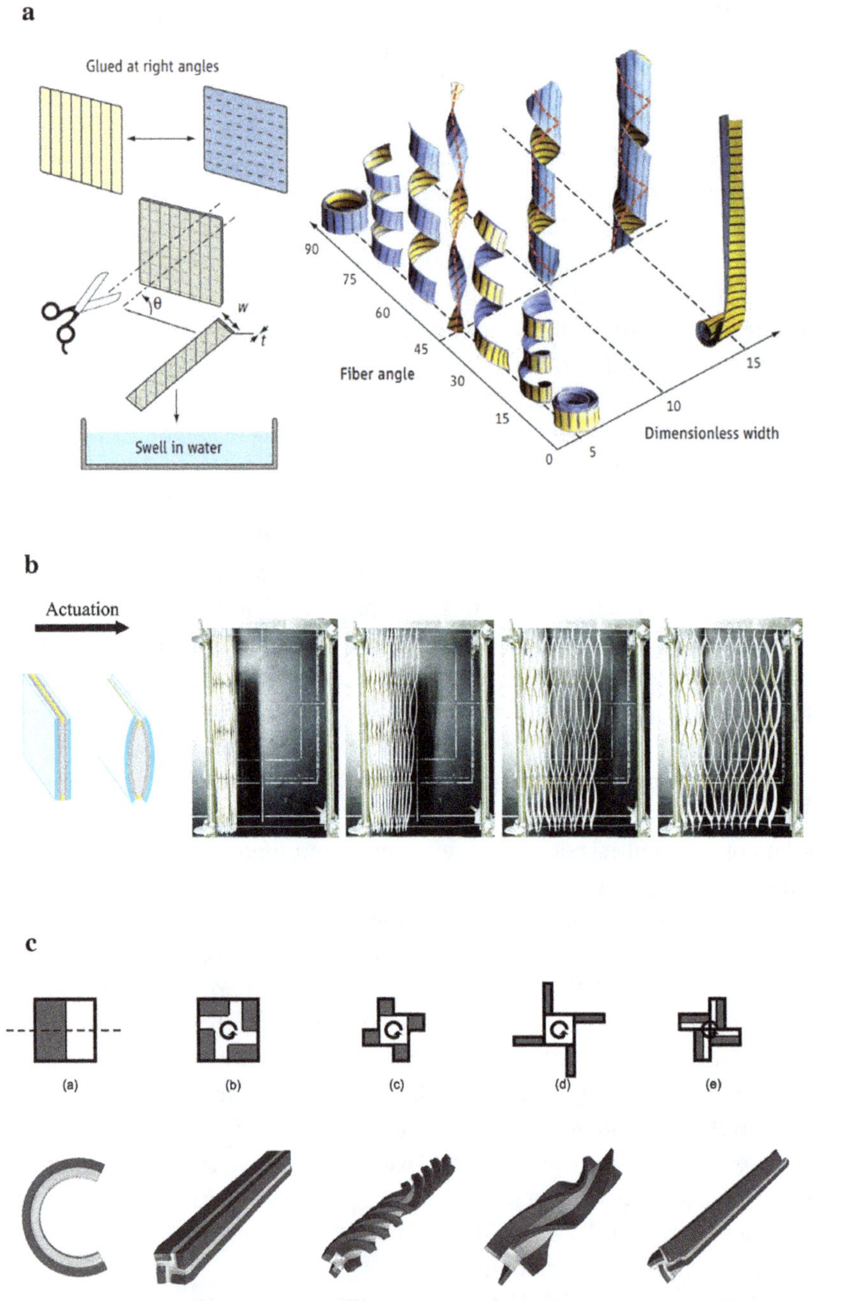

Fig. 2: (a) Exploring different actuation behaviors in a paper bilayer. The fiber orientation within each layer and the dimensionless width of the bilayer structure modulates the hydro-actuated response of the structure. From: [Forterre and Dumais 2011, p. 1716]. Reprinted with permission

2.2 Reconfigurable Mechanical Metamaterials

As we have seen in the previous section, plants have developed simple autonomous devices that, owing to their specific microstructures, harvest changes in environmental stimuli to perform a given function. Very similar strategies are shared by a class of engineering materials called metamaterials.

According to Muamer Kadic, "[Metamaterials] realize extreme or even unheard of effective material properties that go quantitatively and qualitatively beyond (*meta*, Greek) usual material properties of the bulk material they are made of" [Kadic et al. 2013, p. 2]. Due to their regular periodic microstructures with length scales comparable to those of electromagnetic (nm) or acoustic (mm) waves, metamaterials can interact with said waves to create exotic effective material properties, leading to optical applications, such as invisibility cloaks [Schurig et al. 2006; Zhang et al. 2011] and 'lossless' waveguides with suppressed diffraction [Kivijärvi et al. 2016], or to acoustic applications, such as negative refraction index coatings and sound cloaks [Cummer, Christensen, and Alù 2016]. More recently, the metamaterial design paradigm has been leveraged to obtain unusual mechanical properties. For example, origami techniques have been used to structure flat plates into 3D solids whose mechanical properties can be tailored through the imposed folding pattern. For example, the rigidity of a Miura-ori crease pattern (Fig. 3a) can be programmed by introducing metastable 'pop-through' defects [Silverberg et al. 2014]. Tubular structures have been realized by sandwiching together two Miura-ori patterned sheets, thus coupling their folding kinematics. Further stacking a number of these structures by 'zipping' (i.e., by aligning the tubes' main axes) or 'weaving' (in which tubes are placed perpendicularly) results in foldable cellular solids with a variety of different mechanical behaviors. These cellular solids show soft deformation modes that allow them to be flat-folded while at the same time being rigid, depending on the loading direction [Tachi and Miura 2012; Filipov, Tachi, and Paulino 2015] (Fig. 3b). These mechanical metamaterials are not limited to origami techniques and have been based on lattice materials as well. An example of these is the pentamode lattice material that can screen the

Fig. 2 (continued)
from AAAS. (b) The hydro-responsive bilayered honeycomb structure. The schematic representation on the left shows the actuation of a cell with cell walls made up of a responsive (blue) and an inert (gray) layer. The sequential images on the right show a prototypical honeycomb structure buildup of spruce veneer and paper bilayer cells undergoing a unidirectional expansion upon wetting. With permission from: [Guiducci et al. 2016, fig. 9]. License: CC by 4.0 (https://creativecommons.org/licenses/by/4.0/). (c) Various actuation behaviors of structures with different cross-sectional distributions of responsive (dark) and inert (light) materials. Simulated behaviors from left to right: bending, minimal deformation, large twist, small twist, and minimal deformation. Reprinted with permission from: [Turcaud et al. 2011, p. 610, fig. 3].

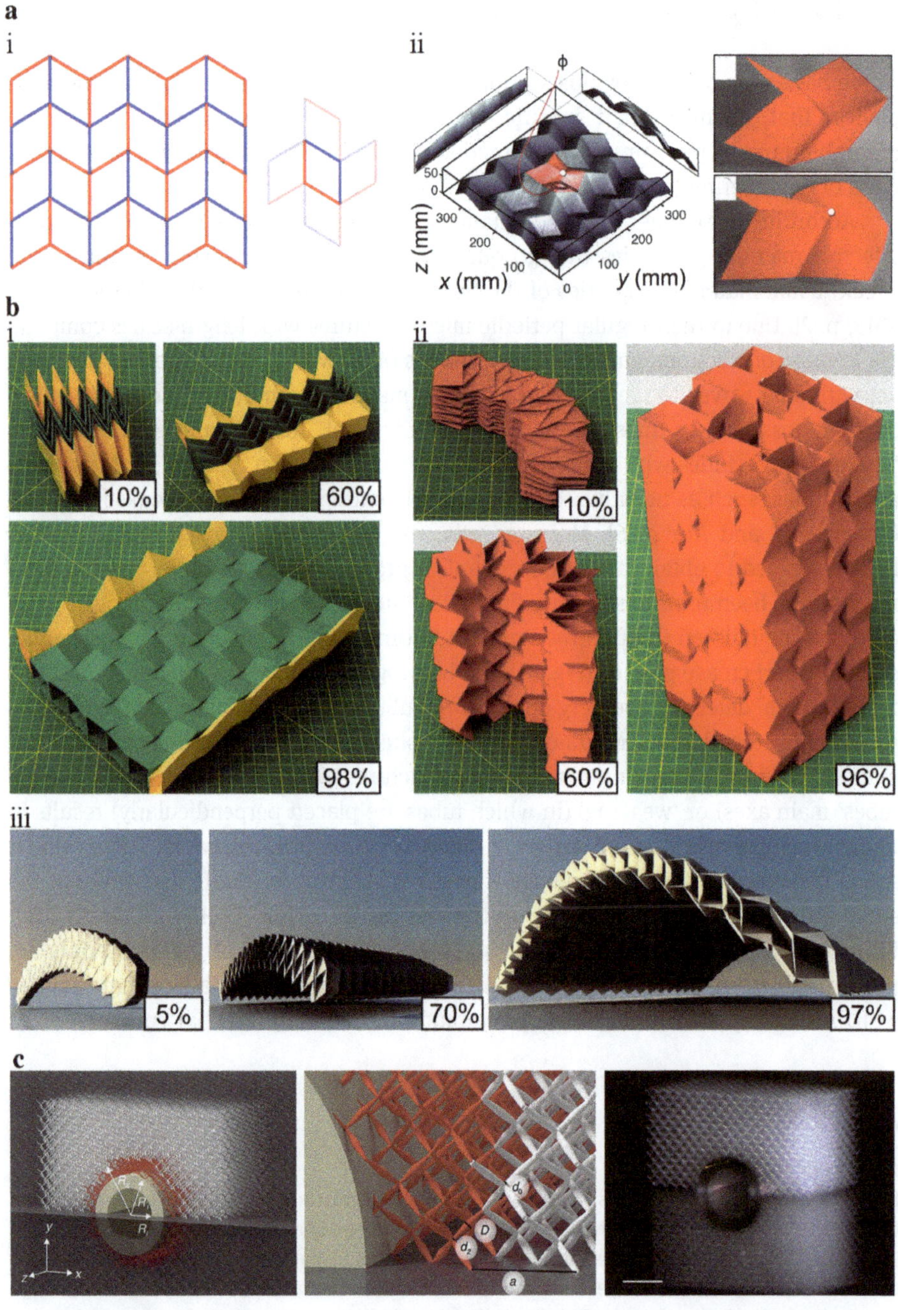

Fig. 3: (a) i) The Miura-ori crease pattern divides a sheet into planar facets hinged by alternating mountain (red) and valley (blue) folds, which join at vertices. ii) Pushing a vertex normally to the sheet eliminates one fold and bends two facets, locally changing the pattern topology and creating a pop-through defect (highlighted in red). This defect is spatially localized, metastable, and

effect of inclusions onto their deformation field and effectively work as mechanical 'unfeelability' cloaks [Bückmann et al. 2014] (Fig. 3c).

The remainder of this section presents some recent examples of mechanical metamaterials that are particularly useful for our scope – that is, they demonstrate a variety of different behaviors that suggest an interpretation as active multifunctional materials.

The reconfigurable mechanical metamaterials presented here in the article of Overvelde et al. are inspired by 'snapology,' a branch of origami invented by Heinz Strobl in which paper ribbons are assembled into complex geometric extruded polyhedra [Overvelde et al. 2016]. In general, these polyhedra can be rigid or flexible depending on whether the facets can rotate about the connecting folding lines. The constituent unit cell of the first reconfigurable metamaterial presented here is built by extruding the faces of a cube to obtain a six-arm cross element; in this particular case, the facets can rotate about the edges as in origami, so that the cross element can deform into radically different configurations (Fig. 4a, b). At extremal configurations (i.e., when the facets of the cross element come in contact), the cross element (state 1) transforms into a hexagonal prism composed of three adjacent rhomboidal channels (state 2), two aligned square channels (state 3), or even a completely flat-folded configuration (state 4).

When the cross element is patterned according to a face-centered cubic lattice, a periodic 3D metamaterial is obtained. In its undeformed state, the metamaterial looks like a 3D network of equispaced square channels aligned with the Cartesian axes. Interestingly, the cube-based metamaterial deforms exactly in the same manner as its constituent unit cell (cross element) so that the reconfigurability and shape-programmability of the unit cell is preserved in the metamaterials as well. Moreover, when the unit cells change configuration, the network of channels defined by them is drastically altered, making it a good candidate for a programmable acoustic waveguide [Babaee et al. 2016]. When the cube-based metamaterial is placed as an acoustic waveguide between an emitter (speaker) and a receiver, the measured acoustic transmittance reveals that each different state results in a completely different acoustic behavior: it can allow 1D sound

Fig. 3 (continued)
reversible but influences the overall rigidity of the sheet, thus allowing to program the sheet's rigidity. From: [Silverberg et al. 2014, figs. 1a,b and 2a–c]. Reprinted with permission from AAAS. (b) Flat-foldable cellular structures based on sandwiched Miura-ori patterned sheets in different stages of deployment. Approximate percentage of extension is shown. i) A bridge structure that resists out-of-plane loading. ii) A structure that interlocks into a fully conforming shape. iii) A deployable architectural canopy with high out-of-plane stiffness for transformable building design. With permission from: [Filipov, Tachi, and Paulino 2015, p. 12325, fig. 6]. (c) A rigid hollow cylinder embedded in a homogeneous 3D pentamode metamaterial environment (white) is covered by a compliant pentamode metamaterial shell (red). Any object can be placed inside of the hollow interior and thereby becomes 'unfeelable'. Reprinted by permission from Springer Nature: [Bückmann et al. 2014. fig. 1].

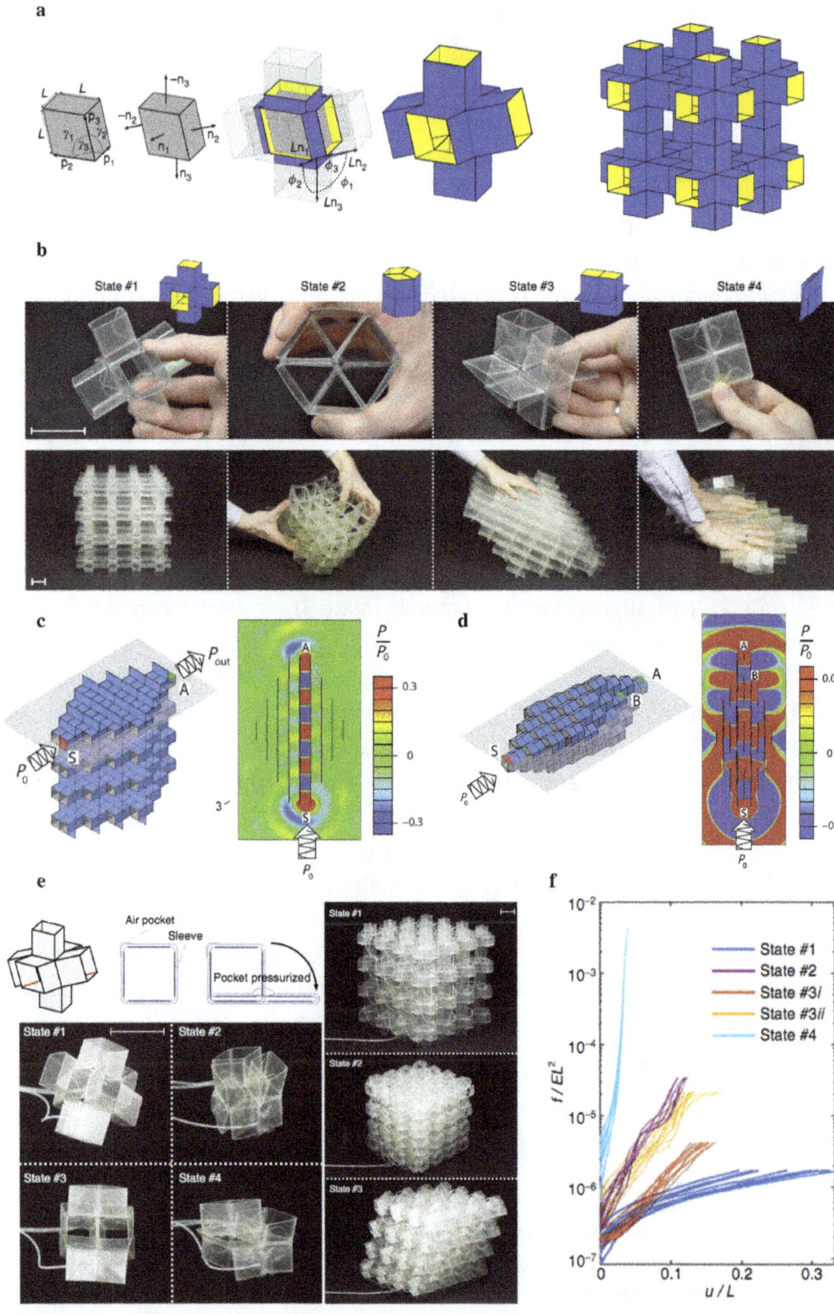

Fig. 4: (a) Deformation modes of the cross-shaped unit cell are characterized in terms of the central rhombohedral angles γ_1, γ_2, γ_3. (b) Extreme values of all admissible γ_1, γ_2, γ_3 combinations identify special states at which facets come in contact (top row: unit cell; bottom row: metamaterial obtained from assembling 4 × 4 × 4 unit cells). Physical prototypes obtained by lamination of PET

propagation along a single straight channel or 2D sound propagation with complex frequency dependence along partially connected parallel channels (respectively, state 3 and state 2 in Fig. 4c, d). Therefore, by simply switching between different mesostructural configurations, the cube-based metamaterial attains radically different acoustic properties.

An interesting aspect of this metamaterial mechanics is that the large deformations bridging between different states (the 'modes') have an energy cost that depends only on the rigidity of the hinges between the facets. On the other hand, its effective mechanical properties depend on the rigidity of the facets and the geometry of the metamaterial at a given state. Since the energy required to rotate the facets about the hinges is in orders of magnitude lower than that needed to deform the facets, configuration changes can be obtained with little energy, allowing the metamaterial to attain a radically different effective stiffness of the metamaterial as a whole. To demonstrate this concept, the authors built a physical prototype of the metamaterial from laser cut polymeric foils and actuated the metamaterial by placing pressurizing air pockets at the folds between the facets (Fig. 4e). As a result, by simply increasing the air pressure, the whole metamaterial could be transformed into one of the aforementioned states. As seen for the acoustic properties, the measured effective stiffness at each state mirrors the drastic configurational changes that the metamaterial undergoes (Fig. 4f).

The metamaterial based on extruded cubes has proven to be an interesting example of a programmable, multifunctional material, in which the architecture of the material – rather than its basic constituents – determines both the reconfigurability and effective material properties. In order to explore the possible design range of metamaterials based on extruded polyhedra, Overvelde et al. extended the method presented earlier to the 28 periodic uniform tessellations of the 3D space, which comprise regular

Fig. 4 (continued)
(polyethylene terephthalate) foils connected by thinner PET ligaments acting as hinges. In all cases, the metamaterial retrieves the same deformation modes as the unit cell. (c, d) Different states corresponding to different networks of channels make the extruded cube metamaterial a programmable acoustic waveguide (arrow indicates exciting acoustic wave): (c) simple 1D channel-like planar wave propagation corresponding to state 3 of the unit cell (red and blue correspond to positive and negative pressure half-waves; side areas external to the excited channel are green, meaning substantial absence of transversal sound propagation). (d) Complex propagation and interferences between waves due to transversally interconnected channels when the unit cell assumes state 2. (e) Schematic representation and physical prototype of a pneumatically actuated unit cell and metamaterial in which reconfiguration can be remotely controlled. (f) The effective stiffness of the metamaterial strongly depends on different conformational states (normalized force–displacement curves are shown). Adapted with permission from: [Overvelde et al. 2016, figs. 2a, 3c,d, 4a,b,d,f, 5c]. License: CC BY 4.0 (https://creativecommons.org/licenses/by/4.0/); and with permission of AAAS from: [Babaee et al. 2016, figs. 3a,b, 4c]. © The Authors, some rights reserved; exclusive licensee AAAS. Distributed under a CC BY-NC 4.0 License (http://creativecommons.org/licenses/by-nc/4.0/).

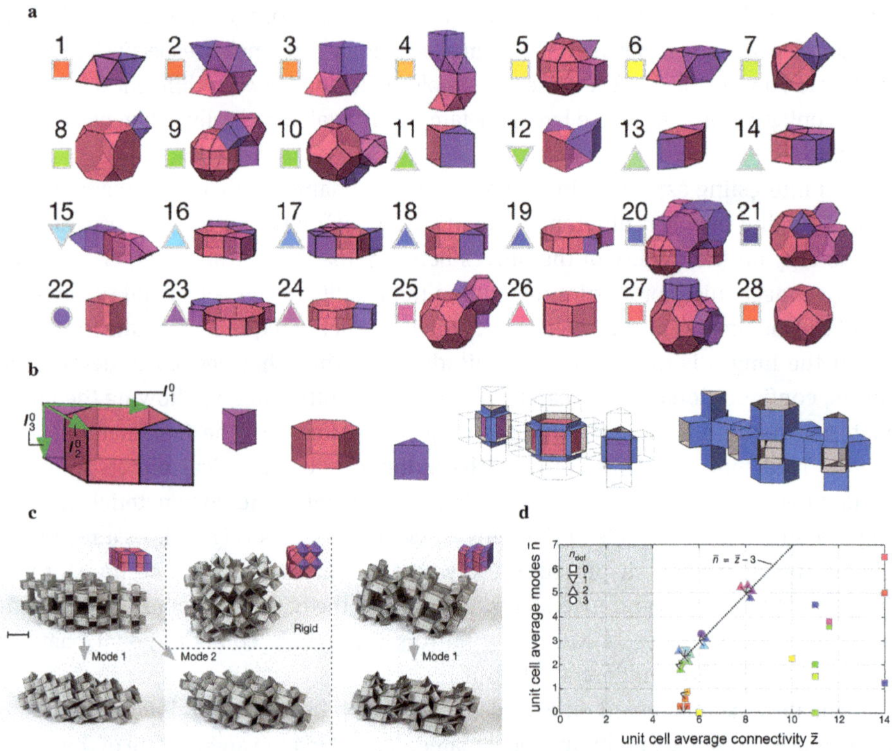

Fig. 5: (a) Design strategy to obtain extruded prismatic metamaterials. Space-filling unit cells composed by convex polyhedra are first expanded, then face pairs are extruded perpendicularly to the polygon faces. Choosing the extrusion length such that the expanded unit cell satisfies periodicity results in well-formed metamaterials in which the polyhedra of the unit cell are connected by prismatic channels. (b, c) Examples of metamaterials based on different polyhedral unit cells show different degrees of reconfigurability: (b) with two degrees of freedom (dof), (c) rigid, and (d) with a single dof. (d) Phase diagram showing that reconfigurability of prismatic metamaterials is proportional to the average number of unit cell modes ($\overline{n}$) and inversely proportional to the average unit cell connectivity ($\overline{n}$). Adapted by permission from Springer Nature: [Overvelde et al. 2017, figs. 2 and 3].

polyhedra, semiregular polyhedra, and semiregular prisms [Overvelde et al. 2017] (Fig. 5a). As for the cube, these metamaterials arise from extruding arbitrary pairs of the polygon faces delimiting the polyhedra of the tessellation unit cell. By enforcing the extrusion to be perpendicular to the polygon faces and by choosing the extrusion length such that the expanded unit cell satisfies periodicity, all space-filling and periodic assembly of convex polyhedra considered could be transformed into well-formed metamaterials (called prismatic architected materials; Fig. 5b).

Unlike the cube-based metamaterial, these metamaterials have a number of deformation modes or degrees of freedom (dof) that differ from those of the expanded unit cell. By analyzing the deformation modes of each metamaterial (Fig. 5c), some empirical rules could be formulated: (1) the higher the connectivity ($\overline{n}$) of a unit cell to its neighbors, the lower the reconfigurability; (2) if the constituent unit cell is rigid ($\overline{n}$), then so is the metamaterial (n_{dof} = 0); (3) the number of modes of the metamaterial is always smaller than or equal to that of the constituent unit cell: $\overline{n}$.

In terms of their stiffness, metamaterials with multiple degrees of freedom are softer and more anisotropic; this is to be expected, since, if a macroscopic deformation 'aligns' with one of the modes, then the stiffness drops to the minimum. Conversely, metamaterials with $n_{dof} = 0$ are fairly isotropic and rigid irrespective of the macroscopic deformation.

Since a major portion of the metamaterials built are rigid (only 13 out of 28 have $n_{dof} \neq 0$), Overvelde explores the design space by extruding only some of the polygonal facet pairs in the periodic unit cell while 'capping' with rigid faces the other pairs. Given the large number of possible face pairs and unit cells to choose from, a total of 0.6×10^6 different extruded metamaterials were obtained. By capping face pairs, fewer connections between polyhedra are formed, resulting in a lower average connectivity of the unit cell; therefore, these extruded prismatic metamaterials have more DOFs (are more reconfigurable) than their constituent unit cells. Moreover, the deformation modes in most of the cases result in internal rearrangements rather than macroscopic changes in volume or strain, thus making it possible to program the properties of the material by changing its internal mesostructure without changing its bulk shape.

Given the large number of different metamaterial designs, the possibility to shift between different geometric configurations, the strong dependence of their properties on said configurations, and the essentially scaleless nature of their designs, these metamaterials have huge potential as active multifunctional materials in a variety of applications, including structural, acoustic, and optical applications.

2.3 Responsive 2D Material Systems

Going to smaller scales, macromolecular sensors can be considered – as with the two aforementioned cases – as a set of active materials in which an elaborate material structure enables the system to undergo a functional transition in response to specific input. Relevant to this work is a special case of binding activity, namely, macromolecular activity at the nanoscale. In molecular and macromolecular systems, interactions between different atoms and molecules which can cause the overall systems energy to increase or decrease, enabling transition between various equilibrium states at the nanoscale.

The equilibrium changes of the atoms and molecules are often accompanied by the release or gain of energy to or from the environment, which, on the one hand, can influence chemical and electrical properties of the system and, on the other, can potentially lead to configurational/geometrical changes at angstrom scale or even at the nanoscale, which in turn can influence the physical or chemical behavior of matter at the molecular level.

One typical example of such macromolecular systems is graphene, a single-atom-thick sheet of covalently bonded carbon atoms with hexagonal arrangement which can have an equilibrium state in a truly 2D conformation on a flat substrate.

The deformation and rolling of 2D graphene sheets into carbon nanotubes as 1D nano-objects at equilibrium is a well-known example of how such configurational/geometrical changes can be induced due to energetically favorable environmental conditions [Biswas and Lee 2011; Calvaresi et al. 2013]. Such configurational changes can be elaborated to create responsive material systems, such as in the case of the pathogen inhibiting behavior of chemically modified graphene oxide sheets.

Graphene oxide (GO), a highly oxidized version of a graphene sheet, can be synthesized within a lateral flake dimension in the range of several micrometers. Thermally reduced graphene oxide (TRGO) is GO with a reduced number of oxygen-containing groups attached to its basal plane. The system includes discrete particles of dendritic polyglycerol (dPG) as a branched type of polymer chain (macromolecule) attached to the reduced graphene oxide (rGO) 2D surface (Fig. 6). Considering this combination as a 2D macromolecular system at nanoscales, the dendritic segments (dPG) can be sulfated by SO_3 end groups (dPGs) on each branch, which gives it the ability to interact with the outer membrane of viruses or bacteria (pathogens) [Gholami et al. 2017].

The energy gained by electrostatic interaction between the dPG or dPGs discrete particles on the 2D molecular sheets and the glycoprotein sites present over the viral envelope is consumed by the 2D sheet to overcome its bending stiffness. The high number of dPG units on the planar sheet of TRGO provides enough multisite adherences and further maximizes the number of interactions with the pathogen membrane – attachment of the sites at each stage would bend the sheet slightly and bring the next attachment sites into the vicinity of the membrane and further increase the efficiency of interactions, gradually entrapping and inhibiting the viral entities (Figs. 6a, b).

This molecular system is tunable in its size and can be specialized to certain pathogens. But studies have shown that the bending of the 2D sheets is not always a straightforward process, and both the density of dPG/dPGs sites and the surface area of the 2D sheets affect the inhibition efficiency of these nano-sensors. For instance, only 2D sheets with a size close to or smaller than the size of the virus would be able to interact with it efficiently [Ziem et al. 2017].

Hence, the function of the system as a pathogen inhibitor can be understood, analyzed, and tuned in regard to all these elements, namely, the type, density, size, and shape of the branches and their end groups, the pathogen type, the size of the 2D sheet, and so on.

Another example of how 2D materials can function as responsive systems derives from their unique electrical properties. One of the properties of graphene as a 2D material is the unrestricted charge mobility in planar direction that can lead to

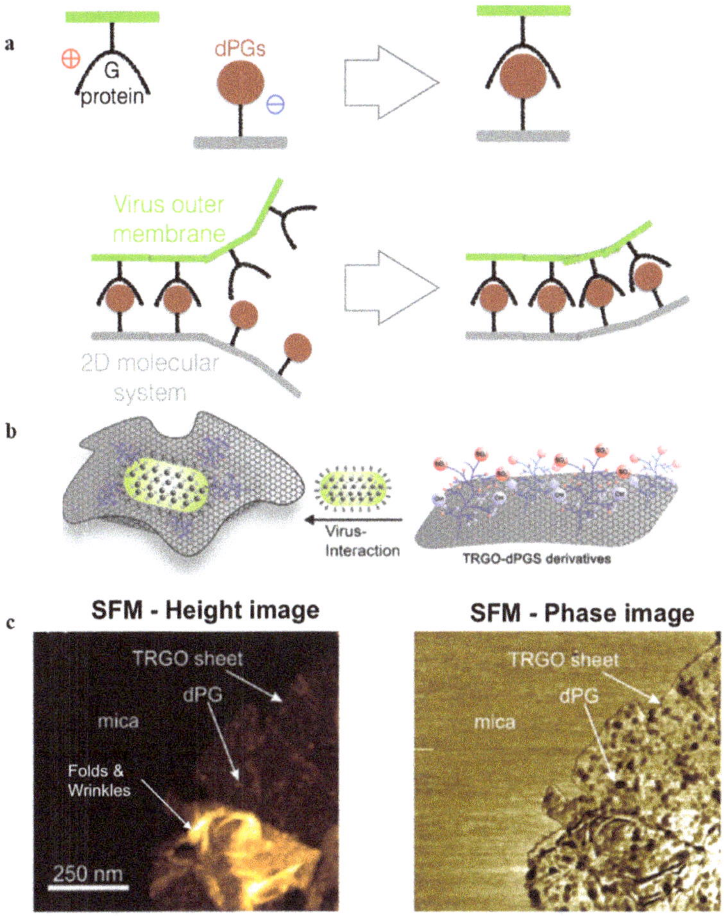

Fig. 6: (a) Representation of the 2D molecular system and dPGs electrostatic interaction as the key for equilibrium change in the conformation of the functionalized 2D TRGO sheets. © Graphic by the authors M.F.G./L.G./S.J./K.R. (b) Schematic representation of the TRGO- and dPG-based viral inhibition molecular system. © Graphic by the authors M.F.G./L.G./S.J./K.R. (c) Scanning force microscopy (SFM) height image of the functionalized TRGO-dPG sheets with their folds and wrinkles deposited onto atomically flat mica crystal surface. SFM phase image demonstrating the locations of the dPG sites with dark circles. (b) and (c) are partly adapted with permission from [Ziem et al. 2016, figs. 1c, d and 2]. © 2016 Wiley-VCH Verlag GmbH & Co. KGaA, Weinheim.

very high conductivities [Bolotin et al. 2008]. Graphene oxide (GO), on the other hand, shows more insulating electrical properties due to the presence of the oxygen-containing groups on its planar structure. By reducing the number of oxygen-containing groups in GO, one can create reduced graphene oxide (rGO) sheets with chemical and physical properties in between the two.

Based on the electrical properties and the ease of synthesis of these rGO, Robinson et al. have managed to create electronic circuitry involving single layer rGO sheets as sensors that are sensitive to the part per billion range simulants of chemical warfare gases. It was observed that the conductivity of rGO can vary with respect to the type of analyte adsorbed onto an rGO layer coating a Si wafer [Robinson et al. 2008, pp. 3137–3140]. This emergent conductivity change is indeed due to the structure of the rGO film and its new equilibrium energy states when an analyte molecule is adsorbed onto it. The structure of the rGO film is based on the hexagonal carbon atom arrangements with random presence of small quantities of oxygen-containing groups. When an analyte is adsorbed onto rGO, the mobility of the electrons within

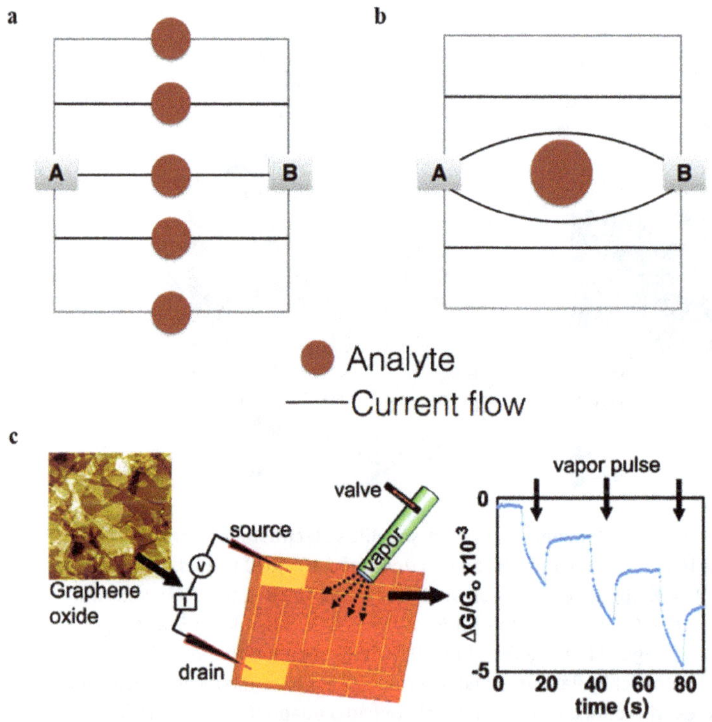

Fig. 7: (a) Considering the sensor elements in parallel configuration between A and B potential difference points. © Graphic by the authors M.F.G./L.G./S.J./K.R. (b) If analyte is not distributed homogenously over all the sensor elements, disruption of current flow, which is the output signal, would be insignificant due to current flow redistribution around the analyte affected region. © Graphic by the authors M.F.G./L.G./S.J./K.R. (c) Scanning force microscopy height image of a few layers of GO deposited onto a SiO_2 layer of a Si wafer substrate overlapping each other and being exposed to certain gas vapors while being connected to a potential difference source. The graph further demonstrates the variation of the overall conductance with respect to time. Reprinted with permission from: [Robinson et al. 2008]. © 2008 American Chemical Society.

the hexagonal carbon structure is disrupted; thus, the conductivity change is a function of the quantity and type of the analyte adsorbed onto the rGO thin film (Fig. 7).

Furthermore, since the oxygen-containing groups within the carbon aromatic rings enhance electron movement disruption, the sensor sensitivity can be tuned by chemically adjusting the rGO reduction state. These properties demonstrate the capacity of rGO as a macromolecular sensor.

3 Analysis of the Case Studies in Relation to Activity

In what sense can the given examples be considered active? Where can commonalities be observed between the examples?

The main commonality among the examples discussed above is the role of the material structure in defining and enabling a function. In all these cases, we have different material structures at various length scales, enabling a regulated response of the system as a whole. These systems can be understood as tailored material architecture that enables a predefined functional activity due to a differential response of the structural units (or, in other terms, the building blocks) to the relevant environmental stimuli. Here, we have a so-called 'hierarchical' structure: materials build a certain functional structure at one length scale, and these structures serve as building blocks to build up higher functional structures at larger length scales, and so on. In this view of activity, we propose that the concept of 'unit' should be rethought and understood in regard to function. Whereas the current active matter discourse takes units as merely similar, discrete, and autonomous physical particles, we argue here that the 'units' within the active system are a set of material structures described and defined by their specific functional role within the function of the system as a whole.

In relation to this different conception of units, one can also question the distinction between internal and external sources of energy, which has been taken as a crucial point in the active matter discourse. The defining characteristic of active matter is that particles convert chemical energy into active movement at the microscopic scale (i.e., at the level of the single particle/unit) – thus, energy is used 'internally.' This is different from other emergent collective behaviors, such as the Rayleigh-Bénard instability (in which convection cells emerge as a new dynamic equilibrium of all equal, passive fluid molecules due to external heating) or turbulence (in forced flowing fluids at large Reynolds numbers). In these dissipative structures, energy is provided externally from the environment to the system. In the examples discussed here, however, by taking this functional understanding of units into account, the boundaries of the system start to be less clear in regard to the source of the energy and where it is processed. It is only in its interrelation with the environment that the activity of the system can be understood, thus making the issue of external/internal source of energy less relevant.

In summary, the material properties and structure of each functional unit and their interrelation within one scale and through the hierarchy of the system, on the one hand, and the interaction of these different levels of the architecture with the environment, on the other, define the functional activity.

Take, for example, the activity discussed in the plant hydro-actuated movement. The rationale behind these environment-sensitive actuation systems is to make the control and regulation of the response to a certain stimulus as passive as possible, reducing the reliance of the system on the active expenditure of energy and on central processing and regulation by embedding and distributing the mechanisms of this regulated response at the material level and through different scales. So, 'active' in these cases can be understood as reactive: a structure at equilibrium carrying the potential to respond to a specific change in the environment and undergo a structural change to fulfil a certain function. Such active systems (natural or synthetic) require less central regulation of the desired response, thus externalizing different aspects of the responsiveness and embedding them in the material structure and the environmental conditions themselves. In the case of the ice plant or pine cone, for example, we have plant material that is so to speak 'dead' *and* active. Responsiveness, in this sense, can be rethought as a concept that we would like to term 'sense-action' in order to emphasize that different aspects of acquiring new relevant information, processing of it and regulation of a relevant response are indeed one inseparable activity integrated within a material system.

Moreover, taking the proposed understanding of unit in relation to the functional roles, one can discuss the activity from different perspectives and at different length scales. By zooming in on interactions at the molecular level, for instance, one can take different macromolecules as units of the system with different affinity to water; by looking at the aspect of the geometrical arrangement of differentially swelling elements, one can describe the activity of the system by the orientation of the non-swelling units (cellulose fibers) within the swellable matrix; by considering the geometrical features of the plant cell and their arrangement, one can talk about cells as units, or, in the case of the bending bilayer, one can discuss the arrangement of the differentially responding layers as the units of the system. Here again, boundaries start to blur. Although it is easy to see water as an external trigger absorbed from the external environment into the cell walls and fueling the activity, the enthalpic/entropic energetic gain responsible for the actuation is realized within the molecules interacting inside the system, making such internal/external distinctions misleading. It is even harder to draw such sharp categorical divisions in the case of the Venus flytrap, where we have, in addition, the living agency playing a role in one stage of the actuation. Although the change in water potential inside the cell and the consequent water inlet and inflation of the cells is a biologically regulated act triggered by the presence of an insect on the leaves, the snapping closure as the final functional activity is a direct consequence of a passive energy release from the structural instability regulated by the specific geometry of the leaf.

Similar to the plant actuation system discussed above, in reconfigurable prismatic metamaterials, chemical variety is replaced by structural complexity; a careful design of their mesostructure provides a huge design space for achieving various properties (rigid or soft, with few or many deformation modes, propagating sound in 1D or 2D, and so on). Here as well, the concept of unit cells and their rules of interaction play a central role. Slight differences in the geometric design of the folding lines hinging together the facets result in dramatically different properties and behaviors at the macroscopic scale. In addition, one can consider that these material structures have the intrinsic potential of becoming autonomous by integrating a responsive phase into their unit cells. For example, instead of actuating the hinges between the facets through an external pneumatic source (as in the extruded cube metamaterial), one could design the hinges as layered structures that would autonomously fold in response to an external stimulus (as with the responsive bending bilayers described in the plant hydro-actuated systems) – in the latter case, the metamaterial would show an active, albeit simple, behavior (reactivity to a stimulus). It therefore seems of little relevance whether such reconfiguration happens autonomously (in response to an environmental stimulus) or not (e.g., when controlled externally by the pneumatic actuator). So, here again, we would emphasize that defining activity based on how and where the energy is processed in the system is limiting, in the sense that it ignores the aforementioned fuzzy boundaries of the structure and their functions.

One can also refer the activity of the responsive macromolecular systems to the commonalities discussed above. The notion of activity at the nanoscale and macromolecular level is not a new concept and has notable examples in many chemical [Weber et al. 2015; Lee et al. 2014; Liese et al. 2017] and biological processes [Reich et al. 2004; Bélteky et al. 2019], as well as in dealing with catalysis [Smith 2002; Haridas, Gupta, and Sreenivas 2008; Bentley, Kang, and Unwin 2019; Parajuli et al. 2007], enzymatic activity [Liese et al. 2017; Cha et al. 2015], DNA translations [Kulkarni et al. 2019; Li et al. 2019; Shao et al. 2020], and so on. However, during the last two or three decades, access to the tools and knowledge of designing macromolecules from their building blocks has allowed researchers to design structures that use such energetical equilibrium changes to fulfil specific functions. In the case of the pathogen inhibiting behavior of the chemically modified graphene oxide sheets, the equilibrium changes from the open 2D sheets to folded conformation around the interacting pathogen is perceived as the activity of the system. At the molecular level, for example, the reactive end groups of the dendritic branches attached to the graphene sheet can be considered as units defining the interaction with specific pathogens. At a macromolecular level, the density of such dendritic dPG/dPGs functional sites can also influence the overall function; hence, each branch can be considered as a higher scale unit, whereas at the full scale, the overall lateral size of the TRGO sheet affects the efficiency of interactions with pathogens. Here again, it is the interrelation of such 'hierarchical' material structure with the pathogens and the surrounding buffer solution (playing the role of biologically relevant environment) that brings about the activity of the system. In the case of the macromolecular chemical sensor, the unit of activity can be discussed in regard to all and any of the involved elements of the system. At the atomic scale, the carbon–carbon bonds and oxygen–carbon bonds within the 2D rGO sheet can be defined as the functional units, since their different interaction with the analyte molecules allows for a change in conductivity. Although the absorption of the discrete analyte particles onto the rGO thin film would change the energetically equilibrated original system, as long as the external force of potential difference is not present, the detection of the energy equilibrium transfer is not possible. What this means within our context is that, here again, such classification based on the internal or external energy source is not a criterion for perceiving the 2D rGO thin film as an active material.

In all these examples, one can go beyond a simple reversible transition between two states and realize a more elaborate system with a degree of control over the modulation of a feedback loop and the self-regulation of the response. For instance, combining the class of reconfigurable prismatic metamaterials presented here with an already large body of research on stimuli responsive materials and synthetic homeostatic chemical systems could lead to scale-free designs of multifunctional and adaptive materials with programmable behavior. Just to give an impression, one can suppose the metamaterial's hinges to be composed of temperature-sensitive phases; if a decrease in temperature induces a forward reconfiguration of the system that enhances its exposure to a heat source, then the temperature of the metamaterial will increase, in turn inducing a backward reconfiguration to the initial state. Realizing a feedback loop in this way enables the system to self-regulate, which can be considered as a higher level of autonomy. Such self-regulating processes are well known in homeostatic biological systems and even in synthetic chemo-mechanical systems based on oscillatory reactions [Prigogine and Stengers 1984, pp. 146–148; Maeda et al. 2007; Horvath et al. 2011]. Successful attempts to integrate these processes into programmable and autonomous homeostatic devices have been realized, although only at the micrometer scale and with simple geometrical constructions [He et al. 2012]. This could bring new design paradigms for active materials beyond the scope of the examples discussed in this article.

4 Taking a Stance: A Case for Active Materials

The objective of our chapter was to critically rethink the current active matter concept by examining the research objects introduced above. What unites the activity of the aforementioned examples within the actuation systems in plants, reconfigurable metamaterials, and macromolecular sensors is that it is their specific *material structures that enable a regulated transition from one functional equilibrium state to another*. If a system is capable of responding to certain conditions and actively changes its state to fulfil a specific function *merely due to the design of its material structure*, we argue, it deserves to be considered as an active material.

The emphasis on the changes in the state of the system recalls the familiar notion of phase transition, which refers to the rearrangement of the atomic packing due to an external condition (temperature, pressure, etc.) and the consequent abrupt change in material properties. In our context, however, these 'active materials' undergo changes in the material structure at various length scales. This is different from the application of the term in the thermodynamic sense, and closer to the so-called phase transformations of cellular materials, which are essentially changes in the geometry of the unit cell from one stable configuration (i.e., convex cell shape) to another stable or metastable configuration (reentrant cell shape) without any change in the underlying cellular structure topology [Restrepo, Mankame, and Zavattieri 2015].

This brings us back to our starting point in the introduction. Regardless of terminology ('phase transformation of cellular solids,' 'passive actuation in plants,' or 'reconfigurable metamaterials,' etc. [Restrepo, Mankame, and Zavattieri 2015; Burgert and Fratzl 2009; Bertoldi et al. 2017]), there is an ongoing body of research that converges on the notion of an activity of a system that is based on changes in the material structure, which is where this category of active materials as we define it becomes relevant.

With the analysis of the examples from the previous section in mind, we can now define general aspects and conditions of these active material systems:

- *Multi-scale structures as a continuous physical material architecture.* Active materials possess a 'hierarchical' structure, where physical units are continuously connected together.
- *Units defined by a functional role.* These physical units are material structures defined by their specific functional role within the function of the system as a whole.
- *Interrelation with the environment.* Environment is an integral part of the activity of the system. The boundaries of the system and where the energy is coming from and processed are defined through the interrelation and functional role of the units and environment.

Based on these criteria, we can now compare our proposed definition of 'active materials' with that of 'active matter' and within a broader context of activity in general.

The systems Menon refers to in his definition of active matter are, as he describes them, "either in the continuum or naturally decomposable into discrete units" [Menon 2010, p. 2]. However, the examples that he introduces are of interest as discrete systems. Bacterial suspensions, the marching behavior of ants, or fish schools, for instance, are all discrete systems whose elements on the micro-level are self-propelled and are the driving forces for patterns, movements, and organizations on the macro-level. This seems to be valid for the big bulk of active matter examples. If researchers employ hydrodynamic descriptions [Toner and Tu 1995], they only theoretically suggest a continuous perception of these discrete systems in order to find a mechanical description for the discrete self-propelled elements. The same applies to thermodynamic descriptions, which also superimpose a continuous understanding of the system, even though the material or system itself suggests clearly definable, self-evident elements, such as fish, birds, bacteria, bees, microtubules, cells, copper particles, or iron atoms.

In the case of the examples discussed above, however, the system itself is continuous on its material level. Different functional structures at different scales are in a functional interrelation with each other in the same smaller or larger scale of the system in a continuous sense. Such interrelations throughout the architecture are a crucial aspect of the system that enables the functional activity. This structural 'interwovenness' sometimes blurs the boundaries between what is taken as the functional units at different scales. Even though, as an analytical strategy, it might make sense to describe and define specific units, these units are, eventually, defined by a boundary that we impose on it – according to the functional role of that part in its interrelation with other structures and the activity of the functioning whole. Consequently, this interrelation extends further, blurring the sharp distinction between the system and its relevant environment. This challenges the classical discourse that emphasizes the internal source of energy of the individual units as a distinguishing feature of active matter systems [Menon 2010; Needleman and Dogic 2017]. Here again, although it might be a useful analytical strategy to talk about such internal/external triggers to discuss specific features of a system, in the case we are putting forward for understanding of such systems, emphasis on such dichotomies is irrelevant.

Bacterial biofilms: a fitting example of active materials

Bacterial biofilms provide a perfect example of an active material (according to our definition) and well symbolize the integration between active and passive elements. As we have seen, bacterial suspensions were studied and considered quite early on as a typical active matter system. A bacterial biofilm, on the other hand, is the typical state in which bacteria are found in the outer environment and not in a host organism. Here, bacteria are physically bound to each other by an extracellular matrix (ECM) composed of proteins, amyloid fibers, and exopolysaccharides that the bacteria themselves produce [Serra et al. 2014; Hengge 2020]. Due to the different concentrations of oxygen and nutrients in the immediate surroundings, in a biofilm bacteria soon specialize into different 'phenotypes,' self-organizing in layers with variable composition. All these 'measures' help the bacteria thrive in the strongly variable environmental conditions typical of a sessile state – a biofilm can grow into a centimeter-sized colony hosting billions of bacteria in a few days. Here, the movement or discreteness of single bacteria is not particularly important; rather, it is their physical integration into a 'local environment' (the ECM) – which they are anchored to and actively influence through signaling and synthetic pathways – that suggests a form of activity. The wrinkling patterns that appear as a consequence of bacterial reproduction, subsequent stress generation, and compositional gradients in the biofilm would not arise without a solid substrate. Hence, the distinction here between active and passive, internal and external becomes difficult to assess: the passive ECM plays as much of an active role as the living bacteria producing it.

Now that we have been able to identify a category of 'active materials' out of the scope of active matter, the question is: how has the term 'activity' come to have such an exclusive position in the active matter discourse within the natural sciences?

What seems to have persisted in the active matter discourse from classical mechanics is the legacy of separating machine and force. From the earliest attempts to construct a model of machine and machinery in the early nineteenth century, machines were characterized as being the result of the coupling of three elements: receptor, transmission, and tools – or, to put it in the language of the mechanical machines of the industrial revolution, motor, gears, and working units, which required fuel and information input to be brought in from outside (workers, steam power, etc.) [Schäffner 1996; Poncelet 1844]. Against the background of this conceptual framework stemming from nineteenth-century mechanics and thermodynamics, activity, then, was perceived mainly through the narrow scope of movements of systems of bodies initiated by an external force working on the receptors, transmitted through the transmitters and into the final functional part. The main critics of this three-part conception of the machine in which the parts are conceived in a linear, hierarchical coupling relation to one another suggested instead a more dynamic operative view of pair-elements of machines [Schäffner 1996, Reuleaux 1875, pp. 592–595]. Nevertheless, this was against the background of a static world and an inherently passive matter.

It seems that current concepts of active matter are somewhat bound to this view – not necessarily in their definitions but in the way research examples are chosen and the phenomenon of active matter is conceived. This representational problem can be traced where examples of active matter are mostly introduced through their capacity for self-driven movement.

We claim that this conception of activity should not define and thereby limit the research objects dealing with activity. The dominance and shortcomings of this view of activity have been questioned extensively within the natural sciences and from a philosophical perspective. In the natural sciences, prominent figures such as Erwin [Schrödinger 1992 (1944), pp. 69–71] and later Ilya Prigogine and Isabelle Stengers have proposed that activity in living systems should be understood in terms of nonequilibrium thermodynamics [Prigogine and Stengers 1984]. Similarly, in philosophy, Manuel DeLanda, to give only one example, has observed that a "shortcoming of nineteenth century thermodynamics, to overlook the role of the intensive and stress only the extensive, to concentrate on the equilibrium form that emerges only once the original difference has been cancelled, has today been repaired in the latest version of this branch of physics, appropriately labeled 'far-from-equilibrium thermodynamics'" [DeLanda 2000, pp. 36–37]. In this view, far-from-equilibrium is a world where activity is cut loose from its inherent relation to passivity as well as from its relation to the living world.

Our proposed conception of active materials, however, suggests an understanding of activity that does not necessarily fall within these extreme cases. If we neither reduce activity to the aforementioned classical mechanical notion of movement nor bind it to nonequilibrium or far-from-equilibrium thermodynamics, then other forms of activity of materials can be noted: namely, changes in the state of the system as articulated above, where *specific material structures enable a regulated transition from one functional equilibrium state to another*. It is the change in the conditions that triggers the transition between two equilibrium states – and this change of functional state, this transition, is a form of activity on the material level that is not activity in terms of far-from-equilibrium states, and nor does it present itself as the clear separation of motor, gear, and units, in the sense of the classical mechanics and model of machines. *If a material system is capable of responding to certain conditions and actively changes its state (shape, mechanical, chemical, optical properties, and so on) to fulfill a specific function merely due to the design of its material structure, we argue, it deserves to be considered as active material.*

5 Conclusions

In this contribution, we have addressed the notion of activity in materials in various types of phenomena from different fields of the natural sciences that deserve to be considered in the active matter discourse. By going through and analyzing the mechanisms behind the activity of a few categorical examples – namely, actuation systems in plants, reconfigurable prismatic metamaterials, and graphene-based materials – we have tried to find the common aspects and criteria of these seemingly different phenomena and bring them together under a new definition of active materials.

In the process of forming a new category of active materials, we were confronted with some aspects of the existing notions of active matter and activity in general, which led to the questioning of some widely held paradigms. The characteristics of what is coined as activity in these systems challenge conventional dichotomies such as system/environment, internal/external energy source, discrete/continuum units, active/passive, or natural/artificial.

The category of active materials that we are putting forward can embrace contemporary efforts to build active, responsive, or smart devices and systems through materiality [Schäffner 2016a; 2016b], complementing the approaches that introduce these traits by programming a centrally controlled behavior via computers, software, and so on [Tibbits 2017].

We argue that our definition of active materials is supposed not only to give an idea of novel developments but also to invite researchers from different natural sciences (chemistry, mathematics, physics, biology, materials science, and engineering) as well as within the humanities (history of science, cultural history, media theory and information philosophy etc.) to work further on these topics. The goal might be to expand the field, critically analyze the accompanying discourses, and design new materials for current challenges and pressing issues such as sustainability.

Bibliography

Babaee S, Overvelde JTB, Chen ER, Tournat V, Bertoldi K. Reconfigurable Origami-Inspired Acoustic Waveguides. Science Advances 2016, 2 (11), e1601019, DOI: 10.1126/sciadv.1601019.

Bélteky P, Rónavári A, Igaz N, et al. Silver Nanoparticles: Aggregation Behavior in Biorelevant Conditions and its Impact on Biological Activity. International Journal of Nanomedicine 2019, 14, 667–687, DOI: 10.2147/IJN.S185965.

Bentley CL, Kang M, Unwin PR. Nanoscale Surface Structure – Activity in Electrochemistry and Electrocatalysis. Journal of the American Chemical Society 2019, 141 (6), 2179–2193, DOI: 10.1021/jacs.8b09828.

Bertoldi K, Vitelli V, Christensen J, Hecke M van. Flexible Mechanical Metamaterials. Nature Reviews Materials 2017, 2, 17066, DOI: 10.1038/NATREVMATS.2017.66.

Biswas C, Lee YH. Graphene Versus Carbon Nanotubes in Electronic Devices. Advanced Functional Materials 2011, 21 (20), 3806–3826, DOI: 10.1002/adfm.201101241.

Bolotin KI, Sikes KJ, Jiang Z, et al. Ultrahigh Electron Mobility in Suspended Graphene. Solid State Communications 2008, 146 (9–10), 351–355, DOI: 10.1016/j.ssc.2008.02.024.

Bückmann T, Thiel M, Kadic M, Schittny R, Wegener M. An Elasto-Mechanical Unfeelability Cloak Made of Pentamode Metamaterials. Nature Communications 2014, 5, 4130, DOI: 10.1038/ncomms5130.

Burgert I, Fratzl P. Actuation Systems in Plants as Prototypes for Bioinspired Devices. Philosophical Transactions of the Royal Society A 2009, 367 (1893), 1541–1557, DOI: 10.1098/rsta.2009.0003.

Calvaresi M, Quintana M, Rudolf P, Zerbetto F, Prato M. Rolling up a Graphene Sheet. ChemPhysChem 2013, 14 (15), 3447–3453, DOI: 10.1002/cphc.201300337.

Cha S-H, Hong J, McGuffie M, Yeom B, VanEpps JS, Kotov NA. Shape-Dependent Biomimetic Inhibition of Enzyme by Nanoparticles and Their Antibacterial Activity. ACS Nano 2015, 9 (9), 9097–9105, DOI: 10.1021/acsnano.5b03247.

Cummer SA, Christensen J, Alù A. Controlling Sound with Acoustic Metamaterials. Nature Reviews Materials 2016, 1 (3), 16001, DOI: 10.1038/natrevmats.2016.1.

Dawson C, Vincent JF, Rocca A-M. How Pine Cones Open. Nature 1997, 390, 668, DOI: 10.1038/37745.

DeLanda M. Deleuze, Diagrams, and the Genesis of Form. Amerikastudien / American Studies 2000, 45 (1), 33–41.

Filipov ET, Tachi T, Paulino GH. Origami Tubes Assembled Into Stiff, yet Reconfigurable Structures and Metamaterials. Proceedings of the National Academy of Sciences 2015, 112 (40), 12321–12326, DOI: 10.1073/pnas.1509465112.

Forterre Y, Dumais J. Generating Helices in Nature. Science 2011, 333 (6050), 1715–1716, DOI: 10.1126/science.1210734.

Forterre Y, Skotheim JM, Dumais J, Mahadevan L. How the Venus flytrap Snaps. Nature 2005, 433 (7024), 421–425, DOI: 10.1038/nature03185.

Friedman M, Krauthausen K. Inspirierte Mechanik. Active Matter als Maschine und Struktur. In: Doll N, Bredekamp H, Schäffner W, eds. +ultra. gestaltung schafft wissen. Leipzig, E. A. Seemann, 2016, 171–176.

Gholami MF, Lauster D, Ludwig K, et al. Functionalized Graphene as Extracellular Matrix Mimics Toward Well-Defined 2D Nanomaterials for Multivalent Virus Interactions. Advanced Functional Materials 2017, 27 (15), 1606477, DOI: 10.1002/adfm.201606477.

Guiducci L, Dunlop JW, Fratzl P. An Introduction into the Physics of Self-Folding Thin Structures. In: Friedman M, Schäffner W, eds. On Folding: Towards a New Field of Interdisciplinary Research. Bielefeld, transcript, 2016, 175–210.

Guiducci L, Razghandi K, Bertinetti L, et al. Honeycomb Actuators Inspired by the Unfolding of Ice Plant Seed Capsules. PLoS One 2016, 11 (11), e0163506, DOI: 10.1371/journal.pone.0163506.

Haridas D, Gupta V, Sreenivas K. Enhanced Catalytic Activity of Nanoscale Platinum Islands Loaded onto SnO 2 Thin Film for Sensitive LPG Gas Sensors. Bulletin of Materials Science 2008, 31, 397–400, DOI: 10.1007/s12034-008-0062-9.

Harrington MJ, Razghandi K, Ditsch F, et al. Origami-Like Unfolding of Hydro-Actuated Ice Plant Seed Capsules. Nature Communications 2011, 2, 337, DOI: 10.1038/ncomms1336.

He X, Aizenberg M, Kuksenok O, et al. Synthetic Homeostatic Materials with Chemo-Mechano-Chemical Self-Regulation. Nature 2012, 487 (7406), 214–218, DOI:10.1038/nature11223.

Hengge R. Linking Bacterial Growth, Survival, and Multicellularity – Small Signaling Molecules as Triggers and Drivers. Current Opinion in Microbiology 2020, 55, 57–66, DOI: 10.1016/j.mib.2020.02.007.

Horvath J, Szalai I, Boissonade J, De Kepper P. Oscillatory Dynamics Induced in a Responsive Gel by a Non-Oscillatory Chemical Reaction: Experimental Evidence. Soft Matter 2011, 7 (18), 8462–8472, DOI: 10.1039/c1sm05226h.

Howse JR, Jones RAL, Ryan AJ, Gough T, Vafabakhsh R, Golestanian R. Self-Motile Colloidal Particles: From Directed Propulsion to Random Walk. Physical Review Letters 2007, 99 (4), 048102, DOI: 10.1103/PhysRevLett.99.048102.

Kadic M, Bückmann T, Schittny R, Wegener M. Metamaterials Beyond Electromagnetism. Reports on Progress in Physics 2013, 76 (12), 126501, DOI: 10.1088/0034-4885/76/12/126501.

Kivijärvi V, Nyman M, Shevchenko A, Kaivola M. An Optical Metamaterial with Simultaneously Suppressed Optical Diffraction and Surface Reflection. Journal of Optics 2016, 18 (3), 035103, DOI: 10.1088/2040-8978/18/3/035103.

Kulkarni JA, Witzigmann D, Chen S, Cullis PR, Meel R van der. Lipid Nanoparticle Technology for Clinical Translation of siRNA Therapeutics. Accounts of Chemical Research 2019, 52 (9), 2435–2444, DOI: 10.1021/acs.accounts.9b00368.

Lee C-L, Liebig T, Hecht S, Bléger D, Rabe JP. Light-Induced Contraction and Extension of Single Macromolecules on a Modified Graphite Surface. ACS Nano 2014, 8 (12), 11987–11993, DOI: 10.1021/nn505325w.

Li L, Xing H, Zhang J, Lu Y. Functional DNA Molecules Enable Selective and Stimuli-Responsive Nanoparticles for Biomedical Applications. Accounts of Chemical Research 2019, 52 (9), 2415–2426, DOI: 10.1021/acs.accounts.9b00167.

Liese S, Gensler M, Krysiak S, et al. Hydration Effects Turn a Highly Stretched Polymer from an Entropic into an Energetic Spring. ACS Nano. 2017, 11 (1), 702–712, DOI: 10.1021/acsnano.6b07071.

Maeda S, Hara Y, Sakai T, Yoshida R, Hashimoto S. Self-Walking Gel. Advanced Materials 2007, 19 (21), 3480–3484, DOI:10.1002/adma.200700625.

Menon GI. Active Matter. In: Krishnan JM, Deshpande AP, Kumar PB, eds. Rheology of Complex Fluids. New York, Springer, 2010, 193–218.

Nédélec FJ, Surrey T, Maggs AC, Leibler S. Self-Organization of Microtubules and Motors. Nature 1997, 389, 305–308, DOI: 10.1038/38532.

Needleman D, Dogic Z. Active Matter at the Interface Between Materials Science and Cell Biology. Nature Reviews Materials 2017, 2, 17048, DOI: 10.1038/natrevmats.2017.48.

Overvelde JTB, Jong TA de, Shevchenko Y, et al. A Three-Dimensional Actuated Origami-Inspired Transformable Metamaterial with Multiple Degrees of Freedom. Nature Communications 2016, 7, 10929, DOI: 10.1038/ncomms10929.

Overvelde JTB, Weaver JC, Hoberman C, Bertoldi K. Rational Design of Reconfigurable Prismatic Architected Materials. Nature 2017, 541, 347–352, DOI: 10.1038/nature20824.

Parajuli O, Gupta A, Kumar N, Hahm J. Evaluation of Enzymatic Activity on Nanoscale Polystyrene-Block-Polymethylmethacrylate Diblock Copolymer Domains. The Journal of Physical Chemistry B. 2007, 111 (50), 14022–14027, DOI: 10.1021/jp075909j.

Paxton WF, Kistler KC, Olmeda CC, et al. Catalytic Nanomotors: Autonomous Movement of Striped Nanorods. Journal of the American Chemical Society 2004, 126, 13424–13431, DOI: 10.1021/ja047697z.

Poncelet JV. Traité de mécanique industrielle. Liège, Leroux, 1844.
Popkin G. The Physics of Life. Nature News 2016, 529 (7584), 16–18, DOI:10.1038/529016a.
Prigogine I, Stengers I. Order out of Chaos: Man's New Dialogue with Nature. New York, Bantam Books, 1984.
Razghandi K, Turcaud S, Burgert I. Hydro-Actuated Plant Devices. In: Kim AH, Guyer RA, eds. Nonlinear Elasticity and Hysteresis: Fluid-Solid Coupling in Porous Media. Weinheim, Wiley-VCH Verlag, 2014, 171–200, DOI: 10.1002/9783527665068.ch8.
Reich S, Gössl I, Reuter M, Rabe JP, Krüger DH. Scanning Force Microscopy of DNA Translocation by the Type III Restriction Enzyme EcoP15I. Journal of Molecular Biology. 2004, 341 (2), 337–343, DOI: 10.1016/j.jmb.2004.06.031.
Restrepo D, Mankame ND, Zavattieri PD. Phase Transforming Cellular Materials. Extreme Mechanics Letters 2015, 4, 52–60, DOI: 10.1016/j.eml.2015.08.001.
Reuleaux F. Theoretische Kinematik: Grundzüge einer Theorie des Maschinenwesens. Vol. 1. Braunschweig, Friedrich Vieweg und Sohn, 1875.
Reynolds CW. Flocks, Herds, Schools: A Distributed Behavioral Model. ACM SIGGRAPH Computer Graphics 1987, 21 (4), 25–34, DOI: 10.1145/37402.37406.
Robinson JT, Perkins FK, Snow ES, Wei Z, Sheehan PE. Reduced Graphene Oxide Molecular Sensors. Nano Letters 2008, 8 (10), 3137–3140, DOI: 10.1021/nl8013007.
Schäffner W. Technologie des Unbewußten. In: Balke F, Vogl J, eds. Gilles Deleuze: Fluchtlinien der Philosophie. München, Fink, 1996, 211–229.
Schäffner W. Immaterialität der Materialien. In: Doll N, Bredekamp H, Schäffner W. +ultra. gestaltung schafft wissen. Leipzig, E. A. Seemann, 2016a, 27–35.
Schäffner W. Neuer Strukturalismus. Eine Geistes- und Materialwissenschaft / New Structuralism: A Field of Human and Materials Science. Graz Architecture Magazine 2016b, 12, 10–31.
Schrödinger E. What Is Life? The Physical Aspect of the Living Cell. With Mind and Matter and Autobiographical Sketches. Cambridge, New York, Cambridge University Press, 1992 [1944].
Schurig D, Mock JJ, Justice BJ, et al. Metamaterial Electromagnetic Cloak at Microwave Frequencies. Science 2006, 314 (5801), 977–980, DOI: 10.1126/science.1133628.
Serra DO, Hengge R. Stress Responses Go Three Dimensional – The Spatial Order of Physiological Differentiation in Bacterial Macrocolony Biofilms. Environmental Microbiology 2014, 16 (6), 1455–1471, DOI: 10.1111/1462-2920.12483.
Shao Z, Flynn RA, Crowe JL, et al. DNA-PKcs has KU-Dependent Function in rRNA Processing and Haematopoiesis. Nature 2020, 579 (7798), 291–296, DOI: 10.1038/s41586-020-2041-2.
Silverberg JL, Evans AA, McLeod L, et al. Using Origami Design Principles to Fold Reprogrammable Mechanical Metamaterials. Science 2014, 345 (6197), 647–650, DOI: 10.1126/science.1252876.
Smith JK. History of Catalysis. Encyclopedia of Catalysis 2002, DOI: 10.1002/0471227617.eoc098.pub2.
Tachi T, Miura K. Rigid-Foldable Cylinders and Cells. Journal of the International Association for Shell and Spatial Structures 2012, 53 (4), 217–226.
Tibbits S, ed. Active Matter. Cambridge MA, MIT Press, 2017.
Toner J, Tu Y. Long-Range Order in a Two-Dimensional Dynamical XY Model: How Birds Fly Together. Physical Review Letters 1995, 75 (23), 4326–4329, DOI: 10.1103/PhysRevLett.75.4326.
Turcaud S, Guiducci L, Fratzl P, Bréchet YJM, Dunlop JWC. An Excursion into the Design Space of Biomimetic Architectured Biphasic Actuators. International Journal of Materials Research 2011, 102 (6), 607–612, DOI: 10.3139/146.110517.
Vicsek T, Czirók A, Ben-Jacob E, Cohen I, Shochet O. Novel Type of Phase Transition in a System of Self-Driven Particles. Physical Review Letters 1995, 75 (6), 1226–1229, DOI: 10.1103/PhysRevLett.75.1226.
Wani OM, Zeng H, Priimagi A. A Light-Driven Artificial Flytrap. Nature Communications 2017, 8, 15546, DOI: 10.1038/ncomms15546.

Weber C, Liebig T, Gensler M, et al. Light-Controlled 'Molecular Zippers' Based on Azobenzene Main Chain Polymers. Macromolecules 2015, 48 (5), 1531–1537, DOI: 10.1021/ma502551b.

Zhang B, Luo Y, Liu X, Barbastathis G. Macroscopic Invisibility Cloak for Visible Light. Physical Review Letters 2011, 106 (3), 033901, DOI: 10.1103/PhysRevLett.106.033901.

Ziem B, Azab W, Gholami MF, Rabe JP, Osterrieder N, Haag R. Size-Dependent Inhibition of Herpesvirus Cellular Entry by Polyvalent Nanoarchitectures. Nanoscale 2017, 9 (11), 3774–3783, DOI: 10.1039/C7NR00611J.

Ziem B, Thien H, Achazi K, et al. Highly Efficient Multivalent 2D Nanosystems for Inhibition of Orthopoxvirus Particles. Advanced Healthcare Materials 2016, 5, 2922–2930, DOI: 10.1002/adhm.201600812.

Mohammad Fardin Gholami is a polymer engineer and a macromolecular physicist who works and researches at the edge of the lower limits of observable scale using scanning probe microscopy. Currently Fardin is a research associate at the 'Cutting' project in Cluster of Excellence 'Matters of Activity. Image Space Material' of the Humboldt-Universität zu Berlin. Gholami's research is oriented towards the development and characterization of active polymeric structures such as one dimensional dendronized polymeric chains and two dimensional graphitic functional systems. Currently he works directly with designers, computer scientists, anthropologists and neurosurgeons on the development of new active tools for simultaneous sensing & cutting inspired by the scanning probe microscopy.

Lorenzo Guiducci is a biomedical engineer with a PhD in physics, science of biomaterials. At present, he is co-leader of the 'Material Form Function' project in the Cluster of Excellence 'Matters of Activity. Image Space Material' of the Humboldt-Universität zu Berlin. His research is generally oriented towards shape changing materials in a variety of forms and contexts: from plants seed capsules, to bacterial biofilms to man-made architected solids, he studies them by applying theoretical and simulation tools from engineering disciplines (structural mechanics, finite element analysis). In particular he is interested in how shape changes at different length scales can augment the properties, behaviors and functions of a material, it being biological or artificial. Currently he is nurturing interdisciplinary collaborations with surface and textile designers: together they explore the design space of self-shaping 3D printed textiles and laser cut shape-programmable auxetic surfaces, with the aim of discovering new design rules and structure function relationships in these material platforms.

Susanne Jany is since 2019 the coordinator at the Federal Chamber of German Architects. She was till 2019 postdoctoral researcher at Cluster of Excellence 'Matters of Activity. Image Space Material' of the Humboldt-Universität zu Berlin. In 2013, she was a visiting scholar at Harvard University, and from 2015 to 2018 she was a research associate at the Cluster of Excellence 'Bild Wissen Gestaltung. Ein interdisziplinäres Labor' of the Humboldt-Universität zu Berlin. Her research focuses on media and cultural studies of architecture, material epistemology, and theories of design. In 2019 her monograph *Prozessarchitekturen: Medien der Betriebsorganisation* (1880-1936) was published at the Konstanz University Press.

Khashayar Razghandi is a material scientist and engineer working at the intersection of the natural sciences, the humanities and design disciplines, between Max Planck institute of Colloids and Interfaces, 'Biomaterial' department, and the Cluster of Excellence 'Matters of Activity. Image Space Material' of the Humboldt-Universität zu Berlin. Coming from a diverse educational background, and backed by a broad interdisciplinary work experience, his research interests cover a range of interdisciplinary research and education projects under diverse yet converging themes of: materials science-design (interdisciplinary processes, methodologies and education), materials philosophy (analog intelligence, information philosophy, and new materialism) and materials sustainability (transient materials, structures, functions and contexts).

Ianis Lallemand

Matter of Agency: Active Materials in Digital Design Research

Digital design projects, once criticized for their lack of materiality, have within the last decade undergone a shift from issues of representation toward questions of material performance and manufacturing. Nurturing new relationships with disciplines such as physics, biology, engineering, and materials science, digital design fully participates in one of the defining trends of the current period: a radical redefining of our relationship with materiality, which has its origins in the growing proximity of computational and material issues. Recently, developments in digital design research have introduced a new set of conceptual and practical challenges into the field, which I propose to situate here in the interdisciplinary context of active matter. As a growing number of designers experiment with the production of emergent structures through bespoke fabrication processes, often involving 'non-calibrated' materials (such as clay and polymers), the dominant understanding of digital manufacturing – based on the idea of an identity of the digital model and its physical realization – is increasingly being called into question. In stark contrast with digital design's industrial beginnings, recent experiments hint toward a participatory understanding of digital production, in which the role of the designer shifts from top–down production of geometric models toward the preparation of a form of openness to active material behavior. This chapter aims to qualify this approach from both theoretical and empirical perspectives, arguing for its recognition as one of digital design's main contributions to the field of active matter, before the use of 'smart' and 'responsive' materials in design projects.

1 Introduction

Digital technologies, once envisioned as the agents of a radical, global dematerialization of our physical environment – an idea perhaps best captured in Nicholas Negroponte's 1995 essay, *Being Digital* [Negroponte 1995] – are today playing an ever-increasing role in the creation of new material structures and behaviors. Far from removing us from material concerns, computation – through its various manifestations, from simulation to

Acknowledgments: The author wishes to acknowledge Andrea Graziano, Marco Palma, Bruno Demasi, Alessio Erioli, and Olivier Dauchot for their collaboration in the projects discussed in this chapter, as well as the following institutions for their financial and material support: EnsadLab; PSL University; the Arts & Sciences chair of the École polytechnique, EnsAD – PSL, and the Daniel & Nina Carasso foundation; digifabTURINg. The author would also like to thank the editors for their fruitful comments and remarks on this chapter.

Open Access. © 2021 Ianis Lallemand, published by De Gruyter. BY-NC-ND This work is licensed under the Creative Commons Attribution-NonCommercial-NoDerivatives 4.0 International License.
https://doi.org/10.1515/9783110562064-012

manufacturing and assembly – has created new modes of interaction with the physical world, by putting complex material behavior within the reach of science, engineering, and design.

As a consequence of this global and interdisciplinary transformation of our relationship with materiality, the field of digital design[1] is undergoing profound changes. From the mid-1990s, when critics voiced concerns about the lack of materiality in digital projects,[2] to the developments of the last decade, digital design's trajectory reveals a clear inflexion from issues of representation toward questions of material performance and manufacturing. These transformations have resulted in the growing proximity of digital design to disciplines such as physics, biology, materials science, and engineering. The creation of new materials, from the tailoring of properties at the microscopic level to the packing of functional building blocks in so-called metamaterials [Kadic et al. 2013], whose properties go beyond what can be found in nature, has thus become a major topic of interest for some designers, as exemplified by the recent successes of initiatives such as MIT's Self-Assembly Lab and Tangible Media laboratories. But perhaps more strikingly, a global rethinking of the relationship to 'traditional' materials – such as clay, steel, timber, and concrete – has developed out of the bridging of computational design methods and digital fabrication processes. The recent works of architects Ronald Rael and Virginia San Fratello,[3] designer Olivier van Herpt,[4] artist Jonathan Keep,[5] and designer collective Co-de-iT,[6] among others, have for instance developed a new aesthetic language based on the emergent behavior of clay, in a process merging additive manufacturing, scripting, and the material's inner plasticity and tendency to buckling (Fig. 1).

The evolution of digital design research's experimental output, made more visible during the past 6 years, reveals an increased interest in the production of emergent

1 This chapter assumes an interdisciplinary position within the field of digital design research, acknowledging that today's context is more adequately defined by a fluid circulation of ideas, concepts, and workflows rather than by strict boundaries between disciplines. I shall therefore understand here the term 'design' in the broader sense of a process of 'conception,' rather than as referring to a specific practice (like architecture of industrial design) implementing such a process.

2 One can mention here, for instance, the critiques expressed by architectural theorist Kenneth Frampton in his 1995 *Studies in Tectonic Culture* concerning the recent tendency to the "dematerialization of building" [Frampton 1995, p. 381] observed in the architectural profession. Although Frampton describes a broader cultural phenomenon and does not explicitly target the influence of computers, his attempt to reassert the centrality of architecture's relationship with structure and construction – in an age where digital technologies began to infuse all aspects of life – has been discussed and criticized by early advocates of a 'virtual' architecture. On this matter, see for instance: [Mitchell 1998]. For a nuanced discussion of Kenneth [Frampton's viewpoint on computers within architecture, see: [Frampton, Allen, and Foster 2003, pp. 55–56].

3 https://www.rael-sanfratello.com/ (accessed November 4, 2019).

4 http://oliviervanherpt.com/ (accessed November 4, 2019).

5 http://www.keep-art.co.uk/ (accessed November 4, 2019).

6 http://www.co-de-it.com/ (accessed November 4, 2019).

Fig. 1: A clay vessel robotically printed by Co-de-iT as part of the inFORMed matter research project. © Bruno Demasi (reproduced with authorization).

material structures, whose qualities challenge the very assumption of an identity of digital form and its material realization. The parametrically modeled, smooth geometries that came to define the mid-1990s and early 2000s in digital architecture and product design, as exemplified by Greg Lynn's and Ross Lovegrove's works,[7] have given way to a vocabulary of discreteness and discontinuity, fracture and roughness; the "pliant [and] supple" [Lynn 1993] geometries of the first digital age are today losing ground to highly textured formations, bearing the volatile marks of unpredictable material behavior.

Architecture schools, once at the forefront of the digitization and 'virtualization' of design practices (the earliest and most famous example being the Paperless Studio experiment, led at Columbia University in the mid-1990s under the supervision of Bernard Tschumi) are among the most prominent experimenters of this new relationship between form and matter in digital design: one can mention, for instance, the Bartlett School of Architecture's Design Computation Lab,[8] as well as the Architectural Association's Design Research Lab,[9] both based in London. Within these environments, bespoke digital fabrication processes, involving the emergent behavior of materials such as wax,[10] ABS,[11] and polyurethane foam,[12] are being experimented at the scale

7 See, for instance, Greg Lynn's tea and coffee service for Alessi (http://glform.com/shop/alessi-tea-coffee-towers/, accessed August 16, 2018) or Ross Lovegrove's water bottle for Ty Nant (http://www.rosslovegrove.com/index.php/custom_type/ty-nant/, accessed August 16, 2018).

8 https://designcomputationlab.org/ (accessed November 4, 2019).

9 http://drl.aaschool.ac.uk/ (accessed November 4, 2019).

10 One can mention here the robotic wax deposition experiments led at the Southern California Institute of Architecture (SCI-Arc) by Brian Harms and Nicholas Barger: http://nstrmnt.com/#/buoyant-depositions (accessed November 2, 2019).

11 On the development of innovative 3D printing processes involving ABS plastics, see for instance the research developed at the Bartlett School of Architecture's Design Computation Lab: Retsin, Garcia 2016.

12 On the development of custom manufacturing processes involving polyurethane foam, see for instance: [Colletti 2013] .

of architectural components or furniture. Firmly rooted in design research, these experiments coexist with more applied programs, which approach the development of new manufacturing processes through the creation of architecture-scale structures – the most notable examples being the series of pavilions[13] jointly built in Stuttgart by the Institute for Computational Design and Construction[14] (ICD) and the Institute of Building Structures and Structural Design (ITKE).[15]

Beyond differences in aesthetics or purpose, these experiments share a common commitment to reinvesting the morphogenetic potential of active material behavior within contemporary digital design, a project sometimes linked to such historical precedents as Frei Otto's use of analog "form-finding" models [Menges 2015, p. 10]. Indeed, it is most often by abandoning the prospect of total digital control, and embracing the opportunities offered by "non-calibrated" materials [Morel 2014, p. 84], that today's designers are discovering new territories of aesthetics and structural performance. As a consequence, digital design workflows are increasingly moving toward a 'participatory' vision of production, in which the designer's authority and aesthetic intent is brought to a form of negotiation with active material processes.

This chapter intends to situate these recent developments of digital design research in the interdisciplinary context of active matter[16] and highlight some potential specificities of design's contribution to this field. The relevance of the notion of active matter in design, as I shall argue here, does not materialize first in applications of 'smart' or responsive materials but rather in the growing adoption of design methods allowing for an openness to matter's "inherent capacity for the generation of form" [DeLanda 2002, p. 134] to be incorporated in the project itself.

The remainder of this chapter is divided into three sections. The next section analyzes the recent transformations of digital design workflows as a shift from a representational to a performative understanding of design, fostering the field's growing interest in active materials. Sections 3 and 4 then introduce two recently completed design experiments: Unspecified Clay, a retroactive production system focusing on emergent clay

13 See for instance [Doerstelmann et al. 2015].

14 https://icd.uni-stuttgart.de/ (accessed November 4, 2019).

15 https://www.itke.uni-stuttgart.de/ (accessed November 4, 2019).

16 The term 'active matter' should be understood here primarily in reference to the interdisciplinary research field developing within natural sciences for approximately the last 20 years which focuses on the emergence of collective motion in living or inert systems composed of multiple interacting entities – such as a flock of birds, or a colloidal solution. For a general introduction to this field, see for instance: [Popkin 2016]. The Manoeuvres project, discussed in this chapter, directly relates to this field through its development in collaboration with active matter physicist Olivier Dauchot. This chapter will also discuss material systems (such as clay extrusions) that, while not falling strictly into the active matter field defined in the natural sciences, are investigated from a *design* point of view as self-organizing systems. These systems may conversely be related to a wider notion of *material activity* that translates within the design field as a questioning of materials' traditionally supposed passivity in fabrication processes.

behavior; and Manoeuvres, an interactive installation producing complex light patterns through the collective motion of physical units.[17]

2 From Representation to Material Performance

Since the advent of the modern architectural discipline in the Renaissance, most notably after Filippo Brunelleschi's innovations,[18] the notion of design has been tied to the assumption of a fundamental divide, both temporal and social, between the world of the project and the world of matter [Boutinet 2002, p. 224]. At the same time, as it separates the designer from craft professions, and the architect from construction workers, design is also envisioned as the prime method through which the intellectual and physical worlds may be reunited [Picon 2004, p. 120]. This dialectic has traditionally defined design as an activity which, although separate and anterior to the concrete process of fabrication, claims a complete authority on the qualities of the latter's output.

Leon Battista Alberti's *De re aedificatoria* (*On the Art of Building*) [Alberti 1988], published in 1485, is the first architectural treatise to build upon this modern conception of design. Alberti's insistence on the role of design is expressed in his distinction between the concept of *lineaments* and the notion of *structure*, or construction: as noted by anthropologist Tim Ingold, "[w]hat Alberti [...] calls 'lineaments' (lineamenta) comprise a precise and complete specification of the form and appearance of the building, as conceived by the intellect, independently and in advance of the work of construction (structura)" [Ingold 2013, p. 50]. This distinction, central to Alberti's architectural theory, is laid out at the very beginning of the first book of *On the Art of Building*:

---

17 Both projects have been developed in the context of a PhD thesis undertaken at EnsadLab, the laboratory of the École nationale supérieure des Arts Décoratifs (EnsAD), from October 2013 to December 2017, under the supervision of Antoine Picon and Samuel Bianchini. This research was situated within the context of the Reflective Interaction group, led by Samuel Bianchini, as well the SACRe doctoral program of PSL University.

18 Filippo Brunelleschi, architect of the dome of the Santa Maria del Fiore cathedral in Florence, is known for his early experiments in linear perspective. The most famous of these experiments aimed at demonstrating the correctness of a perspective painting of Florence's Baptistery of San Giovanni, by means of a small hole drilled into the vanishing point of the painted panel and a mirror. For a global overview of Brunelleschi's life and inventions, see for instance [King 2000]. On Brunelleschi's relationship with perspective, see [Damisch 1995]. In respect to Brunelleschi's role in the history of the architectural discipline, scholar Jean-Pierre Boutinet considers him as the inventor of the architectural *project* in its modern sense. According to Boutinet, Brunelleschi's expertise in perspective drawing allowed him to propose for the first time "a methodology of the *disegno*, or in other words a methodology of the anticipation of the work to be carried out" [Boutinet 1993, p. 10 (translation I.L.)]. For more on the epistemological relationship between the architectural project and the broader concept of design, see: [Vial 2014].

> [t]he whole matter of building is composed of lineaments and structure. All the intent and purpose of lineaments lies in finding the correct, infallible way of joining and fitting together those lines and angles which define and enclose the surfaces of the building. It is the function and duty of lineaments, then, to prescribe an appropriate place, exact numbers, a proper scale, and a graceful order for whole buildings and for each of their constituent parts, so that the whole form and appearance of the building may depend on the lineaments alone.
>
> [Alberti 1988, p. 7]

Alberti's description of the lineaments' purpose clearly corresponds to what is referred to, in modern terms, as the design phase of a building: that is, a stage independent from and chronologically anterior to the building's construction that, however, intends to prescribe every detail of the final object. Noting that lineaments do not "have anything to do with material" [ibid.], Alberti continues: "[i]t is quite possible to project whole forms in the mind without any recourse to the material, by designating and determining a fixed orientation and conjunction for the various lines and angles. Since that is the case, let lineaments be the precise and correct outline, conceived in the mind, made up of lines and angles, and perfected in the learned intellect and imagination" [ibid.] (see Fig. 2).

This definition of the lineaments as a geometric specification ("made up of lines and angles" [ibid.]) of the building's final form, carried out in the intellectual space of the project and not within the material space of construction, has led architectural historian and theorist Mario Carpo to qualify Alberti's writings as a "notational method of design" [Carpo 2011, p. 77]. For Carpo, the concept of lineaments conveys an approach to architecture "predicated on the notational sameness between design and building, implying that drawings can, and must, be identically translated into three-dimensional objects" [ibid., p. 26]. The concept of notation, used by Carpo in reference to Nelson Goodman's book *Languages of Art* [Goodman 1968], expresses the new prescriptive and authorial power that Alberti's treatise ascribes to architectural drawings: with Alberti, drawings become *plans* in the modern sense, or, in Tim Ingold's terms, "full geometrical pre-specification[s] of the intended work" [Ingold 2013, p. 55].

This understanding of drawing as the privileged medium for communicating the forms conceived in the architect's mind relates directly to the Renaissance concept of the *disegno*, defined most notably by Giorgio Vasari,[19] which refers both to the idea of design and its visual expression in the form of a drawing [Jack 1976, p. 4]: indeed, for Ingold, drawing in the Albertian theory of architecture can be understood as the "visual projection of an idea already fashioned in the intellect" [Ingold 2013, p. 55]. This understanding of the architect's role in relation to the production of the *disegno* sets architecture apart from craft professions, and their handling of the physical realities of the building site. Alberti's treatise thus stands at a "pivotal juncture" [ibid., p. 49]

19 The concept of *disegno* is central to Vasari's *Lives of the Artists* [Vasari 1998], first published in 1549–1550.

in the history of the architectural discipline, and "looks forward to a time when the architect would prescribe only the formal outlines of the building, leaving its actual construction in the skilled and capable hands of workmen" [ibid.].

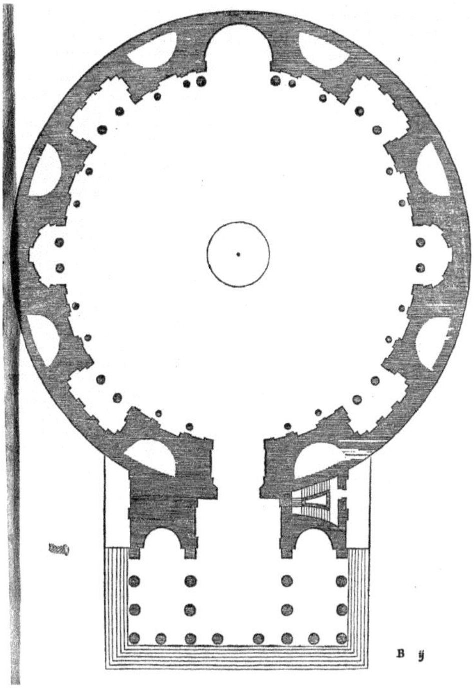

Fig. 2: Sebastiano Serlio [Serlio 1540, p. vii], plan view of the Pantheon. © Bibliothèque nationale de France (reproduced with authorization).

It should be noted, of course, that the professional distancing of architecture with building practice did not imply that construction issues would cease to matter to architects. Alberti's treatise itself includes extensive considerations about the material details of building,[20] and historical examples abound of decisive architectural gestures owing their existence not to a planned decision but to contingencies happening during the physical construction process. One can mention here, for instance, the famous *béton brut* (raw concrete) finishes of Le Corbusier's Unité d'habitation de Marseille: confronted with material limitations, the architect made the choice of not attempting to uniformize the building's surfaces, which bore the marks and

20 Following the introduction of the concept of lineaments, the second book of *On the Art of Building* is thus devoted to the question of materials.

imprecisions of the wood formworks employed during construction. Le Corbusier famously justified this move in aesthetic terms, explaining how one could truly envision concrete as a new type of raw material, worthy to be shown for its own qualities [Boesinger 1991, p. 190].

This impossibility of entirely reducing an architectural work to its plans and perspective renderings has in fact been understood as one of the idiosyncratic qualities of architecture's relationship with its "notational system" [Goodman 1968, p. 221]. Indeed, as noted sharply by architect and historian Robin Evans, the idea that architectural drawing could provide a "complete determination in advance" [Evans 1997, p. 156] of a building, while certainly untrue, has traditionally operated as a sort of defining and enabling fiction within architectural practice. According to Evans, the very possibility of architectural design may thus very well rest on a form of "suspension of critical disbelief" [ibid., p. 154] with respect to drawing's capacity to anticipate every aspect of the final object.

This generative relationship with the ambiguities and limits of design representation, however, appears at odds with the extensive precision and descriptive powers offered by today's digital design environments. Computer drawings, models, and renderings have offered an unprecedented capability to document and author every geometric, material and organizational aspect of a project, from design to fabrication. The growing popularity of design documentation environments, or Building Information Modelling software (BIM), which aims at organizing and quantifying every aspect of the production and life cycle of a building, emblematizes the industry's attempt to rationalize all aspects of material production thanks to digital representation technologies.[21] This agenda all but results from the expanding capacities of digital design and fabrication tools, whose early developments can be situated in the late 1940s.

Modern digital workflows, in the post-war period, developed out of a need to provide greater control over industrial processes, a goal which was approached by enforcing the use of reference geometric data (2D or 3D curves and surfaces) at all levels of the design to manufacturing pipeline. One can think, for instance, of Pierre Bézier's account of the development of UNISURF, the first interactive computer-aided design and manufacturing (CAD/CAM) system, which he led at Renault factories in the mid-1960s. According to Bézier, UNISURF intended to ground the whole production process of a car body – from styling to manufacturing – on a single digital model, with the objective of ruling out the imprecise analog model making and

21 A BIM model is organized around a detailed 3D representation of a building, complete with all its technical systems and subcomponents. This 3D geometric model is supplemented with a range of data, such as manufacturers' specifications, material costs, environmental performance metrics, and parametric relationships between components. This integrated and information-rich 3D model constitutes a single authoritative reference on the basis of which standard documents (such as 2D construction plans) can be generated as needed. For more details on BIM, see for instance: [Kensek 2014].

measurement methods then commonplace in the industry [Bézier 1998]. Even though UNISURF's CAM capacities were mostly intended for prototyping and 3D visualization purposes, rather than the direct production of car body parts, the tool carried within it promise of enforcing a complete conformity between the fabricated cars and their original design.

Ultimately, progresses in the integration of CAD/CAM systems and industrial automation have led to a full realization of this promise in the concept of 'file to factory' processes, a notion that refers to the possibility of a seamless transition from design intent to automated manufacturing. Within this paradigm, any digital design asset is thought to contain, from the moment of its creation, all the information required to set in motion a full product production chain. This idea, although envisioned as a possibility as far back as the 1970s,[22] still constitutes the guiding principles of today's perspectives for the future of manufacturing: the emerging production paradigm known as 'Industry 4.0,' whose conceptual basis lies in the realization of networks of cloud-connected, sensor-enabled manufacturing units – so-called 'cyber-physical' systems [Rajkumar et al. 2010] – aims at creating agile production pipelines able to respond, in near real-time, to any change in design specifications.[23] The most recent advances in industrial manufacturing thus aspire, in a sense, to a quasi-ideal realization of the notational project underpinning modern design, by providing a framework for controlling all material aspects of production within the 'intellectual-digital' space of the project.

This agenda seems so deeply ingrained in CAD/CAM workflows, in fact, that the idea that digital prototyping alone could represent a revolution in design practices should be regarded with caution. For all the transformations brought about by the diffusion of a "making," or "digital DIY" [Anderson 2012, p. 21] culture in design, the greater access to fabrication capacities more often than not befits traditional design thinking, in which geometric form – encoded in a digital file – exerts full control over passive matter. One needs simply to think, for instance, about the very concept of online objects libraries, such as the platform Thingiverse, which provide a collection of user-contributed 'things,' ready to download and 3D print. Here, design is fully identified with the production of geometric form, and material qualities are relegated to a secondary role: the material used in the 3D printing process, as it seems, only matters so much as it proves compliant enough to reproduce the downloaded objects' geometric features.

---

22 See, for instance, the reflections presented by Alvin Toffler on the potentials of industrial automation in his 1970 bestseller *Future Shock*: [Toffler 1970, p. 236].

23 Several key European initiatives, such as the German government 'Industrie 4.0' program and the CREMA (cloud-based rapid elastic manufacturing) project, funded by Horizon 2020 Framework Programme, identify 'Industry 4.0' as a strategic priority for realizing the vision of an "European Industrial Renaissance" called for by the European Commission in 2014 (http://eur-lex.europa.eu/legal-content/EN/TXT/?uri=CELEX:52014DC0014, accessed August 17, 2018).

My goal, of course, is not to take position against the existence of such platforms – which may be home to genuinely inventive projects – but rather to highlight the conservative vision of design they embed. Similarly, in the practice of many accomplished digital designers, materialization is still envisioned as a purely predictable and passive process, in contrast to the 'active' design phase, where the defining features of the object are created. Fabrication, it is often thought, is not able to alter or create form – not because of a technical limitation but rather as a consequence of a conceptual, almost moral impossibility: all material processes resulting in a deviation from the designer's original form are considered erroneous and are thus rejected.

The new design approach gathering strength in schools, laboratories, and practices, of which we have already seen a few examples in this chapter's introduction, is fundamentally more open to material processes yielding unpredicted formal output. Indeed, in projects such as Olivier van Herpt's works with clay, what is at stake is less the intentional production of 'errors' than the recognition of a fundamental form of material activity as a key morphogenetic driver within the design project. Beyond a mere visual interest in exploring 'discontinuous' or 'fractured' form, this emerging trend of design research thus hints at the fact that contemporary designers may be slowly coming to terms with a notational understanding of design that, with some notable exceptions,[24] has prevailed since the Renaissance. As we shall see below, for designers operating within this framework, computation is increasingly used to design fabrication *processes* – which may trigger more or less predictable material responses – rather than the direct visual features of an object.

One key ingredient in this evolution is the fact that today's design workflows are increasingly attentive to the properties of materials and to the degrees of freedom offered by fabrication processes. The approach known as 'material computation,' investigated for instance by MIT researcher Neri Oxman [Oxman 2010], or in much of the research carried out by Achim Menges and his partners at Stuttgart's ICD, has gathered strength since 2010 around the idea that "[m]aterial properties, characteristics and behaviours can now be employed as active design generators" [Menges 2012, p. 16]. This approach has resulted in object or building-scale prototypes exploiting active material behavior, such as the hygroscopic bending of wood veneer panels in Achim Menges' and Steffen Reichert's HygroScope project [Menges

24 One can think here of the form finding experiments led by the German architect Frei Otto, already mentioned above. These experiments were based on dynamic physical models (such as soap films and wool threads) subjected to environmental forces. The 'equilibrium' configurations reached by these models would provide formal solutions to design problems. For more details about Frei Otto's architectural work, see: [Glaeser 1972]. On the generative use of physical models within architectural design processes, one should also mention the use of hanging chain models by Antoni Gaudí to determine the catenary curves used in vaulted ceilings. On this topic, see: [Huerta 2006].

and Reichert 2015], part of the Centre Pompidou permanent collection. Yet, with the exception of a few of such examples, material computation workflows more often materialize as physics-based simulations, which are incorporated into design programs to capture some aspects of 'real-world' dynamics. In this context, the design process' prime objective remains to provide a complete geometric description (in the form of a 3D model) of the objects to be manufactured. While the use of physics-based simulations can certainly produce geometries more akin to comply with the materials' physical properties, the relationship between form and material still follows here a top–down dynamic. Somewhat counter-intuitively, given the name 'material computation,' form is not 'computed' by the actual *performances* of the material itself but by a geometric computer model *representing* the properties of this material.[25]

A more relevant notion to the transformation currently observed in design research may be the idea of "design hacking" [Witt 2011], which envisions the development of bespoke fabrication processes as a key part of the design project. Benefiting from the development of tools such as the electronics prototyping platform Arduino,[26] a growing number of designers are today involved in the building and scripting of their own machining logics, often eschewing traditional assumptions in respect to how a CAD/CAM pipeline should operate. As an example, several researchers have proposed to generate machine toolpaths directly from the design phase, whereas most off-the-shelf CAM software computes toolpaths after the design process per se, once a full geometric model of the target object can be provided. In Manuel Jiménez Garcia's and Gilles Retsin's work with the Bartlett's Design Computation Lab, continuity-based modeling is, for instance, replaced by a voxel-based approach [Garcia and Retsin 2015], in which machine toolpaths are directly generated according to local rules. The full fabrication toolpath, allowing for the manufacturing of the object by robotic polymer extrusion, is then built up from the aggregation of these unit trajectories: geometric data emerges here from fabrication logics, and not the reverse.

By vowing to engage with the design of fabrication processes, rather than the design of full geometric descriptions of the objects to be fabricated, such methods appear ideally suited to the creative and playful exploration of material behavior, beyond the limitations of digital description and simulation. Through 'design hacking' and related approaches, digital design thus appears on the verge of operating a 'performative turn,' in which the formative powers of active material behavior are

25 Notable exceptions to this statement include, as discussed above, projects such as Menges' and Reichert's HygroScope, which are able to dynamically 'compute' their shape through the mobilization of active material properties (such as hygroscopic wood bending). In this case, the shape configuration of the bending wood panels at any given time is not dictated by the 3D model defined in the project's design stage but results from an interaction between wood's physical properties and its environment.

26 https://www.arduino.cc/ (accessed November 4, 2019).

given full recognition at all levels of the design project. It is in this challenging of design's comprehension as a notational, geometry-led practice that one may find digital design's most important contribution to the field of active matter – a field in which the very notion of design, in its most general sense, is increasingly understood as a fundamental and interdisciplinary concept [Schäffner 2010].

The prime issue, for today's digital designers, thus lies in the preparation of a form of openness to material behavior in the design process itself, in a context where conventional tools and CAD/CAM workflows enforce a linear, geometry-oriented approach to object production. The project Unspecified Clay, discussed next, aims at addressing some aspects of this question in a framework of clay-based additive manufacturing.

3 Retroactive Production: Unspecified Clay

3.1 Context

Unspecified Clay is a research project initiated in 2016 in collaboration with the Italian group Co-de-iT, a research and education-focused network investigating the impact of computational techniques within architecture, engineering, design, and craft. Developed primarily with Andrea Graziano, Marco Palma, Bruno Demasi, and Alessio Erioli, the project was carried out in the joint context of Co-de-iT's inFORMed Matter research project, and of the Responsive Matter program, led by the Reflective Interaction group of EnsadLab in Paris. Unspecified Clay was developed as an experimental production system within the framework of these two research initiatives, both undertaking investigations on the topic of active matter from an arts and design-oriented perspective.

As noted in the introduction to this chapter, Co-de-iT had previously led experiments involving the production of emergent material structures in a clay-based additive manufacturing context, through the mobilization of buckling and deformation effects. While this research made an extensive investigation into emergent material behavior, the latitude of the material effects it focused on was in fact entirely set up in the design phase, as no feedback from the material to the digital domain was implemented. When I approached Co-de-iT in 2016 to propose the initial concept of the Unspecified Clay project, we thus agreed on aiming to challenge the linearity of the design to manufacturing pipeline that still characterized these previous experiments, which we identified as the main bottleneck toward implementing a greater openness to material behavior in design workflows.

Accordingly, Unspecified Clay was posited as a retroactive production setup, in which feedback processes would allow a robotic manufacturing unit to build artefacts through repeated cycles of deposition, scanning, and computation. This setup was primarily designed with the intent of exploring the generative potentialities of a reciprocal process of interaction between clay's self-organizing properties and computational logics. Instead of aiming at the materialization of pre-defined digital models, our motive was rather to develop a production system able to discover and foster aesthetic and structural opportunities, emerging from an 'autonomous expression' of material behavior. By this term, we want to convey the idea that the material's main physical properties (such as its density and plasticity) would become generators of form, rather than constraints in a manufacturing process whose output would be determined in advance. In other words, we aimed to let the material organize itself according to its own laws, rather than constraining it to achieve an a priori defined geometric objective (as would be the case, for instance, in a clay molding process). The prime content of the research thus focused on designing a computational system able to negotiate and collaborate with this autonomous expression of clay behavior, with the objective of developing emergent material structures. Our approach to this question centered on implementing sensing and machine vision processes that would allow our digital system to be informed at each fabrication step of the material's configuration. In the remainder of this section, we describe our approach to implementing such a retroactive production process and discuss our empirical investigations and results.

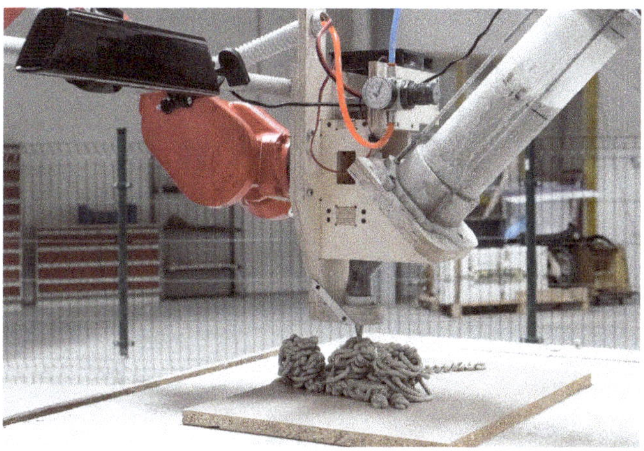

Fig. 3: Unspecified Clay's experimental setup. © Ianis Lallemand.

3.2 Experimental Setup

This initial concept led us to develop the setup shown in Fig. 3. A custom clay extruder, developed by Co-de-iT, is mounted on a six-axis industrial robotic arm (COMAU NS12), which also has a Microsoft Kinect sensor, an infrared-based technology able to compute depth maps of close-range scenes [Zhang 2012]. The robot control program, as well as the system's other computational procedures, including data processing and sensor calibration, are implemented within a single application (Fig. 5), developed with the programming environment processing. Following the idea of a closed feedback loop

Fig. 4: Two successive steps in Unspecified Clay's fabrication process. © Ianis Lallemand.

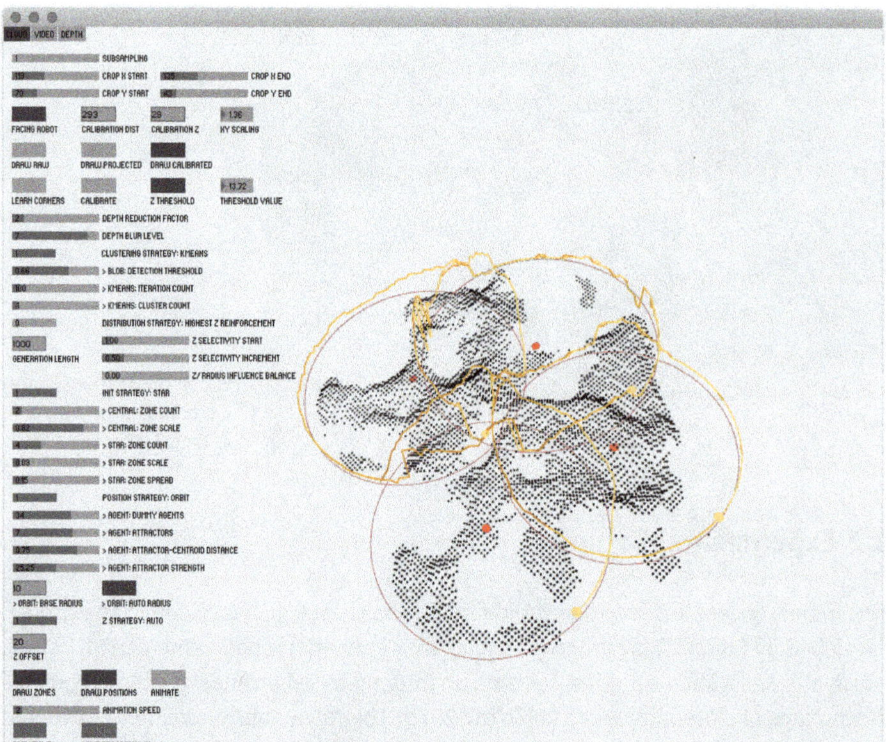

Fig. 5: Processing application developed for the Unspecified Clay project. © Ianis Lallemand.

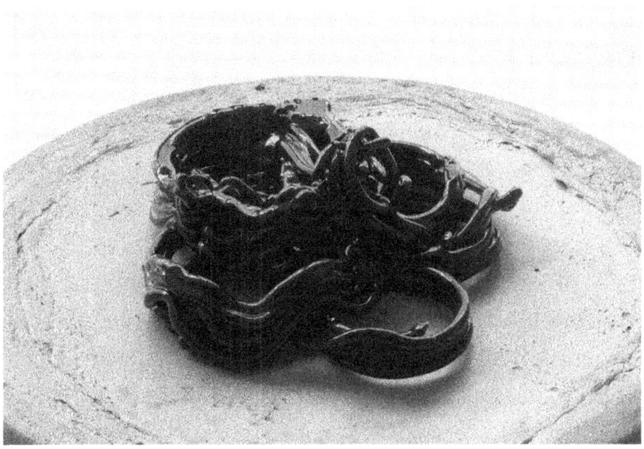

Fig. 6: A finished, glazed ceramics piece. © Ianis Lallemand.

between fabrication outcomes and generated robot toolpaths, each fabrication session is organized as follows:

1. Perform a first series of clay depositions at heuristically selected locations of the working area;
2. Capture the obtained clay structures with the Kinect sensor and compute a 3D point cloud representation of the acquired depth map;
3. Generate robot trajectories based on point cloud data and perform a new series of clay depositions;
4. Repeat steps 2 and 3.

Figure 4 shows two successive steps of the fabrication process, while Fig. 6 shows an example of a finished ceramics piece produced in our experiments, which has been fired in a kiln and coated with a black glaze. With the exception of the first iteration of the fabrication process (step 1 in the list above), Unspecified Clay's output can be described as essentially emergent: no global 3D model specifying the shape of the final object is ever handled by the system, and all form-generating processes – whether computational or material – are realized in 'real-time' within the fabrication phase. This latter characteristic inscribes Unspecified Clay within a background of related design research, which we will briefly discuss below.

3.3 Related Work

Within the set of design research work eschewing the production of traditional 3D models in favor of a direct generation of machine control codes, a small number of approaches have explored the possibility of performing an online adaptation of robotic toolpaths in response to material behavior. As with our approach, most of these projects have resorted to sensing capabilities and feedback mechanisms, either to channel emergent material behavior into the realization of a given shape target, or to develop it for generative or aesthetic purposes.

The project Remote Material Deposition [Dörfler et al. 2014], developed in the context of the research laboratory of architects Gramazio & Kohler, at the ETH in Zurich falls within the first of these two categories. The project aims at tackling the scale issue intrinsic to robotic fabrication processes, which are usually limited by the work envelope of robots and therefore ill-suited to full-scale architectural applications. In this context, Gramazio & Kohler's team developed a robotic fabrication method based on the throwing of cylindrical clay loam units, with the purpose of building proto-architectural structures (walls) over a distance. To handle the uncertainty introduced by the units' deformation on landing, the method makes use of machine vision to adjust the throwing trajectories to previously formed aggregations, thus allowing for a channeling of material behavior toward the realization of a selected floor plan.

Other research[27] investigates the use of feedback mechanisms from a more generative viewpoint, in the respective contexts of plastics, wax, and polyurethane foam-based manufacturing processes. In these works, feedback does not aim at comparing the fabrication's current state to a pre-defined reference but rather at enabling the system to 'go on' from whatever emergent state it has reached previously.

While Unspecified Clay appears closer to these latter examples in terms of intent and conceptual background, its adoption of clay as a working material forces it to face a different set of morphogenetic issues. In contrast to rapid-hardening materials such as polyurethane foam or wax, clay will typically maintain a plastic behavior throughout the whole duration of manufacturing. This exposes our setup to the effect of indeterminate material deformations, both during and after printing (as the addition of new clay creates extra compressive stress on existing structures). As we mentioned earlier, Gramazio & Kohler's Remote Material Deposition research also featured similar deformation events; yet, the staging of these in a generative, open-ended workflow constitutes, to our knowledge, a novel research issue, which Unspecified Clay is thus first to explore.

3.4 Hypothesis and Methods

Unspecified Clay makes the hypothesis that the maintained plasticity of deposited clay can be computationally channeled into the production of materially differentiated artefacts, exhibiting variable properties of density, texture, and articulation. This hypothesis was experimentally tested through the implementation of two main sets of procedures.

27 See, for instance, Matias del Campo et al.'s Autonomous Tectonics [Del Campo et al. 2013], Ryan Luke Johns' Augmented Materiality [Johns 2014] and Roland Snook and Gwyllim Jahn's Stigmergic Accretion [Snooks and Jahn 2016].

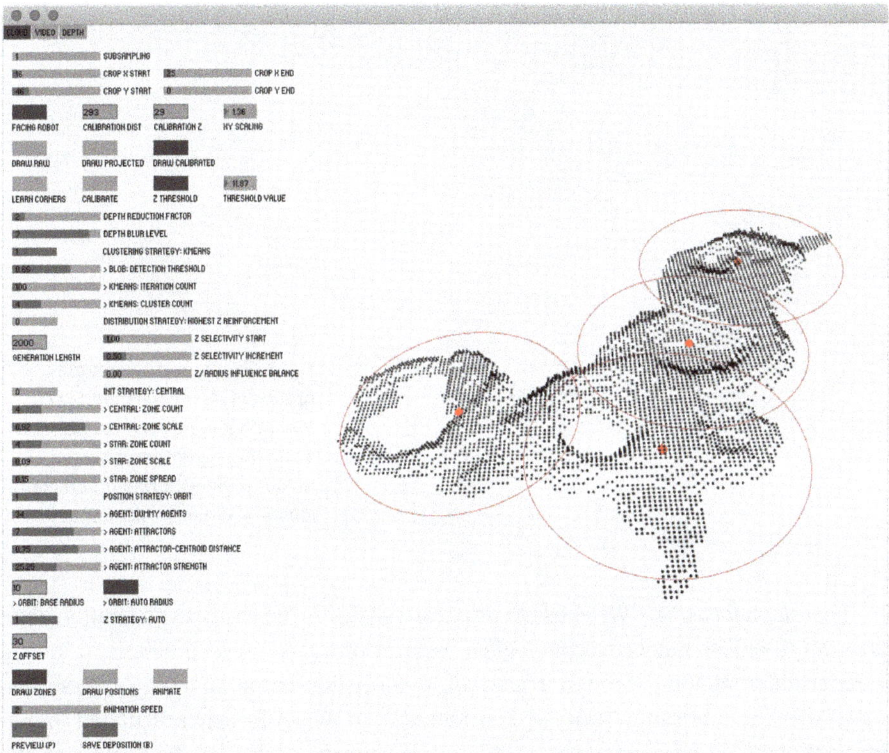

Fig. 7: Example of candidate deposition areas detected by the Unspecified Clay software. © Ianis Lallemand.

Sensing – To generate new deposition trajectories from the scanning of previously deposited structures, the system first computes a set of candidate deposition areas from the 3D point cloud morphology by using a K-means algorithm (Fig. 7). In each detected cluster, an agent-based simulation is instantiated and configured given the cluster's properties (area and centroid height). The generated agent positions are then automatically translated into robot control code, using so-called linear motion instructions (i.e., while keeping the extruder's vertical orientation constant). The implemented algorithm allows for the generation of a wide range of trajectories, from dense and intricate paths derived from Craig W. Reynolds' swarming simulation model [Reynolds 1987] (Fig. 8), to orbit-like revolutions around cluster centers.[28]

---

28 The swarming simulation model is here taken as a heuristic approach to the generation of deposition toolpaths. More specifically, the characteristics of the model enable the generation of spatial curves with a controlled degree of redundancy and self-overlay – managed through the tuning of the simulation settings – while adopting ever-changing morphologies. These two characteristics are suitable in the context of the discussed experiment, as they allow for a building up of clay in

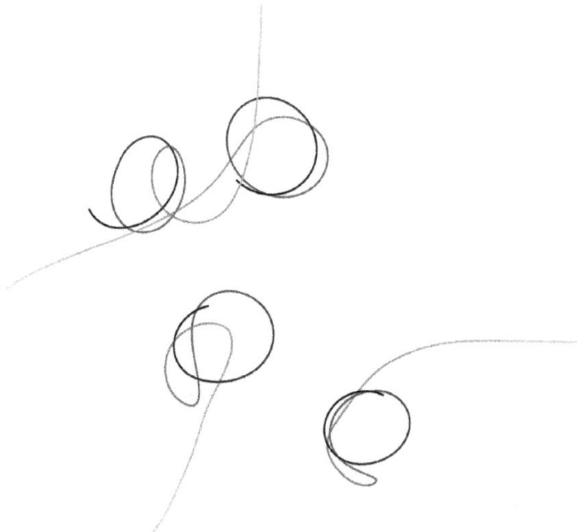

Fig. 8: Example of trajectories produced by Unspecified Clay's toolpath generation procedure. A swarming algorithm is here instantiated within four detected clusters. © Ianis Lallemand.

Deposition focusing – We chose to progressively focus the toolpath generation algorithm on the most salient morphological features of the print, as a means to induce material differentiation through a gradual, deformation-based shaping of these features. After the fabrication process' first iteration, in which loosely articulated formations are obtained from simple geometric forms (such as circles arranged in a star-like configuration), the system thus proceeds to identify, select, and reinforce the highest material clusters. The algorithm implementing this behavior acts on two physical parameters: the allocation of extruded material between detected clusters, and the vertical distance between the extruder's nozzle and the extrusion's target surface (the main parameter controlling material organization). A reward mechanism progressively drives the material allocation procedure toward favoring higher clusters, while the extruder's vertical offset relative to the deposition surface is slightly decreased after each iteration. The result is an increasingly selective deposition behavior obtained by: (i) allocating more material to the 'fittest' emergent structures, thus increasing the effects of clay deformation, and (ii) achieving greater organizational density in these areas through the progressive focusing of deposited mass, induced by the reduction of the extruder's vertical offset.

This algorithm is implemented using a simple reward mechanism based on a power function of the form $r = z^{\alpha}$ where r is the reward allocated to a given deposition zone, $0 < z \leq 1$ is the normalized height of the zone's centroid, and $\alpha \geq 1$ is the

specific locations while keeping a certain degree of indeterminacy in the printing toolpaths, suitable for exploring new formal opportunities.

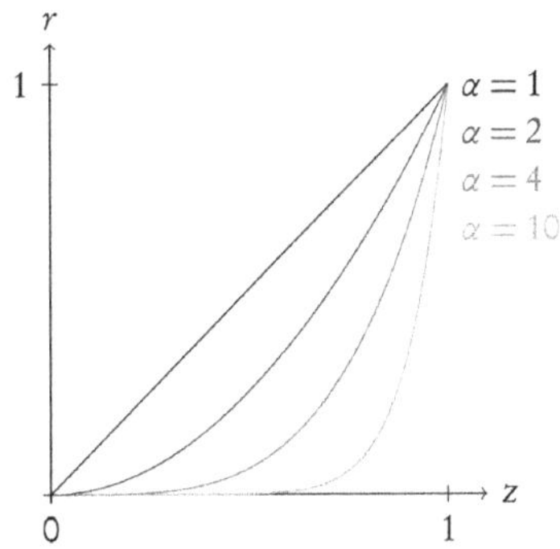

Fig. 9: Evolution of the power function used in Unspecified Clay's reward procedure under changes of the α parameter. © Ianis Lallemand.

reward mechanism's selectivity parameter. After computing r for each deposition zone, the system divides the obtained values by the sum of all attributed rewards, yielding the proportion of the total number of deposition points to allocate to each centroid. Initially set at 1, α is incremented by a fixed value after each iteration – an operation which, by increasing the convexity of the reward function (Fig. 9), provides a simple way to make the reward mechanism increasingly selective toward higher clusters. After enough iterations, this process would ultimately result in a single cluster, winner-takes-all focusing strategy.

In addition, the extrusion's vertical offset is decreased at a constant rate after each iteration, typically shifting from 50 mm to 10–20 mm. When computing new deposition trajectories, this offset is locally added to the height information provided by the point cloud representation of previously deposited structures. This allows agents from the trajectory generation algorithm to maintain a constant vertical offset relative to prior topography, reducing the risk of collisions between the robot and already present material.

3.5 Results and Discussion

Three material samples resulting from different fabrication sessions are presented in Figs. 10–12. They have been obtained with different agent simulation parameters, while maintaining the same settings otherwise. This process allowed us to evaluate the adaption of the implemented deposition focusing procedure to different material morphologies, induced by several modes of toolpath generation.

Overall, the performed experiments tend to validate our initial hypothesis of being able to channel, through computational means, emergent material behavior into differentiated structures. In particular, the samples highlight our approach's capacity to achieve an increase of organization along the extrusion's vertical axis. In each of the samples shown in Figs. 10 and 11, the initial, loosely structured mass has progressively

Fig. 10: Experimental sample produced as part of the Unspecified Clay project. © Ianis Lallemand.

Fig. 11: Experimental sample produced as part of the Unspecified Clay project. © Ianis Lallemand.

Fig. 12: Experimental sample produced as part of the Unspecified Clay project. © Ianis Lallemand.

been re-shaped into a few higher density clusters, through the addition of increasingly targeted depositions. In the highest clusters, further differentiation has been achieved as the algorithm drove the extruder to operate in direct contact with the clay, resulting in characteristic flat-topped structures. Figure 12 exhibits a more layered morphology, induced by the use of orbit-like agent trajectories. This example suggests that the fabrication can be intentionally steered toward different structural typologies by means of high-level parameters, while achieving local organizational details through the autonomous interaction of algorithmic and material behavior.

In its current form, the system provides a proof of concept of a retroactive, clay-based production system, intended primarily for generative explorations. The project's technological developments, however, are susceptible to applications in the broader context of sensor-enabled robotics. Retroactive reward strategies, as implemented in Unspecified Clay, allow fabrication algorithm designers to use actual material behavior as a validator of previous robotic actions, making it possible to work with a broader class of material processes, regardless of the latter's propensity to be digitally simulated.

As a conclusive note, here, we should acknowledge the two main limitations of the project in its current state, which will be the subject of future work. First, the implemented clustering strategy (K-means algorithm) requires manually setting the numbers of clusters to detect. While the gradual increase in convexity of the reward function $r = z^{\alpha}$ leads to the automatic pruning of non-relevant clusters over time, our system may still fall prey to an overestimation of the cluster count in the first iterations of the fabrication process, where the reward function's selectivity is virtually nonexistent. Tackling this issue would require developing a new clustering strategy involving a form of morphological analysis of the point cloud data, so as to better adapt the search procedure to the real features of the deposited clay.

A second limitation of our current approach rests in its handling of possible collisions between the extruder and previously printed structures. While each generated toolpath makes use of the local topographic information provided by the point cloud, no check is currently performed to ensure that the extruder's body does not collide with structures adjacent to the toolpath. Embedding the effector's geometry into the trajectory generation routine would provide an immediate solution to this issue, yet the problem might be better approached by rethinking how the robot moves in space, in order to allow the use of a greater range of motion commands (and not simply linear ones). This approach would require implementing an inverse kinematics solver into the application, which would signal a deeper integration of manufacturing logics into the system's design.

4 Manoeuvres

4.1 Context

As stated before, one of the most salient manifestations of the increasing relevance of 'active matter' in design lies in the field's increased proximity with other research areas, such as physics and materials science. The Manoeuvres project, discussed in this section, provides a concrete example of how this conceptual proximity might evolve into design research collaborations.

Initiated amid the growing involvement of EnsadLab in the Arts & Sciences movement,[29] and benefiting from this laboratory's long tradition of exchanges with its neighboring scientific institutions, Manoeuvres has been developed in collaboration with physicist Olivier Dauchot.[30] The project finds its origins in one of the team's experiments, initiated in 2003 to explore the collective motion of self-propelled disks, or 'walking grains.' Situated within the field of active matter, this research aimed at creating a model experiment demonstrating the same type of behavior as Tamás Vicsek et al.'s influential theoretical model, published in 1995.[31]

With the aim of achieving precise experimental control, EC2M's setup was designed to eliminate all forms of external influence, like biological or chemical factors, in the production of the grains' motion [Deseigne, Dauchot, and Chaté 2010, p. 1]. To this end, EC2M thus elected to build a fully designed and artificial system, which they describe as the first "well controlled experiment" [ibid.] able to create large scale collective motion – or a "fluctuating, collectively moving ordered phase of the type frequently observed in simple numerical models" [ibid.]. To achieve this result, EC2M used purely mechanical means to set up each grain into directed motion, in an attempt to produce the local alignment behavior central to Vicsek et al.'s model. The experiment employs specially designed grains, equipped with a mechanically 'active' structure, and a homogeneous source of energy in the form of a vibrating plate on which the grains stand. The grains' design (Fig. 13) consists in a 4 mm diameter, micro-machined copper–beryllium disk, bearing two 'legs' on its bottom side: a copper–beryllium tip, machined from the same stock as the disk, and a rubber skate glued to the disk's surface. This structure, composed of two standing points having different mechanical

29 EnsadLab's mother institution, EnsAD, is a founding member of the first Arts & Sciences chair in Europe, along with the École polytechnique and the Daniel & Nina Carasso foundation: http://chaire-arts-sciences.org/ (accessed August 24, 2018). The SACRe doctoral program, in the context of which Manoeuvres was developed, is also a key driver of Arts & Sciences projects within EnsadLab.

30 Director of the Collective Effects and Soft Matter team (EC2M) of the laboratory Gulliver UMR 7083 at the École supérieure de physique et de chimie industrielles de la ville de Paris (ESPCI).

31 Vicsek's model was introduced to study the behavior of nonequilibrium systems [Vicsek et al. 1995, p. 1226]. More specifically, in the original article in which it was first presented, the model was used to demonstrate the existence of a kinetic phase transition (characterized by the alignment of particle velocities) in "systems of particles with biologically motivated interaction" [ibid.]. Although theoretical and based on computer simulations, the article indeed compared its results to the collective behavior of biological systems such as "schools of fish, herds of quadrupeds, or flocks of flying birds" [ibid.]. From a practical viewpoint, the model consists in a simple velocity update rule: at each time step of the simulation, the new direction of motion of each particle is set to "the average direction of motion of the particles in its neighborhood of radius r with some random perturbation added" [ibid.].

response under vibration, creates a polar asymmetry which allows the grain to achieve directed motion, in the rubber skate to tip direction. The type of ordered formations produced by the experiment is visible in Fig. 14, showing a zenithal view of the vibrating plate (cropped to its circular boundary).

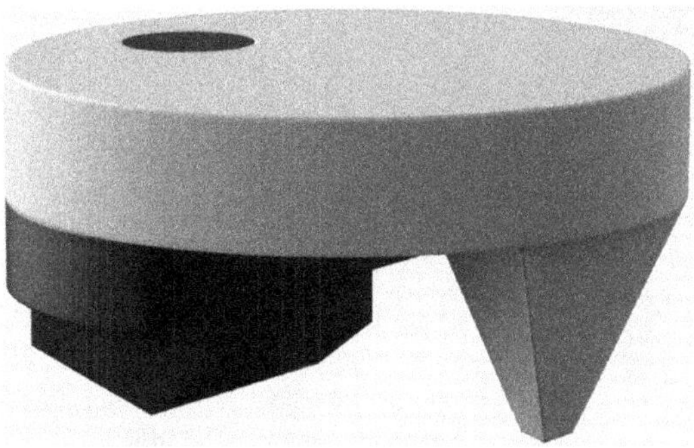

Fig. 13: View of the grain design used in EC2M's original experiment. © Ianis Lallemand.

From a designer's point of view, one of the most striking features of this experimental setup undoubtedly resides in its 'analog' nature: the system appears completely devoid of digital control, at the exception of the simple function generator plugged into the vibrating plate. The local alignment behavior followed by the grains, and the resulting collective motion achieved by the group, both arise here, in fact, as purely emergent outcomes of physical collisions, whose dynamics are not controlled by any kind of programmed rule. This original approach, which struck us upon our first visit of the laboratory, appears to be in direct contrast to the direction most commonly assumed in the field of design research, where the topic of physical swarming systems has been subject to several investigations in recent years. Physical realizations of multi-agent simulations – which have been extensively used in the digital design community over the last two decades[32] – have indeed, up to now, been primarily investigated from a robotics-oriented viewpoint, in projects such as the drone-based flight-assembled architecture, realized by Gramazio & Kohler's research lab in collaboration with Raffaello D'Andrea's team at the ETH Institute for dynamic system and control [Augugliaro et al. 2014]. In such projects, the adopted approach – coincident

32 The use of such methods in digital design was certainly favored by the field's close relationship with the animation community (to which Craig W. Reynolds belonged) in the mid-1990s, as famously illustrated by Greg Lynn's use of animation software in his early work (see for instance [Lynn 1999]). Current uses of multi-agent algorithms include the generation of complex geometries [Snooks 2012] or the tackling of optimization tasks such as, for instance, the determination of an adequate tessellation of an architectural shell given aesthetic as well as digital manufacturing constraints [Schwinn and Menges 2015].

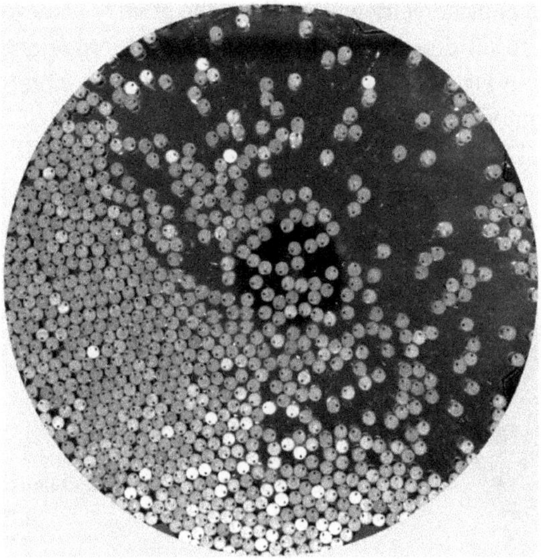

Fig. 14: The 'walking grains' original experiment (view from above). © Olivier Dauchot, Gulliver laboratory UMR CNRS 7083, ESPCI Paris (reproduced with authorization).

with current robotic research in collective behavior[33] – focuses on the programming of rules at the unit level. Alignment behavior is then produced through a combination of digital sensing and motor actuation, which ensures the proper response of every unit to these built-in rules.

While this research direction offers the advantage of lending itself to the realization of many types of applications, including large-scale construction tasks, its rule-based approach requires setting up a technological apparatus that might not fit all types of design projects. In particular, the field of interactive or 'practicable' installations – image, sound, or shape-production devices calling for a physical engagement of their viewers[34] – has recently witnessed a growing interest for the use of passive[35] actuators based on responsive materials, which offers a lower level of perceived complexity than digital systems. The Lotus Dome installation, by the Dutch artist Daan Roosegaarde, takes for instance the form of a hemispheric assemblage of heat-responsive 'smart foils,' at the interior of which rests a motorized stage light: when a person approaches the dome, the light rotates and illuminates the viewer, heating and 'opening' the smart foils on the beam's path (Fig. 15). Besides its evident poetic qualities, this project also offers distinct advantages in terms of maintenance

33 See, for instance, the Kilobot project, led by Harvard University's Self-Organizing Systems Research Group: [Rubenstein et al. 2014].

34 For a more thorough definition of the notions of interactive and practicable artwork, as well as a discussion of their relevance from an art-historical perspective, see, in particular: [Bianchini and Verhagen 2016].

35 We use here the term 'passive' to describe an actuator not drawing energy from an electrical power supply.

over motor-based installations: the foils, devoid of internal mechanics and not dependent on a digital controller, are much less likely to fail during the work's operation. Far from being trivial, this feature constitutes a crucial argument for the use of material-based actuation mechanisms in contemporary art and design contexts.

In light of these reflections, we envisioned EC2M's experiment as a unique opportunity to approach the design of an agent-based, interactive production system, allowing for the creation of form through physical – rather than simulated – collective behavior. This realization led us to the development of Manoeuvres, of which we will now present the main concept and design principles.

Fig. 15: View of the Lotus Dome installation by Dutch artist Daan Roosegaarde. © Roosegaarde Studio (reproduced with authorization).

4.2 Concept

The opportunity of designing an interactive system in dialog with EC2M's research was approached through the formulation of a light projection concept: we proposed to translate the collective motion of self-propelled units, or 'grains,' into a moving image, by equipping each unit with a rod-shaped structure, and casting the shadow of the latter on a surface. The projection was to be calibrated so that the shadows could reach about 1.8 m in height, creating a landscape of abstract yet human-scaled figures. Figure 16 shows a preparatory drawing of this concept, as well as a digital render of the expected light projection effect. By reference to EC2M's research, we envisioned to build a vibration-based system, at times robust and visually simple, with the objective of inviting the viewer to engage in a direct interaction with the system's behavior. Initially left open, the question of the viewer's mode of interaction became evident once the project had reached its first functional stages: thanks to the analog nature of the motion, and the natural 'grip' provided by the grains' rods, we realized that the viewer could easily manipulate the grains by hand, like pieces in a game of checkers – with the exception, of course, that the pieces would here move in an autonomous fashion. This 'analog' physical interaction ultimately suggests a blurring of boundaries between the viewer's and the grains' behaviors, an idea reflected in the production of a human-sized image out of the grains' shadows.

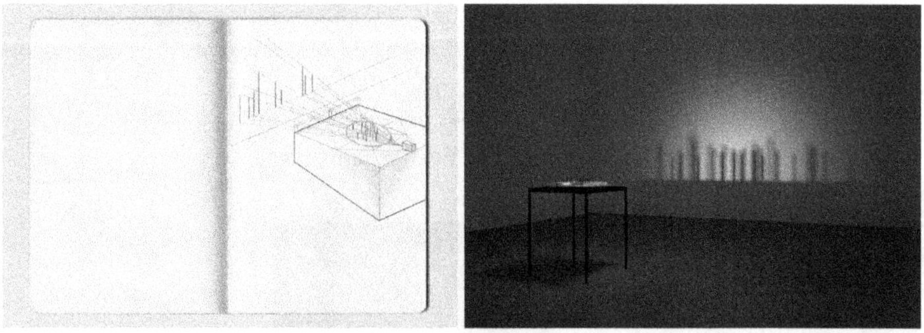

Fig. 16: Initial concept drawing and digital rendering of Manoeuvres. © Ianis Lallemand.

The ambition of inducing a form of proximity between the system and its viewers arose from the zoomorphic or even anthropomorphic associations that most observers would use when describing EC2M's original experiment, whether they were highly familiar with it – as in the case of the team's own personnel – or being confronted with it for the first time. Although being completely inert, and lacking any zoomorphic or anthropomorphic features on their own (due to their simple, disk-like shapes), the grains where indeed frequently compared to insects, animals, or even human beings, when commented on in the context of their collective behavior. While many of these comparisons would hold from a scientific viewpoint – in the same way as Vicsek et al.'s work can be said to model the principles behind some natural systems' behavior, such as bird flocking – we found most promising, from an artistic point of view, to approach the grains' motion as the model behavior of a human crowd. In this context, we imagined Manoeuvres as staging an interplay between two projection processes: the mental projection of a human behavior into the motion of inert, non-figurative objects, and the light projection of these objects' contours into a symbolic, shadow-formed 'crowd.'

4.3 Design

Manoeuvres' design was naturally inspired by EC2M's original experiment, as required for the production of the same type of collective behavior which it originally displayed. However, beyond general principles (such as the use of an 'active' material structure in order to drive the grain's motion), EC2M's experiment could not be reproduced 'as is.' Given the small original grain diameter (4 mm), larger grains were in fact needed to carry shadow-casting rods, of a sufficient section, in a stable way; this in turn implied designing a larger vibration system than the one originally used by EC2M. Due to the complexity of the coupling between the grains and the plate, this sizing up of the system's core elements meant that the existing design for the grains legs could not simply be reused in a scaled form, and that a new design, fitting our system's properties, had to be developed through experimental testing. Although our system could have, in theory, made use of the same vibration production mechanism as EC2M's experiment (an industrial, electromagnetic servo-controlled shaker), convenience and cost considerations drove us to develop an alternative solution. Our discussions with Olivier Dauchot, revealing a mutual interest for approaching the problem through a 'hacking' mindset, led us to adopt inexpensive and quickly iterable manufacturing processes, like 3D printing, and off-the-shelf components, such as loudspeakers or surface transducers, to develop a more economical, lighter vibration system. Although not offering the level of control ex-

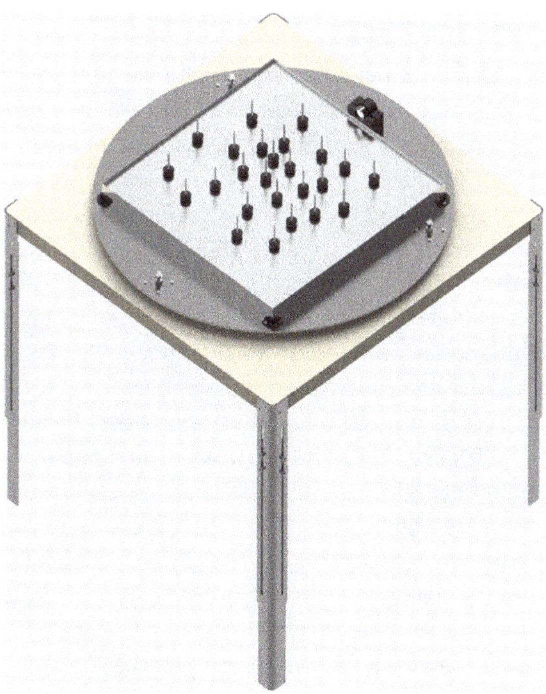

Fig. 17: Rendering of Manoeuvres' final setup. © Ianis Lallemand.

pected from a scientific apparatus, this approach constituted a viable solution in the design-oriented context of Manoeuvres.

The resulting setup, shown in Fig. 17, is composed of five main components: a table, a disk-shaped support frame, a square-shaped vibration unit, an LED lamp, and a set of walking grains. Figure 18 shows a detailed view of the frame's and

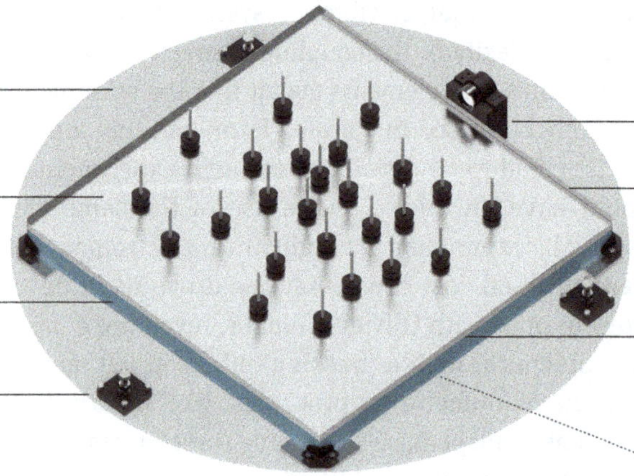

Fig. 18: Detail of Manoeuvres' vibration system. The vibration system is pictured in a cutaway view to reveal its internal structure. © Ianis Lallemand.

vibration system's components. The frame, waterjet-cut from a 5 mm thick aluminum sheet, has a 750 mm diameter circular profile. Its surface bears a 500 × 500 mm cut, inside which the vibration plate is inserted. Four support aluminum tabs, located at each corner of the cut, allow the vibration plate to rest on the frame with minimal contact. The frame is placed on top of a custom-made table, equipped with four waterjet-cut steel legs that can be adjusted in length, so as to allow for the tuning of the light projection's height. The contact with the table is realized by three height adjustment screws, which are used to calibrate the frame's (and thus the vibration plate's) horizontal inclination. Illumination is provided by a 1,000 lumen LED lamp (Cree XM-L T6), placed in a 3D printed case attached to the frame's surface.

The vibration system's internal structure (visible in Fig. 18) is composed of an assemblage of three materials: a 30 mm thick extruded polystyrene board, a 10 mm thick foam board, and a 1 mm thick aluminum covering on top. The vibration is provided by four surface transducers secured to the bottom face of the polystyrene board by threaded rods embedded in the material. We used commercial transducers (Visaton EX 60 S) designed for sound and music applications, which we fed an 80–125 Hz sine wave produced by a Pure Data program running on a Raspberry Pi Linux micro-computer, processed through a 25 W Hi-Fi amplifier (Dynavox CS-PA1). Because of their location, the transducers are not visible when the setup is observed from a normal viewpoint. For aesthetic reasons, the polystyrene board's and foam board's side faces are also hidden by a square-shaped aluminum 'fence,' attached to the frame with 3D printed connectors, and rising 10 mm above the vibration system's top surface. This fence performs the additional function of preventing the grains from falling onto the frame when moving.

Figure 19 shows an exploded view of the grain design. The assembly starting point is a 20 mm diameter 3D printed PLA disk, the bottom side of which includes an

Fig. 19: Exploded view of Manoeuvres' grain design. © Ianis Lallemand.

extruded triangular profile which constitutes one of the grain's legs. The second leg consists in a 5 × 5 mm skate, cut from a 1 mm thick rubber vibration damping pad (3 M 0500.54 G.70). As in EC2M's original experiment, the different mechanical responses exhibited by the two legs' materials (PLA and rubber) induce a polar asymmetry that allows the grain to achieve directed motion; as we realized soon, however, this behavior can only be observed if the grain has a higher mass than given by its PLA disk alone. To provide the requisite mass, we thus designed a 10 mm diameter threaded cylinder on the top side of the disk, which allows securing an M10 nut to the grain.

A crucial feature of the grain design is the vertical rod, required by Manoeuvres' light projection concept. As the rod has to rise by about 50 mm over the PLA base to create large enough shadows, the structure becomes prone to the development of energy-dissipating vibrations, which can harm the grains' motion. This effect can be limited by reducing the mass of the vertical rod. We thus used a 5 mm diameter, 50 mm long nylon threaded rod, inserted through the PLA disk and secured to it by two M4 nuts, which help keep the rod in a perfectly straight position – a requirement otherwise difficult to meet given the grain's thin profile. A hollow cylindrical ABS cover completes the design by hiding the grain's internal components, resulting in a simpler external appearance.

4.4 Results and Discussion

Manoeuvres was first presented to the public as part of a performance organized at the Espace Pierre-Gilles de Gennes in Paris, on June 23, 2016, following the completion of the project's first working prototype. After a short discussion of our system's relationship with EC2M's original experiment, viewers were invited to freely experiment its behavior and engage with the grains' collective motion. The project was then exhibited at the Gaîté lyrique, in Paris, as part of our PhD defense's exhibition, which ran from December 2 to 5, 2017. These two contexts, differing both in terms of duration and settings – the Gaîté lyrique exhibition featured other works and a specially designed scenography – provided two complementary testing scenarios of the

Fig. 20: View of the exhibition at the Gaîté lyrique, Paris, December 2–5, 2017. © Jean-Marc Lallemand (reproduced with authorization).

Fig. 21: A light projection produced by Manoeuvres. © Ianis Lallemand.

project's concept and design implementation. Figure 20 shows a view of the Gaîté lyrique exhibition, and Fig. 21 presents a light projection produced by the system.

Although Manoeuvres' setup, by its analog nature, follows a different approach than explored in Unspecified Clay, the project may also be envisioned as an experiment in designing feedback mechanisms, for the purpose of enabling reciprocal interactions between a production apparatus and its material environment: just as Unspecified Clay's retroactive algorithms allowed the system to build on emergent external conditions, so could the material principles of motion built into Manoeuvres negotiate the emergent behavior of the public. During the performance at the Espace Pierre-Gilles de Gennes, for instance, a surprisingly efficient (albeit rather erratic) motion mechanism was experimented by some viewers, who discovered that the agents could still 'walk' when laid on their side, their shadow-casting rod touching the vibration plate – an unplanned configuration which was easily supported by our system. Far from being anecdotal, this example illustrates the intrinsic playfulness and discoverability built into the project, by means of its analog design principles. Such qualities may motivate the use of analog design methods as an alternative to traditional, digitally oriented approaches to interaction design, which have in the past few years tied our existences to an increasing amount of opaque, black box like devices [Haque 2007, p. 60] – from our phones to the growing population of 'smart' assistants inhabiting our homes.

On another level, experimenting with Manoeuvres led us to envision the very notion of material activity from a new perspective. We indeed observed that the system's self-organizing behavior had the capacity to permeate the organization of the public itself: groups of viewers would form, structured both by direct interactions with the device, and by indirect, interpublic interactions – such as, for instance, the commenting of light projection patterns created by others, or the discussion of the grains' motion principles, experienced by observing other viewers play with the system. Manoeuvres' active 'grain material' appeared to us, in this respect, as equally structured and structuring, as the reach of its performances extended up to the human crowd. Such an observation ultimately calls for an understanding of material activity as a true form of agency within the creative project, in a stronger sense than usually envisioned in design-oriented applications of 'smart' or responsive materials.

5 Conclusion

This chapter has introduced two complementary approaches for the exploration of material activity within a research-oriented, digital design practice. The discussed projects, Unspecified Clay and Manoeuvres, follow our initial call for design processes achieving a form of openness toward material behavior, by adopting a performative, rather than representational approach to form making.

In Unspecified Clay, digital techniques are mobilized to set up a closed feedback loop between a robotic 3D printing system and the physical behavior of clay, letting form emerge from this reciprocal process of interaction as it unfolds over time. Beyond the setting up of the experiment's technical apparatus, the design approach focuses here primarily on the creation of computational processes, seeking to channel the interplay of robotic, digital, and physical behavior by means of implicit methods rather than geometric prescription.

Manoeuvres, based on an existing experiment by Olivier Dauchot and the EC2M research team, investigates the concept of active matter through the question of collective motion – an object of ever-increasing attention in the physics community since the publication of Vicsek et al.'s theoretical behavior model in 1995. The project takes the form of an interactive production system, whose output – a light projection – is obtained through the collective motion of active grains, following a behavior-to-form translation process inspired by the digital method of multi-agent systems. Derived from EC2M's initial experiment, Manoeuvres' design is fundamentally driven by an openness to material behavior: in an alternative approach to sensor and motor-based approaches, collective motion is here achieved by tapping into the physical qualities of a specially designed material unit (the grain), which is left to interact, without any form of programmed control, with the behavior of its fellow partners.

These two instances of open-design processes ultimately illustrate two contrasting traits of digital design research's relationship with the notion of active matter, of which we have mentioned several other examples throughout this chapter. On the one hand, more digital processes today implement sensing and data analysis mechanisms (such as the 3D scanning and feedback procedures built into our Unspecified Clay project) that allow them to operate in 'greater proximity' with the physical world. Instead of attempting to fully predict the properties of physical materials through digital simulations, new production strategies instead implement feedback mechanisms into manufacturing systems, enabling the continuous update of computational models with real-world data. On the other hand, digital designers are increasingly drawn toward the production of analog components, transposing methods and ideas from the digital world into the new landscapes of physical behavior opened by today's active materials.

While these two approaches may at first seem to embody divergent research trajectories – the first of them focusing on the expansion of digital systems' capacities, and the latter advocating for the development of new analog production methods – what is at stake in both cases is an evolution of design practice toward a performative – rather than representational – approach to materiality. In contrast to representational design approaches, which intend to geometrically define form in its totality before physical realization, the two case studies discussed in this chapter demonstrate how one can approach the production of form as a dynamic process, happening in the 'here and now' of the physical mobilization of material performances. As materials become active participants in the definition of form, designers invest increasingly in the development of dynamic production systems, often based on similar ideas regardless of their digital or analog nature – such as the notion of agent-based systems, developed by both Unspecified Clay and Manoeuvres projects. Far from being antagonists, analog and digital approaches thus appear as complementary directions for the investigation of performative design methodologies and processes in the era of active materials.

Bibliography

Alberti LB. On the Art of Building in Ten Books. Trans. Rykwert J, Leach N, Tavernor R. Cambridge, MIT Press, 1988.

Anderson C. Makers: The New Industrial Revolution. New York, Crown Business, 2012.

Augugliaro F, Lupashin S, Hamer M, et al. The Flight Assembled Architecture Installation: Cooperative Construction with Flying Machines. IEEE Control Systems Magazine 2014, 34 (4), 46–64, DOI: 10.1109/MCS.2014.2320359.

Bézier P. A View of the CAD/CAM Development Period. In: Blomenthal J, Barnhill RE, Barsky BA et al. Graphics Remembrances. IEEE Annals of the History of Computing 1998, 20 (2), 35–51, here 37–40.

Bianchini S, Verhagen E, eds. Practicable: From Participation to Interaction in Contemporary Art. Cambridge, MIT Press, 2016.

Boutinet JP. Psychologie des conduites à projet. Paris, PUF, 1993.

Boutinet JP. Projet. In: Barus-Michel J, Enriquez E, Lévy A, eds. Vocabulaire de psychosociologie. Toulouse, Érès, 2002, 222–230.

Carpo M. The Alphabet and the Algorithm. Cambridge, MIT Press, 2011.

Colletti M. An Example of [En]coding Neo Materialism: ProtoRobotic FOAMing. In: Werner LC, ed. [En]Coding Architecture. Pittsburgh, Carnegie Mellon University, School of Architecture, 2013, 57–65.

Damisch H. The Origin of Perspective. Cambridge, MIT Press, 1995.

Del Campo M, Fure A, McGee W, Manninger S, Flexer A. Autonomous Tectonics: A Research into Emergent Robotic Construction Methods. In: Gengnagel C, Kilian A, Nembrini J, eds. Rethinking Prototyping: Proceedings of the Design Modelling Symposium Berlin 2013. Berlin, Universität der Künste, 2013, 417–432.

DeLanda M. Philosophies of Design: The Case of Modeling Software. In: Salazar J, ed. Verb Processing. Barcelona, Actar, 2002, 130–143.

Deseigne J, Dauchot O, Chaté H. Collective Motion of Vibrated Polar Disks. Physical Review Letters 2010, 105 (9), 098001, DOI: 10.1103/PhysRevLett.105.098001.

Doerstelmann M, Knippers J, Koslowski V, et al. ICD/ITKE Research Pavilion 2014–15: Fibre Placement on a Pneumatic Body Based on a Water Spider Web. Architectural Design 2015, 85 (5), 60–65, DOI: 10.1002/ad.1955.

Dörfler K, Ernst S, Piškorec L, et al. Remote Material Deposition: Exploration of Reciprocal Digital and Material Computational Capacities. In: Voyatzaki M, ed. What's the Matter: Materiality and Materialism at the Age of Computation. Barcelona, ETSAB – UPC, 2014, 361–377.

Evans R. Translations from Drawing to Building. In: Evans R. Translations from Drawing to Building and Other Essays. Cambridge, MIT Press, 1997, 153–193.

Frampton K. Studies in Tectonic Culture: The Poetics of Construction in Nineteenth and Twentieth Century Architecture. Cambridge, MIT Press, 1995.

Frampton K, Allen S, Foster H. A Conversation with Kenneth Frampton. October 2003, 106, 35–58, DOI: 10.1162/016228703322791016.

Garcia MJ, Retsin G. Design Methods for Large Scale Printing. In: Martens B, Wurzer G, Grasl T, Lorenz WE, Schaffranek R, eds. Real Time: Proceedings of the 33rd eCAADe Conference. Vol. 2. Vienna, Vienna University of Technology, 2015, 331–339.

Glaeser L. The Work of Frei Otto. New York, Museum of Modern Art, 1972.

Goodman N. Languages of Art: An Approach to a Theory of Symbols. Indianapolis, Bobbs-Merrill, 1968.

Haque U. The Architectural Relevance of Gordon Pask. Architectural Design 2007, 77 (4), 54–61.

Huerta S. Structural Design in the Work of Gaudí. Architectural Science Review 2006, 49 (4), 324–339, DOI: 10.3763/asre.2006.4943.
Ingold T. Making: Anthropology, Archaeology, Art and Architecture. London, Routledge, 2013.
Jack MA. The Accademia del Disegno in Late Renaissance Florence. The Sixteenth Century Journal 1976, 7 (2), 3–20, DOI: 10.2307/2539556.
Johns RL. Augmented Materiality: Modelling with Material Indeterminacy. In: Gramazio F, Kohler M, Langenberg S, eds. Fabricate: Negotiating Design and Making. Zurich, gta Verlag, 2014, 216–223.
Kadic M, Bückmann T, Schittny R, Wegener M. Metamaterials Beyond Electromagnetism. Reports on Progress in Physics 2013, 76 (12), 126501, DOI: 10.1088/0034-4885/76/12/126501.
Kensek KM. Building Information Modeling. London, Routledge, 2014.
King R. Brunelleschi's Dome: How a Renaissance Genius Reinvented Architecture. New York, Walker & Company, 2000.
Le Corbusier PJ. Œuvre Complète. Vol. 5. 1946–1952. Boesinger W., ed. Zurich, Les Éditions D'Architecture, 1991.
Lynn G. Architectural Curvilinearity: The Folded, the Pliant and the Supple. Architectural Design 1993, 102 (8), 15.
Lynn G. Animate Form. New York, Princeton Architectural Press, 1999.
Menges A. Material Computation: Higher Integration in Morphogenetic Design. Architectural Design 2012, 82 (2), 14–21, DOI: 10.1002/ad.1374.
Menges A. Fusing the Computational and the Physical: Towards a Novel Material Culture. Architectural Design 2015, 85 (8), 9–15, DOI: 10.1002/ad.1947.
Menges A, Reichert S. Performative Wood: Physically Programming the Responsive Architecture of the HygroScope and HygroSkin Projects. Architectural Design 2015, 85 (5), 66–73, DOI: 10.1002/ad.1956.
Mitchell WJ. Antitectonics: The Poetics of Virtuality. In: Beckmann J, ed. Virtual Dimension: Architecture, Representation, and Crash Culture. New York, Princeton Architectural Press, 1998, 205–217.
Morel P. Computation or Revolution. Architectural Design 2014, 84 (3), 76–87, DOI: 10.1002/ad.1761.
Negroponte N. Being Digital. New York, Alfred A. Knopf, 1995.
Oxman N. Material-Based Design Computation. PhD thesis. Cambridge, MIT, 2010.
Picon A. Architecture and the Virtual: Towards a New Materiality. Praxis 2004, 6, 114–121.
Popkin G. The Physics of Life. Nature News 2016, 529 (7584), 16–18, DOI: 10.1038/529016a.
Rajkumar R, Lee I, Sha L, Stankovic J. Cyber-Physical Systems: The Next Computing Revolution. In: Proceedings of the 47th Design Automation Conference. ACM, 2010, 731–736.
Retsin G, Garcia MJ. Discrete Computational Methods for Robotic Additive Manufacturing: Combinatorial Toolpaths. In: Posthuman Frontiers: Data, Designers, and Cognitive Machines. Proceedings of the 36th Annual Conference of the Association for Computer Aided Design in Architecture (ACADIA). Ann Arbor, University of Michigan Taubman College, 2016, 332–341.
Reynolds CW. Flocks, Herds and Schools: A Distributed Behavioral Model. ACM SIGGRAPH Computer Graphics 1987, 21 (4), 25–34, DOI: 10.1145/37401.37406.
Rubenstein M, Ahler C, Hoff N, Cabrera A, Nagpal R. Kilobot: A Low Cost Robot With Scalable Operations Designed for Collective Behaviors. Robotics and Autonomous Systems 2014, 62 (7), 966–975, DOI: 10.1016/j.robot.2013.08.006.
Schäffner W. The Design Turn: Una Revolución Científica en el Espíritu Del Diseño. Kepes 2010, 7 (6), 61–75.
Schwinn T, Menges A. Fabrication Agency: Landesgartenschau Exhibition Hall. Architectural Design 2015, 85 (5), 92–99, DOI: 10.1002/ad.1960.

Serlio S. Il terzo libro di Sabastiano Serlio bolognese, nel qual si figurano, e descrivono le antiquita di Roma e le altre che sono in Italia, e fuori d'Italia. Venice, Francesco Marcolini da Forlì, 1540.

Snooks R. Volatile Formation. Log 2012, 25, 55–62.

Snooks R, Jahn G. Stigmergic Accretion. In: Reinhardt D, Saunders R, ed. Robotic Fabrication in Architecture, Art and Design 2016. Cham, Springer, 2016, 398–409.

Toffler A. Future Shock. New York, Random House, 1970.

Vasari G. The Lives of the Artists. Trans. Bondanella JC, Bondanella P. Oxford, Oxford University Press, 1998.

Vial S. De la spécificité du projet en design: une démonstration. Communication et Organisation 2014, 46, 17–32, DOI: 10.4000/communicationorganisation.4699.

Vicsek T, Czirók A, Ben-Jacob E, Cohen I, Shochet O. Novel Type of Phase Transition in a System of Self-Driven Particles. Physical Review Letters 1995, 75 (6), 1226–1229, DOI: 10.1103/PhysRevLett.75.1226.

Witt A. Design Hacking: The Machinery of Visual Combinatorics. Log 2011, 23, 17–25.

Zhang Z. Microsoft Kinect Sensor and its Effect. IEEE Multimedia 2012, 19 (2), 4–10, DOI: 10.1109/MMUL.2012.24.

Ianis Lallemand is a Paris-based designer, researcher, and educator whose work is situated at the intersection of computation, generative design, and digital manufacturing. He is an assistant professor at the École supérieure d'Arts et de Design, Tours Anger Le Mans (TALM), an associated researcher in the Reflective Interaction group at École nationale supérieure des Arts Décoratifs (EnsAD) Paris, and a Co-de-iT member. He received his PhD in design from Paris Sciences et Lettres University (PSL)/EnsAD Paris in 2018. For further information: https://www.ianislallemand.com/.

Essays: Part 2: Historical and Philosophical Perspectives

Sylvie Kleiman-Lafon, Charles T. Wolfe

Unsystematic Vitality: From Early Modern Beeswarms to Contemporary Swarm Intelligence

"Cette grappe est un être, un individu."[1]

1 Materialism, Vital Materialism, and Self-Organization: Enter the Beeswarm

The eighteenth century was the century of self-organization but also that of materialism, inasmuch as it was then that certain thinkers proclaimed themselves to be materialists (rather than just being labeled as such by enemies of various sorts).[2] If one seeks to read these two features together – one hesitates to call them 'facts' or 'events' – one arrives rather quickly at an influential metaphor, the beeswarm. But a metaphor of, or for, what? Irreducible organic unity, most broadly – spelled out in the vocabulary of the period in terms of synergy, sympathy, and sensibility, but also of cohesion, consensus, and conspiration: individual bees have their characteristics, their intentions, and their own purposes, but they also 'conspire,' 'cohere,' and 'consent' in the name of a larger living unity, the swarm, although this leaves open further questions such as the exact nature of the order or organization yielded or enacted by the swarm (bottom-up? top-down? emergent? etc.). Bees and their modes of organization have fascinated many observers, and have served as inspiration and/or as metaphors for a variety of ideas, most prominently of the social variety. That is, bees – their hive structure for instance, or the existence of a queen – have served as exemplars of social order (although it turns out that only a fraction of bee species are actually social), and of an allegedly innate 'mathematico-architectural' potential in these

1 "This cluster is a being, an individual" [Diderot 1975–, vol. 17, p. 120] ("grappe," i.e., "cluster" is a term Diderot uses for the swarm). Unless stated otherwise, all the translations from French and Spanish were done by the authors (S.K/C.W.).
2 On the eighteenth century as the century of self-organization, see: [Sheehan and Wahrman 2015]; on eighteenth-century materialism, see: [Bloch 1997; and Wolfe 2014].

Acknowledgments: The authors wish to thank Stéphane Schmitt for his suggestions. Charles Wolfe wrote his draft while in residence at the Institute for Advanced Study at CEU (Budapest) and is grateful for the pleasant working conditions.

Open Access. © 2021 Sylvie Kleiman-Lafon, Charles T. Wolfe, published by De Gruyter. BY-NC-ND This work is licensed under the Creative Commons Attribution-NonCommercial-NoDerivatives 4.0 International License.
https://doi.org/10.1515/9783110562064-013

animals.[3] By contrast, the phenomenon of the swarm seemed disturbing – less organized, less virtuous – to many such observers. This changed when the beeswarm became a core metaphor for organic, indeed organismic unity and in some cases (notably in Denis Diderot) for the organic or organismic features of living matter itself. It is this story of beeswarms as a mode of organization that we wish to tell.

Materialism has gone through a number of incarnations in its history, a history which itself is a matter of some dispute: are Empedocles, Julien Offray de La Mettrie, and 'new materialists' all part of one history, on one track, one line of development? Materialism has been treated as a particularly nefarious species of 'cold, dead mechanism,' reducing everything in the world – values, feelings, persons, aesthetic experience and so on – to mere 'lumps' of passive matter,[4] mere aggregates or collections of particles, sometimes even invoking the image of a swarm of bees, as when Jonathan Swift mocks the opinion of "choice *virtuosi*" that "the brain is only a crowd of little animals" clinging together "in the contexture we behold [...] like bees in perpendicular swarm upon a tree" [Swift 1801, p. 263–264].[5] But, as one of us has sought to argue elsewhere [Wolfe 2016, ch. 4; 2017a], we would be wise to put at the center of our histories of materialism the existence of a 'vital,' 'embodied' materialism in which life – organic life, biological life, embodiment, flesh, and blood in contrast to abstract mathematical or otherwise purely physicalistically construed entities – is the core issue. This can be shown on various levels and contexts: for instance, the marked 'anti-mathematicism' of authors like Bernard Mandeville and Denis Diderot (i.e., their view that future programs in medicine, natural history, physiology, etc. should not rely uncritically on the privilege or prestige of the mathematical models derived from the successes of the physico-mechanical sciences of the seventeenth century)[6] or

---

3 The interest of early modern and eighteenth-century thinkers in bees, in general, and in the beehive or the beeswarm, in particular, rests upon a long tradition that saw the society of bees as an emblem for human society. From the Renaissance onward, numerous beekeeping manuals based on the observation of nature presented bees as the most profitable insects for man ("What creature for profit can compare to the bee, or the silk-worm?" Purchas asks [Purchas 1657, chap. I, n.p.], while Moffett judges them to be "framed for the nourishment of man" [Moffett 1658, p. 889]) but also "political" insects. On the history of bees as emblems or metaphors, see: [Preston 2005; Woolfson 2010 Prete 1991; Fraser 1931; Quiviger 2003].

4 According to John Hancock, in his 1706 Boyle Lecture attacking the neurophilosophy of Thomas Willis, the brain is a lump of matter "of a clammy and unactive Nature and Substance; [which] seems as far as we can judge of it to be a *meer passive Principle*, as to the Acts of inward Sensation and Intellection" [Hancock 1739, p. 243].

5 Swift refers specifically, in an unambiguously political gesture, to Hobbes' Leviathan as such a degraded form of unity, like a swarm of bees or "carrion corrupted into vermin" [Swift 1801, p. 264].

6 For Mandeville, for instance, "the Scheme of bringing Mathematicks into the Art of Medicine is not of many Years standing yet [...] a great length of time must be required before an entire System can be form'd, that shall be applicable to all Cases, and by the Help of which; Men shall be able to explain all *Phenomena* that may occur" [Mandeville 2017 (1711), p. 110], while for Diderot: "the realm of the mathematicians is an intellectual one: what we take to be rigorous truths absolutely loses this advantage when

their concern with ideas derived from the 'life sciences' (any combination of medicine, physiology, natural history – 'biology' as such did not exist under that name until the late years of the eighteenth century [McLaughlin 2002; Gambarotto 2017; Wolfe 2019]). It is also at least an amusing fact that the earliest usage of the term 'materialist' seems to have meant 'pharmacist,' in the sense of those who prepared and purveyed the *materia medica* [Bloch 1978]. Granted this vital emphasis shifts and arguably disappears by the time materialism is reconfigured in post–World War II Anglophone philosophy as 'physicalism' [Wolfe 2016]. But our interest in what follows is to explore an aspect of this vital materialism that is crucially bound up with its appropriation of a core metaphor from eighteenth-century medical vitalism: the beeswarm.

Beeswarms in the early modern period are objects of discussion in varied and colorful contexts, ranging from questions of social and political order to mathematical considerations concerning the hexagonal pattern of beehives. The development of 'mixed' (applied) mathematics from the beginning of the seventeenth century ('pure' mathematics being geometry and arithmetic) had given rise to a wide range of applications, from mechanics to architecture or optics. Demeter and Schliesser take the diffusion of Newton's *Principia* as a turning point for the "mathematization of all fields of knowledge from natural philosophy through medicine to moral philosophy": the application of principles of geometry to the architecture of the hive is an illustration of this tension between the use of mathematics to understand nature and 'pure' mathematics being necessarily "severed from matter," as Francis Bacon had put it [Demeter and Schliesser 2019].[7] These discussions of beehives in fact combine a mathematical focus with a physicotheological reading of hive architecture, as the complex patterns seem to reflect an innate and higher design, something Georges-Louis Leclerc de Buffon was to sneer at later on. In volume 4 of his *Histoire naturelle* (1753), referring to the great naturalist René-Antoine Ferchault de Réaumur's apiphilia, he stated that a fly (by which he meant any winged insect) should not take up any more room in scientific writing than it does in Nature [Buffon 1753, p. 92 (and more generally, pp. 90–100); Le Menthéour 2009, commentaries; Wallmann 2017].[8] Réaumur had asked the mathematician Samuel König to calculate geometrically how the bees arrive at the maximum number of cells with the minimum amount of wax, and by consequence judged that the bees' optimal choice reflects

it is brought down to our earth [...] mathematics, especially of the transcendent sort, lead to nothing particular without experience; they are *a kind of general metaphysics which strips bodies of their individual properties*" [Diderot 1975–, vol. 9, p. 29, emphasis S.K/C.W.]. For more on anti-mathematicism in the eighteenth century as, not an *anti*-scientific discourse but a *pro*-scientific discourse, see: [Wolfe 2017b].

7 For further reading on 'pure' and 'mixed' mathematics, see [Andersen and Bos 2006; Brown 1991; Roux 2010, 2011]; see also [Mancosu 1996].

8 Le Menthéour notes that Bonnet replied to Buffon defending Réaumur's position on the intelligence of bees, in his *Considérations sur les corps organisés* [Bonnet 1985 (1762), pp. 290–291]. Hoquet suggests that Buffon is less bluntly opposed to Réaumur's position than it appears [Hoquet 2018] – Buffon is 'radicalizing' it instead of opposing it.

divine wisdom [Réaumur 1740, p. 389];[9] his fondness for bees can be seen in his nicknaming them his *petit peuple* ("little people") [Réaumur 1740, p. 475]. For Buffon, in contrast, the complexity of the cell architecture of beehives was not an indication either of the intelligence of the individual bees or of a higher intelligence of a designer, but rather an emergent property of countless "mechanical interactions," themselves often imperfect, of thousands of "automata" [Buffon 1753, pp. 98–99].

There is no single, monolithic notion of 'machine' in the early modern and Enlightenment periods (nor presumably in any other period): the term could be used to refer to the body (as La Mettrie often did), or often, to theatrical machines producing what we might call special effects. Thus the *Dictionnaire de l'Académie* defines "machine" in 1694 as "a set of parts or organs which form a whole, living or not, and produce determinate effects without transmitting a force externally; organism, body."[10] Equally fundamental here is the fascination with the automaton as a kind of intermediate figure, like a machine but closer to something level. Indeed, the machine often functions as a kind of go-between, enabling the interface between a mechanistic ontology (à la Galileo or Descartes, in which everything reduces to the fundamental properties of size, shape and motion) and heuristics, within which actual machines can serve as 'matière à penser,' so to speak. In Georges Canguilhem's elegant phrase: "Essentially, a machine is a mediation or as mechanics say, a relay (*relais*)" [Canguilhem 1965, p. 87]. Interestingly, by the nineteenth century the situation is different, with the *Encyclopédie méthodique* explaining (in 1808) that one should no longer use the expression 'machine humaine' but, instead, for example, 'organism,' because "the term 'machine' seems to refer to a system of causes and effects which belongs wholly to the mechanistic theory."[11]

But in all these contexts, the real object of interest is *organization*: whether in the 'built' form, like their hives, or in the curiously 'emergent' form, like their swarms,[12] the way in which bees seemed to generate a mode, or different modes, of organization – what later theorists of self-organization would call 'order for free.' We first discuss early modern interpretations of the phenomenon of swarming in bees, which are inseparably 'sociopolitical' and 'metaphysical,' inasmuch as they call attention both to the nature of social interaction, order and disorder, and political régimes (if bee societies should serve as inspiration for human societies, it is indeed crucial to decide whether the former are monarchical or republican) and to the nature of unity and disunity themselves, as core

9 The result of the calculation was close to that which had been found by Maraldi [Maraldi 1712]. See also Fontenelle's presentation of Maraldi's work in the *Mémoires de l'Académie Royale des Sciences*, 1712, pp. 5–12.

10 *Dictionnaire de l'Académie*, cited in [Cayrou 1948, p. 530].

11 "Machine," in [Encyclopédie méthodique 1808, p. 310].

12 The distinction between the swarm and the hive is not that clear-cut. Both behave in the same way, although technically the swarm announces the colonization of new hives.

principles (for biology but also for questions of identity more broadly). Thus the editors of a recent, influential collection of essays on current evolutionary-biological perspectives on individuality amusingly observe that while Charles Darwin had written in his notebooks, "He who understands baboon would do more towards metaphysics than Locke," today, it is more an issue of 's/he who understands beehive': "When one replaces 'baboon' by 'beehive,' one starts to appreciate the complexities of biological individuality and why we must explain how, from the many, one can emerge" [Bouchard and Huneman 2013, p. 11]. We then turn to the beeswarm's association with materialism in the early modern context, as a kind of chaotic (Lucretian) multiplicity of forces, and then to the specifically eighteenth-century vitalistic development of this metaphor, in which it occupies center stage (now in an entirely positive, non-pejorative sense), and its philosophical (indeed, materialist-philosophical) reappropriation by Diderot, before concluding with some reflections on swarm intelligence, emergence, and attempts to devise swarm algorithms, in a marked focus on the swarm as a kind of mind versus as a kind of life (as in, e.g., earlier vitalist discussions).

2 Beeswarms: Order or Disorder?

The beeswarm offers the powerful image of a collection of "little bodies," as Samuel Purchas calls them, that have no existence outside the single organism they constitute. Comparing the single bee to "a drop of water" that "hath no power" compared to the seas, or to a "spark of fire" that lacks the destructive power of "the communion of many flames," Purchas describes the swarm as the only beneficial form of organization for the bee: "*una Apis, nulla Apis*, one Bee is no Bee, but a multitude, a swarm of Bees uniting their forces together, is very profitable, very comfortable, very terrible, profitable to their owners, comfortable to themselves, terrible to their enemies" [Purchas 1657, p. 16]. The single bee as a self-sufficient, autonomous part is not to be found in early-modern conceptions of nature. For Margaret Cavendish, the necessary gregariousness of the beeswarm illustrates the continuum of matter in nature, where inanimate and animate matter form "a single, continuous, self-subsistent organism" in which "self-subsistent atoms" [O'Neill 2013, p. 316] cannot exist:

> Nature would be like a beggar's coat full of lice: Neither would she be able to rule those wandering and straggling atoms, because they are no parts of her body, but each is a single body by itself having no dependence upon each other. Wherefore, if there should be a composition of atoms, it would be a body made of parts, but of so many whole and entire single bodies, meeting together as a swarm of bees. [Cavendish 2001, p. 129]

The various theories on the generation of bees that were popular until the second half of the eighteenth century reinforced the idea of a continuity between inanimate and animate matter, as bees were believed, like most crawling or flying insects, to swarm from the carcasses of dead animals or from "putrefaction" [Moffett

1658, p. 897].[13] The collective body of bees, variously described as "a large parcel," a "stock of bees" (John Thorley), or a "lump of Bees" (Joseph Warder) appears as shapeless as the cells of the beehive are regular and ordered. Cavendish, however, sees the swarm as a perfectly defined shape as the bees gather "in a round Figure or Globe, like the world; which shews the round figure is not only the most profitable, having the least waste, and largest compass, but the securest Figure, being the most united, not only by drawing in all loose and wandering parts, but combines them all together with a round Circle Line" [Cavendish 1656, p. 163].[14] Although the image of the harmonious orb is unusual, the physical unity of the swarm mirrors the unity of the community of insects, the close-knit throng functioning as a single-bodied organism. Influenced, no doubt, by Purchas and other authors of beekeeping treatises, Robert Boyle uses the example of the beeswarm to discuss the motion and cohesion of bodies that are firmer than fluids though not entirely solid:

> I have more than once taken pleasure to look upon an heap of swarming Bees, for though they make not up a liquid but coherent body, which may be turn'd upside down without losing its coherence, and which being beheld at a distance, seems to be one entire mass or body; yet it is evident to him that looks at them near enough, that the particular Bees that swarm have most of them their distinct and peculiar motions, and that yet these motions of the particular Bees destroy not the coherency of the heap; because that when one of the more innermost Bees removes, as she lets go her hold from those that she rested on before, and goes away from those that rested on her, so she meets with others on which she may set her feet, and comes under others that in like manner set their feet on her, and so by this vicissitude of mutual supports their coherence and their removes are made compatible; and if instead of Bees, the swarm consisted of extreamly little flies, their particular motions would perhaps be inconspicuous.[15]
>
> [Boyle 1772, p. 399]

However incoherent the movement of each individual corpuscle or bee may seem, the swarm is perceived as a coherent whole, moving in one single direction.

Cavendish also describes the swarm or "commonwealth" as "one body, or rather, all those little bodyes are as one great Head, or rather, as one Minde, and their influence united by a general agreement, as one Minde" [Cavendish 1656, p. 162–163]. John Thorley resorts to an equivalent comparison between the swarm of bees and the united body and mind to describe the uniform volition and concerted actions of the bees: "If Soul and Body are once separated, the Man dies [...] These little Creatures thus firmly knit together in sincere affection, and combination in Power, appear effectually secured from all Assaults and Enemies" [Thorley 1744, p. 14]. As early as the

13 [Worlidge 1676, pp. 4–5] confirms that bees "are sometimes engendered by putrefaction" and quotes Virgil's *Georgics*: "A wonder not to be believed, he sees/From the dissolved Entrails, Swarms of Bees." See: [Tunstall 2016, pp. 207–211; Baine Campbell 2006].

14 Baine Campbell links the circular shape of the bee colony to the round shape of the utopian space, see: [Baine Campbell 2006, p. 621].

15 On bees and particularly swarms in Boyle, Purchas, and Hartlib, see: [Schmidgen 2013, pp. 99–100].

seventeenth century, the fascination for the swarm was expressed not only as an interest for life and matter but also for swarm cognition and intelligence, a topic to which we will return in closing. The cohesion of the swarm is the condition of the bees' survival, for, as Cavendish points out, "they know that if the commonwealth be ruinated, no particular person can be free" [Cavendish 1656, p. 163]. Thorley also describes swarms as bodies of interdependent bees moved by the common good:

> They are a Body Politick, and cannot live separate and alone. A *Bee*, as a solitary Insect, and without her Company, is an insignificant, impotent, helpless, useless Creature; cannot work, nor propagate her species, nor secure herself from numerous Enemies and Evils. But what noble, excellent Purposes do they serve, thus incorporated; and by a social Spirit thus strictly united together jointly pursuing the Publick Profit and Advantage. [Thorley 1744, pp. 12–13]

The collective success of the commonwealth depends, for Thorley as for the majority of these authors, upon its cohesion and consistency.

3 Are Bees Royalists or Republicans?

The beehive itself was an ancient metaphor for political order (at least as far back as Plato, and Aristotle[16]). Indeed, whether used in a scandalous fashion like in Mandeville, or in a virtuous fashion like in many other authors, honeybees (*Apis mellifera*) were most often taken as exemplars of good, useful, productive social interaction – of industry – as was the beaver, and also of altruism: "The bee nation was perceived to cooperate as a group of comrade-members, a commonwealth in which the individual subsumes herself in the collective enterprise and good of the colony, and each bee acts for the general benefit of all the other bees" [Preston 2005, p. 53].

As naturally social insects, bees were systematically held up as an example for the human polity in the early modern context: "A Bee, like a man cannot live alone, if shee be alone, shee dies," writes Purchas who further describes bees as "political creatures" that "destinate all their actions to one common end," their commonwealth being comparable to the body "in the Fable of *Menenius Agrippa*" which died "when the rest of the members to ease themselves wronged the belly" [Purchas 1657, pp. 16–17]. Some authors also resort to a martial metaphor[17] to describe this apian society as a top-down

16 On Aristotle, see: [Aristotle 1965, 488a33 ff.]; cf. [Johach 2007] – although Aristotle also states that human beings are *more political* animals, than bees [Aristotle 1984, 1253a8].

17 On the origin of the bees as war-mongering soldiers, see: [Freeman 1980, pp. 186–188]. The martial metaphor or simile is also frequently used in medical treatises to describe the similar functioning of another form of swarming, that of the animal spirits. See, for example, Thomas Willis, who insists on the unity of the animal spirits in action as they "allways agree mutually between themselves, and intimately conspire together" and adds: "the Spirits contiguous one with another are set like an Army in Array" [Willis 1683, pp. 23–24]; see also: [Willis 1679, pp. 137–138].

organization placed under the authority of a unique leader. Those who were convinced that the swarm was under female rule, described the bees as fierce Amazons or compared them to the "Female troops of *Thrace*" [Dinsdale 1740, pp. 16–17]. Bees have to form "a disciplined army" [Arbuthnot 1728, p. 7],[18] driven by "mutual emulation" to maintain the indivisibility of the swarm, which is perpetually threatened by its own collapse or disunion. For some authors, their "thronging legions" [Dinsdale 1740, p. 23], described as "a numerous army, strongly entrenched, in which every common Soldier was a perfect Hero, that would sooner die than yield," taught by nature "to hate and abhor Riots, Tumults, Treasons, and Rebellions" [Thorley 1744, p. 10], do not form a self-governing commonwealth striving as a "single Minde" for the common good, but a regiment of "submissive bees" acting under the authority of one "Royal Mind" [Dinsdale 1740, pp. 16–17].

If for Cavendish "the general agreement" holds the swarm together, it does not imply that the organization of the commonwealth should be strictly horizontal and egalitarian. In his poem on beekeeping, Joshua Dinsdale reverses the Mandevillian trope from *The Fable of the Bees* and suggests that bees used to form an ideal republic and were imitated by men until men, unlike the frugal bees, became deluded by luxury and lost sight of the common good: "Thus Hives were first with Golden Honey fraught, / And their Republics sharpen'd human Thought, / Inspir'd, with Love of Public Good, Mankind, / 'Till Fraud and Luxury debauch'd the Mind" [ibid., p. 4]. By tradition, however, the political model associated with the commonwealth of bees is that of a monarchy, the anthill being, in nature, the incarnation of the republic. For Cavendish "the Ant and the Bee resemble more in their wise industry than in their government of the Commonwealth, for the Bees are a Monarchical government, as any may observe, and the Ants are a Republick" [Cavendish 1656, pp. 164–165].[19] But she further refuses to assign greater value to one or the other, and concludes: "The monarchical Commonwealth of the Bees is as wise and as happy as the Republic Commonwealth of the Ants" [ibid., p. 165]. For others such as Thorley, who underlines the difference between the commonwealth and the monarchy, only the ants live in "a Democracy, or Common-wealth, having no overseer or ruler" [Thorley 1744, p. 47]; bees, he insists, have instinctively formed into "a most compleat and admirable Monarchy, the original and most natural Form of Government" [ibid., p. 43]. In his *Feminine Monarchie*, Charles Butler stresses their natural abhorrence for both polyarchy and anarchy, and their creation by God to set for men the example of the "perfect monarchie, the most natural and absolute form of government" [Butler 1609, ch. 1, (A3)]. While it is tempting to contrast an eighteenth-century, Rousseauean

18 For the anonymous author of *The Congress of Bees* – Arbuthnot – the bees are more "valiant Warriors" and "puissant Heroes" [Arbuthnot 1728, p. 7] than Swift's Lilliputians or Dryden's Pygmies; see also: [Moffett 1658, p. 891].

19 On commonwealths of insects, and more particularly on ants and bees, see: [Hobbes 2012, pt. II, ch. 17].

reading of bee societies as radically republican and egalitarian, with earlier monarchist visions of the same,[20] in fact tensions between the two readings always existed. After all, Miguel Cervantes in *Don Quixote* had already described the beeswarm as a republic presented as the emblem of Arcadian abundance.[21]

The monarch ensures the unity and cohesion of the commonwealth of bees, but whether the ruler was a king or a queen was still a matter of debate in the eighteenth century, mainly because of conflicting theories regarding the generation of bees. As late as 1795, a curious episode concerning bees took place at the Revolution-era École normale in Paris (to be precise, in year III of the Revolutionary calendar, which spanned 1794–1795). The naturalist Louis Jean-Marie Daubenton was ridiculing Buffon's 'anthropomorphic' presentation of the lion as the 'king of the jungle,' when a particularly misogynistic student named Laperruque (literally, The Wig) jumped up and declared, "much worse than the king is the queen!," referring to the queen bee: "more extraordinary still, a queen within a republic" [Daubenton 1800, p. 92].[22] Lions, he continued, are simply feared by their subjects whereas bees are genuine courtiers, surrounding the queen and protecting her. Daubenton responded that the queen, who he preferred to term the "female bee," is just an egg-layer; the real power lies with the worker bees, who only respect the others (female and male) because they are needed for the reproduction of the species [Daubenton 1800, p. 93]. Daubenton nevertheless emphasized the role of the queen, including in the case of swarms: "the swarm would not subsist if it did not have any queen, that is to say, a female bee [...]. Whenever there is no female bee in a swarm, it soon goes back to its former hive" ("l'*essaim* ne subsisteroit pas s'il ne s'y trouvoit une reine, c'est-à-dire une abeille femelle [...] lorsqu'il n'y a point

20 Thus, Gaukroger notes that bees had traditionally been considered a model monarchical community (referring to influential texts such as [Butler 1609], [Purchas 1657], [Warder 1712], and [Thorley 1744], which we discuss here – but, as we note above, this was not always the case (it is in fact possible to see a tension between monarchical and republican 'readings' of beehives and swarms as always present, in beekeeping texts and in more explicitly political and/or theoretical texts) [Gaukroger 2010, p. 401, footnote 35].

21 "En las quiebras de las peñas y en lo hueco de los árboles formaban su república las solícitas y discretas abejas, ofreciendo a cualquier mano, sin interés alguno, la fértil cosecha de su dulcísimo trabajo"; in Smollett's eighteenth-century English translation, the bee republic became a commonwealth: "In clefts of rocks and hollow trees, the prudent and industrious bees formed their commonwealth, offering without interest, to every hand the fruitful harvest of this delicious toil" [Cervantes de Saavedra 1755, I, p. 54]. On bees in *Don Quixote*, see: [Brewer 2014, pp. 34–36; Cascardi 2012, esp. pp. 63 and 156].

22 Thanks to Stéphane Schmitt for this reference; cf. [Drouin 1992, p. 333]. D. Allen mentions a medieval charm against swarming that refers to the bees as feminine ("victor dames") [Allen 2003, p. 96n.]. Some authors of the period describe the queen as a 'mother bee' (on gender politics within the hive see: [Prete 1991]), while it is startling to see Maraldi, in his impressive report on bees to the Académie des Sciences in 1712, referring to the queen as "le Roy" [Maraldi 1712, p. 9]. As for the link between bees and lions, it is actually biblical (Judges 14:8), as discussed already by Moffett in the seventeenth century (see above).

d'abeille femelle dans un *essaim*, il revient bientôt à l'ancienne ruche" [Daubenton 1966 (1757), 994b]).

Jeffrey Merrick has called attention to the role of the microscope in the evolution of the theory regarding the gender of the apian monarch, as it unraveled the mode of reproduction inside the hive, and with it its "moral and political perversity" [Blum 1988, p. 5; Merrick 1988]. Some authors such as Butler (*Feminine Monarchie*), Warder (*The True amazons: or, the monarchy of bees*), or Thorley (*The Female Monarchy*) displayed their opinions on the subject in a programmatic title, the natural benevolence of the queen bee confirming the "just and kind" exercise of absolute sovereignty while ensuring the remarkable fertility of the commonwealth.[23] For others such as Moses Rusden, bees are governed by a male king who fertilizes the "animable matter" disposed in the cells, while the honeybees "are the Female, but not by virtue of any Copulation or Conception, but because they supply the place and Office of the Female" [Rusden 1685, p. 42]. King or queen, the monarch of the beehive is not presented as a brutal autocratic ruler (however prone to wage war on other insects) but as a benign sovereign, equipped either with an innocuous, poisonless sting, which is to be understood as a symbol of its authority, or with a lethal sting it only uses against enemies or rival princes. The physical characteristics of the monarch are natural elements of distinction that set it apart. Thomas Moffett reports in his seventeenth-century *Theatre of Insects* that according to some "curious searchers into the work of nature," the 'matter' bees are made of conditions their social role and behavior:

> The best and noblest bees are generated and bred out of the Lion, and the Kings and Princes of them do derive their pedigree and descent from the brain of the Lion, being the most excellent part of his body: it is no wonder therefore if they, proceeding and coming from so generous a flock, do assail the greatest beasts, and being endures with Lion-like courage, do fear nothing.
> [Moffett 1658, p. 891]

The "king of the whole swarm," which, for Moffett, is elected by the other bees for its natural superiority, is "always of an excellent shape and twice as big as any of the rest." It is all the more exceptional, since it is "not begotten a little worm at the first, as the Bees are, but presently able to fly" [Moffett 1658, pp. 891, 892]. While Rusden's monarchical hive is ruled by "a natural King, and set apart by Nature herself" [Rusden 1685, p. 17] and monarchy is passed on to the royal heirs, for Moffett the monarch's election guarantees the unity of the *civitas*: "Bees are swayed by Soverainty, not by Tyranny, neither do they admit of a King properly so called by Succession or by lot,

23 For Mary Baine Campbell, treatises on beekeeping were willing "to distort their closely observed accounts of bee sociability in the service of maintaining the power of the bee 'polity' to analogize, and thereby to authorize, prevailing norms such as gender hierarchy in government, the superior usefulness of male labor, or the chastity and monogamy of women" [Campbell 2006, p. 622]. On the gender of the apian monarch, see Prete [1991]; and on the hive as a model for human polity, see: [Campana 2013; Merrick 1988].

but by due advice, and circumspect choice; and though they willingly submit to regall Authority, yet so, as they retain their Liberty; because they still keep their prerogative of election" [Moffett 1658, p. 892]. Purchas offers a discordant vision of the nature of monarchy in the apiary. All the actions of those "political creatures" have "one common end" and they all "aim at the publick welfare" [Purchas 1657, p. 34], but they strive "under one Commander, (who is not an elected Governour" [ibid.]. For Purchas, who also imagines "an Amazonian Commonwealth" governed by a "Queen Bee"), notwithstanding the fact that the swarm is unanimously preoccupied by the common good in all its actions, the bees cannot be entrusted with the choice of their ruler "for the vulgar often want judgement, raising the worst and wickedest to the Throne" [Purchas 1657, p. 16].

However indispensable to the general happiness and productivity of the hive, and more importantly to the political cohesion of the commonwealth, the king or queen can also be the cause of its division. The ruler's untimely death would lead to "the certain and total Destruction of the State" [Thorley 1744, p. 7], overturning the natural order of the commonwealth and threatening the regeneration of the swarm: "But the King being dead, the subjects are perplext, the Drones lay their young ones in the Bees cells, and all things are out of order" [Moffett 1658, p. 892]. Having no king or queen would lead the community of bees to its collapse, but more than one monarch would be equally dangerous. Faced with conflicting sources of authority, the bees would no longer be able to perform their respective duties: "under two or more Kings they will not be, for they will neither work, nor breed, nor be quiet" [Rusden 1685, p. 16]. For Moffett, however, the bees will protect the "Master Bee" at the peril of their lives as the indispensable protector of the indivisibility of the swarm, but only insofar as he is not tempted to yield to his own selfish and tyrannical passions: "much less do they put him to death, unless as tyrants are wont he makes his lust the rule of his Government, or being negligent of the Common-wealth, takes no care of it" [Moffett 1658, p. 893].

4 Swarming

In some cases, the metaphor of the swarm is strongly opposed to that of the hive. Yet as we discuss below with respect to the specifically 'biological' uses of the beeswarm metaphor, both images illustrate the same general idea of organismic unity.[24] Granted, bees swarm when it is time to leave a hive and found a new one, and in that sense, the swarm is a transient state, but that does not mean that it has nothing to do with organismic unity, on the contrary. In the early modern discussions, swarming – the planned or accidental division of the *civitas* into several colonies – is seen as happening either when a younger prince tries to rise to power or when the hive becomes too populous

24 On metaphors of organism overall, see: [Schlanger 1971].

and the threat of sedition arises. Joseph Campana points to the fact that the division of the close-knit commonwealth is prompted neither by gender issues nor by class conflict; the new swarm is formed under the pressure of the younger bees eager to cluster around a new ruler [Campana 2014, p. 72].[25] Dinsdale describes swarming as the bold move of conquerors extending the boundaries of an empire:

> When now the Hive too populous appears, / And the wise Regent a sedition fears, / She strait exhorts the Youth their Lot to try, / [...] And the Foundation of an Empire lay. / [...] Strait little Clangors give the shrill Alarm, / And animate the young advent'rous Swarm; / They flock their new-elected Queen around, / And pant for Glory at the martial Sound.
>
> [Dinsdale 1740, pp. 22–23]

The division of the apian organism through swarming is comparable to cell division as it produces a new organism (a new swarm clustered around a new monarch) and also extends the more shapeless body of bees over an ever-growing territory, extending the apian dominions over some other "Rock or clefted Tree" [Dinsdale 1740, p. 22]. Swarming is a driving force, but it also threatens the commonwealth with partial or complete destruction if the new monarch is lost on the way or if the old monarch is opposed to the departure of some of its subjects (an opposition which for Moffett justifies regicide).

Even when it is ruled by a benevolent monarch, and however inspired by the common good, the industrious swarm as a 'body politick' is paradoxically further threatened by its own parts. The general idea of a harmonious community of selfless bees put forward by all the early-modern authors of bee-keeping manuals hides a more divided social model, which harbors selfish passions and other vices. The swarm is in fact not exactly as uniform as it would seem, and besides the hard-working, disinterested honeybees and their respected king or queen, another kind of bee, the male bee or drone, is an element of potential disruption. Conspicuously absent from Dinsdale's ideal swarm, the drone was, as Campana observes, "a central node of anxiety" for many of the authors we discuss [Campana 2014, p. 71]. For Cavendish, they are human-like bees that do not aim at the common good but take advantage of the work of others: "Men are not like Beasts, to Work for a General Profit, but like Drones, to Rob the Particular Labours of the Commonwealth."[26] Drones seem

25 Campana quotes Thomas Hill, who explains swarming as a necessity for the younger bees: "for such is the nature and propertie of the bees, that assoone as the swarme of the yong bees are bred with the kings, and they be strong & able to flie away, then as disdaining the swarms of the old bees, they seeke the more government. For they be such living things as delight to rule alone, not seeking aide or counsel of the elder bees" [Hill 1608, p. 14].

26 And Cavendish adds: "neither is it amongst Mankind as amongst Beasts, for amongst Beasts there are more Bees than Drones, but amongst Mankind there are more Drones, as I may say, than Bees, that is, there are more Unprofitable, than Good Commonwealths men" [Cavendish 2012, letter 205, pp. 220–221]. For Cervantes also, if the republic of bees in general is a model for the human polity, the drones are the paradoxical apian incarnation of human passions within the hive: "For I would have you know, my friends, that your idle and lazy fellows are the same in a commonwealth as drones in a bee-to form a community of their own, although they are indispensable to reproduction either because their role is, according to some authors, to impregnate the queen or because they heat the hive by their presence alone, allowing the young bees to grow. Rusden presents them as the defenseless victims of the ruthless honeybees but insists on the fact that "no colony or stock can well strive without them" [Rusden 1685, p. 8].[27] While Purchas seems to consider them idle but useful, for the vast majority of authors their usefulness is questionable. For Samuel Hartlib, their presence is endured until swarming takes place, but their supposed role is not even worth mentioning: "There is in every Hive a great number of attendants somewhat larger than the Bees (we call them Drones) which are fed by the labours of the Bees as long as they prepare for Swarming; but as soon as the Bees resolve to send out more Colonies, they fall upon the Drones and kill them" [Hartlib 1655, p. 13].[28] Moffett differentiates between older and less productive honeybees – even those who "can do nothing at all" – and drones: "for they do not as they do, spoyl the Combs and steal the Honey" [Moffett 1658, p. 893]. In the first decade of the seventeenth century, John Day imagined a short play in which the bees express their grievances in the Parliament presided over by Prorex, the Master-Bee. They complain of several insects personifying weaknesses or moral flaws, such as Parcimonious, the "thrifty bee" or Pharmacopolis, the "quacksalver." Various petitions against mortal enemies of the hive are also presented to the magistrate. Among these enemies is "the surly Humble Bee" who "will neither pay / Honey nor Waxe, doe Service, nor obey," and "by Stealth / Makes dangerous inroads on your common-wealth, / Robs the day-labourer of his Golden Prize," but also "the lazie Drone":

> Our native Country Bee, who like the Snaile
> (That bankrowt-like makes his own Shell his Jayle
> All the day long) Ith' evening Plaies the thief,
> And when the laboring Bees have tane reliefe,
> Begone to rest, against all right and lawe
> Acts Burglary, breaks open their house of straw,

hive, that consume the honey which the industrious labourers have made" ("Porque quiero que sepáis, amigos, que la gente baldía y perezosa es en la república lo mesmo que los zánganos en las comenas, que se comen la miel que las trabajadoras abejas hacen" [Cervantes de Saavedra 1755, II, p. 301]).

27 About the drones, he writes: "they have neither stings nor fangs, which is the reason why they are so easily beaten and killed by the Honey Bees, having no Weapons offensive or defensive" [Rusden 1685, p. 7].

28 (This is a collection of letters and notes on bees, only some of which are by Hartlib himself.) Moffett also emphasizes the violence that characterizes life in the orderly hive, in which deformity and imperfection cannot be tolerated even among the workers themselves: "And if the chance to find among young ones any one that is a fool, unhandsome, hairy, of an angry disposition, ill-shapen, or naturally ill-conditioned, by the unanimous consent of the rest, he gives order to put him to death, lest his souldiery be disordered, and his subjects being drawn into faction, should be destroyed" [Moffett 1658, p. 892]. The unity and cohesion of the swarm is also conceived as resting upon the uniformity and homogeneity of its parts.

> And not alone makes pillage of their hives,
> But (Butcher-like) bereaves them of their lives.[29] [Day 1641, n.p.]

As opposed to the humble bee or to other alien foes such as the "cruell waspe," the drone is the enemy from within, like a cancerous cell set on destroying the organism that harbors it. As any living organism would oppose a tumor, the worker bees turn against the drones and kill them to prevent the self-destruction of their commonwealth. The process is thus described by Moffett:

> The Dors also and Drones they kill as often as they want room for their works (for they take up the innermost part of the Hive) and take away from them both their honey and their victuals. As also when their honey fails and there is a dearth then they go pell mell among themselves, and fight as it were for life and bloud, the short bees they fall upon the long, the smaller sort set upon the Drones (as idle and unprofitable). [Moffett 1658, p. 894]

The conclusion of these intestine wars is unpredictable, but for Moffett the annihilation of the drones and the corollary survival of the shorter bees will yield "an excellent swarm" [Moffett 1658, p. 895].

Some of these tensions between republican and monarchical readings of the bee collective also reflect different attitudes toward the question of instinct, which we have not discussed here. For most of the texts surveyed above tend to present the hive as acting instinctively for the common good, but on the other hand it also requires a benevolent and just monarch, otherwise the entire collectivity qua system collapses (implying that queen bees can be 'bad' and that without a chief whip, bees can follow something else than the common good, despite their instinct for what ensures the common good).

In the vast majority of texts on beekeeping and texts that use the commonwealth of bees as a model for human polity, the swarm or hive as a steady and uniform body preoccupied by the common good always contains the ferments of its own ruination or obliteration. To have more than one monarch is a threat, but ambitious princes trigger swarming and therefore the regeneration and extension of the species. Drones are useless thieves, but by their very presence they allow the newly hatched eggs to grow into honeybees. In the *Fable of the Bees*, Bernard Mandeville's "grumbling hive" takes after this long and consistent tradition [Mandeville 1924 (1714, 1729)]. Mandeville uses this rich literature on bees and plays with its internal contradictions. His hive full of knaves works toward to the common good, despite the cracks in its close-knit unity. For all its violence, for all its cheating and honey-gorging individuals, the Mandevillian hive is like the hives described by Moffett and others: a coherent community within which private vices do mean public benefits.

---

29 Fairfax Withington, points to the fact that pre-republican America identified the unproductive drone with British custom officials who came in swarms to "destroy crops" and feed on "the productivity of the colonies" [Withington 1988, p. 44].

5 Disturbing Swarms: Beeswarms and Materialism

We have already seen how beeswarms served as exemplars of different kinds of order and organization and, in authors such as Cavendish, how they could also function as heuristics (i.e., heuristically charged metaphors) for more abstract considerations of unity and disunity, mereology and overall the identity of what one might call 'dynamic wholes,' powered by the constant interaction of their living parts, and the emergent properties thereof. They are never just something merely natural, or "simply there," as Donna Haraway remarks about a very related notion, that of organism: "Politically and historically, I could never take the organism as something simply there. I was extremely interested in the way the organism is an object of knowledge as a system of the production and partition of energy, or as a system of division of labour with executive functions" [Haraway 2006, p. 136]. But swarms could also be seen as disturbing, as is still conveyed in the German expressions *Schwärmen* and *Schwärmerei*. We have already mentioned Swift's sharp dismissal of materialist claims concerning the inherence of mind, soul or self in such an organ as the brain, *using the image of the beeswarm to hammer in the impossibility of such inherence*: the unity of thought, consciousness, selfhood, self-awareness, and so on are irreconcilable with the idea of the brain as a mere "swarm of Bees" [Swift 1801, pp. 263–264].

Samuel Clarke had already argued in familiar post-Cartesian terms that the soul cannot be material because consciousness is indivisible, while matter is divisible, but Clarke also pointed to what we might call the problem of *biological-personal identity*. He suggests that we imagine "three or three hundred Particles of Matter, at a Mile or any given distance one from another; is it possible that all those separate parts should in that State be one individual Conscious Being?" Suppose then that "all these particles" are brought together "into one System, so as to touch one another": will they thereby, "or by any Motion or Composition whatsoever," become "any whit less truly distinct Beings, than they were when at the greatest distance? How then can their being disposed in any possible System, make them one individual conscious Being?" [Clarke 1731, pp. 23–24]. In later eighteenth-century reactions against materialism, this is repeatedly heard in different variations. For the Abbé Nicolas-Sylvestre Bergier, if the brain is a mere "heap of molecules of matter," how could "a simple and indivisible act such as willing be attributed to the brain?" Bergier rejects the key organismic premise of the beeswarm metaphor without mentioning it specifically, stating that a collection of wills or minds can never amount to one will or mind [Bergier 1771, p. 176]. In his very visible and influential Boyle Lecture (the first of the series) against atheism, Richard Bentley had specifically rejected materialist ideas of vital, embodied matter with reference to the beeswarm:

> a great number of... living and thinking Particles could not possibly by their mutual contract and pressing and striking compose one greater individual Animal, with one Mind and Understanding, and a Vital Consension of the whole Body: *anymore than a swarm of Bees, or a crowd*

> *of Men and Women can be conceived to make up one particular Living Creature compounded and constituted of the aggregate of them all.* [Bentley 1693, p. 13 (emphasis S.K/C.W.)]

Unlike Samuel Clarke and Bergier, Bentley is not just worried about chaos and the loss of the unity of the person; he is also concerned, somehow, with what will become a positive, ideologically neutral claim in the medical vitalists we discuss in the next section, and then again a 're-materialized' claim in Diderot (as we discuss subsequently): that each parcel of organic matter (or matter in general) could come to be a 'life' of its own: "Every Stock and Stone would be a percipient and rational Creature. We should have as much feeling upon the clipping off a Hair, as the cutting off a Nerve" [Bentley 1693, p. 13]. Seen from the standpoint of the early twentieth century and beyond, this vision of the beeswarm is ironic, given that it became instead the image – a living argument, as it were – for *organicism rather than materialism*! (We briefly discuss the work on 'social insect collectives' and the idea of 'superorganisms' associated originally with the entomologist William Morton Wheeler in closing.) But the first step in that direction was to reconfigure the beeswarm as a core metaphor for organic unity, in other words, for the identity of organisms as distinct from mere mechanisms, heaps of matter, or other assemblages lacking the newly discovered and discussed features of organic interconnection and thus self-organization.

6 The Beeswarm as Scientific Metaphor: Vitalism

We now come to the crucial passages concerning the beeswarm in a vitalist context, or rather, more broadly, in a context in which it serves the purpose of metaphorically expressing the structural and relational feature(s) of organismic unity. The context shifts in that the swarm is no longer discussed in terms of sociopolitical order (or disorder), nor as a source of anguish regarding materialism and disorder, but rather as a legitimate heuristic for biological and medical thought broadly construed. It is first used in this sense (as *grappe d'abeilles* or *groupe d'abeilles*) by the famous naturalist Pierre Louis Moreau de Maupertuis, Secretary of the Berlin Academy of Sciences, in his *Système de la nature* (1751), then in a series of writings from the medical thinkers we now know as the 'Montpellier vitalists'[30] (notably by Théophile de Bordeu

30 The term 'Montpellier vitalists' is generally used to refer to the group of physicians and professors of medicine (but also anatomy, botany, etc.) at the Faculty of Medicine in Montpellier, beginning in the mid-eighteenth century; the term "vitalist" was applied to this group since at least the 1790s, and indeed served as a self-description during those decades [Rey 2000; Williams 2003; Wolfe and Terada 2008]. Significant figures of this school include Louis de La Caze (1703–1765), Théophile de Bordeu (1722–1776, also known for his appearance as a fictional character in Diderot's *Rêve de D'Alembert*), Henri Fouquet (1727–1806), Jean-Joseph Ménuret de Chambaud (1739–1815),

and Jean-Joseph Ménuret de Chambaud), and finally in Diderot, from the mid-century to the late 1760s. The basic intuition centered around a way of describing the organism (or 'animal economy,' in the period's vocabulary) that did not treat it strictly in mechanical or mechanistic terms, but rather, as a whole formed of parts which have, or *are*, independent lives. With Diderot, the beeswarm is transposed from 'vitalist medicine' to 'materialist philosophy,' in his then-unpublished *Le Rêve de D'Alembert* (1769) [Duflo 2003; Wenger 2012, p. 40].

In his *Système de la nature*, which bears the more informative subtitle *Essai sur les corps organisés,*[31] Maupertuis gives the shortest and also simplest statement of the beeswarm metaphor. His concern is how to illustrate the dynamics whereby organisms are formed out of various organs or parts, and this is where the metaphor comes in. How is it that a body is composed of thousands of elements that somehow 'knew' how to place themselves in the right position in the course of embryonic development? Like in the case of "an army seen from a distance," which "might appear to us as a great animal" (an ancient image, and one which Boyle had also applied to the beeswarm, as we saw above), similarly, "a bee-swarm, when the bees are assembled and united on the branch of a tree, only presents to our gaze a body lacking any resemblance to the individuals which composed it." The outward appearance of a beeswarm leads us to disregard that it is composed of thousands of tiny insects [Maupertuis 1756, pp. 154–155].

Bordeu makes much more extensive use of the image, in his most celebrated work, the *Recherches anatomiques sur la position des glandes et leur action* (also from 1751), in a section revealingly entitled "How to understand the action of all the parts, their departments, and their periodic motions" (§ CXXV). After a long analysis of the relations between the "general" circulation and "particular" (or "specific") circulation, which Bordeu describes in Hippocratic terms as being like "small circles which gradually form a larger one,"[32] as well as between different types of blood vessels, raising issues of 'communication' between parts,[33] Bordeu acknowledges that he has to resort to a metaphor (he initially says a "comparison"): that of a cluster or

the – mainly unacknowledged – author of many important medical entries in the *Encyclopédie*, and, perhaps most famously, Paul-Joseph Barthez (1734–1806) in the later eighteenth century.

31 This text first appeared in Latin in 1751 (supposedly in Erlangen – actually in Berlin) under the title *Dissertatio inauguralis metaphysica de universali naturae systemate*, with the pseudonym Dr. Baumann; Maupertuis translated it into French in 1754, now with a more explicitly 'biological' title, as *Essai sur la formation des corps organisés* (with a pseudonymous translator's name); it was later included in his 1756 *Œuvres* under the title *Système de la nature*.

32 "I have customarily used the term 'circle' to convey that a part, even if it receives blood by means of the general circulation, as occurring in the largest vessels, nevertheless has a particular circulation, depending on whether it is in action or not; the other parts which 'feel' this action, belong to its department, its circle, etc." [Bordeu 1818, I, p. 187].

33 "The least part should be considered as 'a body apart', so to speak. True, it acts by means of the general circulation, but it is as distinct as the system of blood vessels is distinct from the chiliac swarm of bees. As the beeswarm arguably represents the single most condensed expression of the animal economy in Montpellier vitalism, it is worth quoting it at length:

> Might I make use of a comparison which, however rough, may be useful?
>
> I compare the living body, in order to properly estimate the particular action of each part, to a swarm of bees which cluster together [*se ramassent en pelotons*], and hang from a tree like a bunch of grapes; I find the image suggested by an ancient author, that one of the lower organs was an *animal in animali*,[34] to be quite helpful. Each part is, so to speak, not quite an animal, but a kind of independent machine [*machine à part*] which contributes [*concourt*] in its way to the general life of the body.
>
> Hence, following the comparison to a bee swarm, it is a whole stuck to a tree branch, by means of the action of many bees which must act in concert to hold on; some others become attached to the initial ones, and so on; all concur [*concourent*] in forming a fairly solid body, yet each one has a particular action, apart from the others; if one of them gives way or acts too vigorously, the entire mass will be disturbed: when they all conspire to stick close, to mutually embrace, in the order of required proportions, they will comprise a whole which shall endure until they disturb one another. [Bordeu 1818, I, p. 187]

Bordeu then tries to spell out the literal correspondence of the image: the interconnection of the bodily organs, the way each organ has its "district" and its "action"; importantly, he adds that "the relations between these actions, the resulting harmony, is what *makes health*" [ibid., emphasis S.K/C.W.]. Disturbance in this relation between the parts is what constitutes illness, of varying severity. In other texts, Bordeu (as well as other Montpellier vitalists, notably Ménuret and Henri Fouquet) spoke in related terms of the different organs not as mere 'parts' but as 'little lives.' In his work on 'chronic illnesses,' Bordeu does not specifically use the beeswarm image but presents three "theorems" which describe the living body as "an assemblage of several organs, each of which live in their own way, which feel more or less, and move, act or rest in fixed times; for, following Hippocrates, all parts of animals are animate" [Bordeu 1818, II, p. 829]. He also explains that the organs are "expansions of the

vessel system, or as the circulation of the lung and the liver are from what occurs in ordinary large vessels" [Bordeu 1818, I, p. 187].

34 [One might say 'has a life of its own.'] Bordeu notes, regarding the expression *animal in animali*, that the ancients already held that each part of the body had a particular form of action [Bordeu 1818, I, p. 188]. Fouquet refers to Galen on the theme of each organ having its own "life," and adds that other ancient authorities, including Plato, compared the liver, among other organs, to an animal contained in another animal; an image that Van Helmont applied to the uterus [Fouquet 1802, p. 78, n. 4]. Cf. [Rey 1997, pp. 137–138]. Harvey uses the same expression, for the womb, in his work on generation: "as if the Womb were Animal in Animali, one living creature in another; and had a peculiar independent motion of its own" (the expression is also used in the original Latin text) [Harvey 1653, § 68, p. 415]. Diderot also speaks of a particular organ, the eyes, as "un animal dans l'animal" [Diderot 1975-, vol. 17, p. 500] and then states more generally that man can be understood as "an assemblage of animals in which each one retains its peculiar function" [ibid., p. 501], describing the relation between organic parts in terms of 'sympathy.'

nerves," and – in terms already familiar to us – defines what he calls "general life" as "the sum of all the particular lives"; "all of the parts are both causes, principles and final causes" [Bordeu 1818, II, p. 829]. The parts of an organism are constituted by a stable interaction, not just of inanimate parts but of 'lives,' that is, of *organs considered as lives*, like individual bees in the collective swarm.

Ménuret describes movement and sensation as two basic properties which exist in modified forms in every organ; "they give rise to a corresponding number of *particular lives*, the whole of which, in concert [*concours*] and mutual support, form the general *life* of the body"; Fouquet says that "each organ senses or lives in its own way, and the agreement [*concours*] or sum of these particular lives comprises life in general, just as the harmony, symmetry and arrangement of these little lives comprises health" [Ménuret 1765b, *Enc.* XI, 361b; Fouquet 1765, *Enc.* XV, 42b]. In his article "Observation," Ménuret refers to both Bordeu's and Maupertuis' works, noting that they were published in 1751, but that Bordeu's has priority, having been authored in 1749. He mentions the beeswarm and Bordeu in order to emphasize that life in the body occurs, or is best described as, a "connection of actions" (*liaison d'actions*). After criticizing earlier medical commentators for failing to notice the interconnectedness of organic phenomena in the living body, he makes explicit use of our key metaphor, and explicitly praises the two earlier authors for having first introduced it:

> One could, following these authors, compare man to a flock of cranes which fly together, in a particular order, without mutually assisting or depending on one another. [*In contrast,*] The Physicians or Philosophers who have studied and carefully observed man, have noticed this *sympathy* in all animal movements – *this constant and necessary agreement in the interaction of the various parts, however disparate or distant from one another*; they have also noticed the disturbance of the whole that results from the sensory disagreement of a single part. A famous physician (M. de Bordeu) and an illustrious physicist (M. de Maupertuis) likewise compared man, from this luminous and philosophical point of view, to a *swarm of bees which strive together to hang to a tree branch. One can see them pressing and sustaining one another, forming a kind of whole (une espèce de tout), in which each living part contributes in its way, by the correspondence and direction of its movements, to sustain this kind of life of the whole body*, if we may refer in this way to a mere connection of actions (*liaison d'actions*)[35]
>
> [Ménuret 1765a, pp. 318b–319a (emphasis S.K/C.W.)].

Without wanting to read into this text metaphysical considerations which are foreign to it (whether issues of mereology, the idea of an ontology of relations, or that of structural realism), it is the case that in these discussions of "sympathetic relations" between parts (organs), the *materiality* of the relations is recognized. It is not

---

35 In her discussion of the political uses of the metaphor of flocks of cranes, Kalff reminds us that Aristotle had listed, in addition to bees, wasps and ants, cranes as social animals [Kalff 2014, p. 437]. Wallmann notes that in Bodin, bees and cranes are also used together as examples, but in a monarchical sense, more or less opposite to the vitalist usage, while Kalff cites various Renaissance authors for whom the crane was a model of republican equality [Wallmann 2017, p. 147].

just a matter of different entities communicating with one another in the void, so to speak. In this passage in particular, the cranes are presented as merely contiguous: they fly together "without mutually assisting or depending on one another," which is to say, without crucial interconnecting phenomena such as 'sympathy.' Interestingly, Ménuret uses more mechanistic – or mechanism-friendly – language than his peers do, speaking of connections, movements, pressure, support, agreement in relation between parts, and so on. Whether the term used is 'metaphorical,' like the beeswarm, 'technical,' like that of 'organic sympathies,' or somewhere in between the two, like the 'circle of action,' we can see that Ménuret and Bordeu are trying to articulate a structural, relational concept of interaction among living parts ("lives") which does not rely on strictly linear causality – in other words, that is not strictly mechanistic, although (contrary to claims by some earlier interpreters who rely on a somewhat facile opposition between mechanism and vitalism) it is also not *anti*-mechanistic: one might see their focus on 'structure' as a kind of 'expanded mechanism.' This is also shown by the frequent usage of the Hippocratic maxim, "everything concurs, consents and conspires together in the body" [Ménuret 1765b, 363b].[36] The forces and actions of the animal economy are too intimately intertwined to be quantified according to purely mechanical laws of force and motion.

Returning to the *grappe* or *groupe d'abeilles*, we are tempted to ask what kind of concept it is; it is an 'image,' of course, but one which its authors clearly intend as encapsulating their speculative and practical efforts. Bordeu himself is aware of the difficulty and asks "to be allowed a metaphor," when dealing with forces that govern "a thousand singular motions in the human body and its parts," given that "we do not even know which terms to use to describe certain motions in plants or properties of minerals." He admits at the end of this passage that he can only provide "a way of conceiving things [*une manière de concevoir les choses*], metaphorical expressions, and comparisons" [Bordeu 1818, I, p. 163]. The status of metaphors in scientific investigation has gone through considerable changes (most prominently due to Mary Hesse's insistence on their importance); leaving outside the outright denial that metaphors have any 'purchase' on scientific reality, one can summarily distinguish between three views on their role in science, in increasing strengths: (i) they play a weak heuristic role, (ii) they contribute to theory construction and thus play a stronger heuristic role, and (iii) they are equivalent to models, which contain various possible analogies to be investigated.[37] We might say that for Bordeu et al., the beeswarm is a scientific metaphor in sense (ii); for Diderot, it reaches the status of an ontological claim itself, within which further theories can be investigated, hence like sense (iii).

36 'Sympathy' is often used in connection with terms such as 'cohesion,' 'conspiration,' and 'consensus' in these texts.

37 This useful way of distinguishing different senses of metaphor in science is suggested in [Petkov 2015], referring respectively to Ortony, Black, and Hesse.

Ludwig Feuerbach understood Leibniz's theory of monads (as nested individuals) on the model of the beehive – an image which is not used by Leibniz himself. In his 1837 *Darstellung, Entwicklung und Kritik der Leibniz'schen Philosophie* (*Presentation, Development, and Critique of the Leibnizian Philosophy*), he wrote that the body, which the monads "bring together and hold together," is the beehive, while the dominant monad is the queen or mother bee. Building on the kind of organismic intuition we have seen developed in Bordeu, Maupertuis, and Ménuret (none of whom credit Leibniz in those passages – although Bordeu does so elsewhere[38]), Feuerbach adds that the bees "do not live in such a loose connection as the beasts of a herd; they constitute *one* whole; every individual bee is to be seen as just one member of this organism, having only a partial life [*Theilleben*], a particular function, like an organ in my body"; yet every bee is nevertheless "an individual in itself, a particular being that stands on its own legs" [Feuerbach 1837, p. 86, cit. in Smith 2011, p. 140].[39]

Related images of organic unity can be found in non-vitalist authors of the period, but not as sharply focused as the beeswarm in Bordeu and Ménuret. Thus, the well-known Geneva naturalist Charles Bonnet reintroduced sociopolitical language to describe the tree as an autonomous "organic society": "an assemblage of a multitude of subordinate organic productions, tightly connected to each other, all participating in a shared life and needs, yet each of which also has its own life, needs and functions" [Bonnet 1769, p. 164].[40] He continues by describing how all of these individuals work toward the "common good" of this society, that is, the tree, while at the same time seeking out their own personal good – this seems to echo Mandeville, but as we noted earlier, in this respect Mandeville was only taking up, albeit with much more talent and biting irony, a model (private vices/public benefits) that was in fact there in most beekeeping texts from the beginning. Across the Channel, Samuel T. Coleridge wrote, in his unfinished "theory of life," of a "tendency to individuation" characteristic of the way life unites the parts of a body, life being defined "as *the principle of individuation*, or the power which unites a given *all* into a *whole* that is presupposed by all its parts" [Coleridge 1848, pt. III, p. 42], but in fact stresses this "*élan vital*"-like character of striving and unification more than the Montpellier vitalists do.

The understanding of organic individuality in the vitalist authors surveyed above does not treat such individuality (also known as 'specific modes of organization') as a thing but as a 'system,' a dynamic relation between individual vital centers (the little 'lives') which are interrelated by means of 'sympathy,' 'consensus,' 'conspiration,' and so on, that is, various forms of reciprocity, in a 'circle of action.' As Elisabeth Wallmann

---

38 Bordeu named monads (along with Buffon's organic molecules) in his list of the main "hypotheses on the elements of bodies" [Bordeu 1818, II, p. 925]. The "fortunes" of Leibniz's monadology read as a kind of biological theory in the eighteenth century is another story.

39 Thanks to Justin E.H. Smith for this reference.

40 See also: [Citton 2006, ch. 6, on individuation].

notes, while naturalists such as Réaumur or the writers of beekeeping manuals defined the swarm as "the temporary and highly unstable formation of a collective of bees as they searched for a new home," medical theorists such as our vitalists emphasized instead "the way in which the swarm seemed to form a new bodily unity irreducible to the insects that formed it. The swarm-body, they argued, mirrored the economy of the human body, similarly composed of seemingly independent parts that became one with the body as a whole" [Wallmann 2017, p. 117]. But authors like Richard Bentley who feared the materialist implications of the beeswarm – indeed its specifically *vital*-materialist implications – may have felt vindicated in the end because after this vitalist treatment in which the beeswarm lacks disturbing ontological and/or political overtones, its appropriation and transformation at the hands of Diderot, in what many consider to be his masterpiece, *Le Rêve de D'Alembert*, is a key moment of radical materialism.

7 Unsystematic Vitality: Diderot's Vital-Materialist Beeswarm

First, a brief introduction to the context in which our core image appears in Diderot's work. *Le Rêve de D'Alembert* contains, in an unusual, experimental prose form, some of Diderot's most important thinking at the intersection between metaphysics and the newly emerging life sciences. The work remained unpublished for many years after his death, and was given as a gift in manuscript to Catherine the Great. It is, famously, composed of three dialogues, each of which features characters named after real, living figures of the time. The first dialogue, between Diderot and Jean Le Rond d'Alembert, covers traditional philosophical issues such as self and world, matter and thought, the existence of God, sensation and the true properties of objects. The second, longest and central dialogue involves the somnolent D'Alembert, the doctor Bordeu (who, in an earlier draft, Diderot had named La Mettrie), and Mlle de Lespinasse. It contains the image of the beeswarm, which belongs to the part in which Lespinasse reports D'Alembert's apparently incoherent dream utterances. The third dialogue is shorter again, and involves only Doctor Bordeu and Mlle de Lespinasse discussing certain issues from the earlier dream discussions, including monsters considered as both biological and social problems, the relation between matter and sensation, and the nature of biological reproduction with explicit attention to its sexual dimension.

The image of the beeswarm comes after a long description of chemical concepts which Mlle de Lespinasse reprises from D'Alembert's 'ravings.' This description concludes with a reference to a type of "unity" that "only exists in the animal," a type of "action and reaction" which binds the parts together. At this point, D'Alembert cries out (or rather, it is reported that he cried out):

> Have you ever seen a swarm of bees leaving their hive? [...] The world, or the general mass of matter, is the great hive [...]. Have you seen them fly away and form a long cluster of little winged animals, hanging off the end of the branch of a tree, all clinging on to each other by their feet? [...] This cluster is a being, an individual, some sort of animal [...]. If one of these bees decides to pinch somehow the bee it is clinging onto, do you know what will happen? [...] [T]his one will pinch the next one; [...] as many pinching sensations will arise throughout the cluster as there are little animals in it; [...] the whole cluster will stir, move and change position and shape [...]. [S]omeone who'd never seen the formation of a cluster like that would be tempted to think it was a single animal with five or six hundred heads and a thousand or twelve hundred wings.[41] [Diderot 1975–, vol. 17, p. 120]

Diderot had actually used this image, but in much more summary fashion, in his *Pensées sur l'interprétation de la nature*, some thirteen years earlier, in 1753. There, commenting on Maupertuis' *Système de la nature* in the final sections of the work, he reflected on the mechanisms of generation, how 'information' is conveyed in the seminal fluid, how the elements retain a kind of 'memory,' and so forth. Mentioning the equally evocative image of the polyp, Diderot added that these "may be compared to a cluster of infinitely small bees which, as they only retain a living memory of one position, would cling to one another and remain in that situation, in accordance with the position most familiar to them" [Diderot 1975–, vol. 9, p. 80].

Diderot takes over from Bordeu the idea that individual organs 'live' in the organism like tiny animals composing one larger animal, but with a newer meaning, in which the distinction between contiguity and continuity is central. Indeed, as Rudy Le Menthéour observes, the difference between a hive and a swarm for a thinker like Diderot is that the relation between parts in the former is merely contiguous, whereas in the latter it is continuous. François Pépin emphasizes that this new model in which external parts are assimilated together (when the legs of the various bees join together to form "a single animal") is also, crucially, a *chemical* one [Le Menthéour 2009, p. 211; Pépin 2011, p. 141]. The life of the animal, "l'animal entier," is the composite of the life of each organic component, interacting in a relation of "sympathy" (the modernized form of the Hippocratic *sympathia panta*, used in these texts as a technical term for nervous interconnection), which sometimes is not dependent on any center, any 'controller' at all: "these are sensing and living organs, coupling, sympathizing and concurring towards the same goal, without the participation of the whole animal" [Diderot 1975–, vol. 17, p. 501]. (On the political side, Mandeville, too, was not concerned with the role of the monarch as a possible central or centralizing force that keeps the hive together, even if the latter was a monarchical community. For him the collection of private interests is what keeps it from falling apart.)

41 For discussion, see: [Dieckmann 1938, pp. 86–87].

This raises the question of the *unity* of the organism (in the *Rêve*, the unity of the self, which Mlle de Lespinasse worries about – to which the reply, via the character Bordeu, is precisely a doctrine of *organismic* unity, that is, you are yourself because of the individuality of your body or *organisation*). After all, if an organism is a sum of many lives, whether this is an additive or one that involves qualitative shifts, where is the limit? This is another one of the difficult questions that neither Diderot nor Bordeu – both of whom pose it – resolves to anyone's satisfaction, including their own. One recalls that Bordeu introduced the image of the beeswarm as a *metaphor* of organic unity, and Diderot, although he expands on it and adds other metaphors for the nervous system including the spider's web and the harpsichord (for the vibrating 'strings' of the nervous system) does not present it as anything other than that.

Diderot brings together a more mechanistically oriented account of a structural relation between solid parts (from Albrecht von Haller), the more holistic sense of an integrated network of sensibility/sympathy (from Bordeu and others) and various other theories of organic matter concerning what we might call 'vital *minima*,' that is, the minimal constituents of organic life which are themselves "alive" and possessed of animate properties. Contrasting with theories such as Haller's, he collapses any residual dualist distinction between irritability and sensibility (which in Haller and other authors had served to preserve a concept of soul): "Generally, in the animal and in each of its parts – life, sensibility and irritation" [Diderot 1975–, vol. 17, p. 449]. Specifically as regards the beeswarm metaphor, one can see that Diderot is both nudging it in the direction of a different register, a different discursive space, but also, emphasizing holistic properties even further. Where Maupertuis and Bordeu were indeed reflecting on the nature of 'wholes,' Diderot is willing to state that the world itself is the hive (or swarm), in a kind of suspension of boundaries. Diderot would doubtless have endorsed this statement of Henri Bergson, regarding the "things of life": "who can say exactly where individuality begins and ends, whether the living being is one or many... In vain we force the living into this or that one of our molds" [Bergson 1911, p. x].

Up until now, we have treated the beeswarm metaphor primarily in its biomedical context, with some limited comparative reflection on its sociopolitical usage. But, as Diderot's case highlights, this metaphor plays other roles as well. As we saw with Clarke, Bentley and others above, the swarm could be a problematic image inasmuch as it conveyed dispersion rather than unity; as such, it is no coincidence that it was associated with materialism, for a major objection to materialism, in its treatment of life, body, and person, is the seeming absence of any 'center' or 'self' within the system of living parts. Sometimes, this is presented as disturbing, as dangerous Epicureanism; sometimes, as a fascinating feature of organic nature overall, as in Johann Wolfgang von Goethe's reaction: "Countless animals in a drop, that moved among each other with unspeakable agility and shortly gathered themselves together into a thick, swarming cluster."[42] Indeed, as their fascination with the image of the beeswarm shows, a number of materialists – call them 'vital materialists' – are deeply concerned with providing an account of the organism or body as something more than a set of interlocking, solid parts, although this "something other or more" is *not* understood as either 'soul' or 'vital force.' We might say that the issue is not just organismic unity but also biological individuality.

Additionally, this opens onto what contemporary researchers will call 'swarm intelligence,' that is, the question again of unity in multiplicity, but posed no longer in terms of life but of mind; it is actually not easy to historically demarcate arguments for organic unity in terms of *life* (cohesion, sympathy, chemistry, etc.) from arguments in terms of *mind*, which do become more visible in the nineteenth century and beyond.[43] Kate Tunstall has suggested that Diderot's usage of the beeswarm illustrates "the way in which body parts cooperate and perform actions without the need for an immaterial soul" [Tunstall 2016, p. 218]. That is, it is not just that the body can think – a more garden-variety materialist claim – but that what she calls, loosely referring to some ideas in cognitive science, "embodied thinking" takes place throughout the body "and in the relationships between bodies": swarm intelligence is "extended" as well as "embodied" [ibid., p. 205]. A similar description, in this case taking eighteenth-century vitalism (very broadly construed) as its object, is given by Catherine Packham: vitalism "dethrone[ed] the mind from its assumed role of reasoned governance of the body, and by envisioning, instead, a body capable of automatic and autonomous, if unconscious and instinctual, self-direction and self-preservation" [Packham 2012, p. 19].[44]

While it is interesting, and surely fruitful to conceive of the beeswarm in Diderot as a model of self, of distributed cognition, of plural selfhood, it seems to us that what Stephen Gaukroger referred to as the "unsystematic vitality" [Gaukroger 2005] of a beeswarm is being used as a true model of natural processes (rather than as a mathematical one).

---

42 Goethe's notes on infusoria experiments (1785–1786), May 11, n° 9, in *Schriften zur Naturwissenschaft*, I.10, p. 39, quoted in [Goldstein 2011, p. 9].

43 The case of Leibniz seems unusual, in that he explicitly distinguished aggregates from organisms ("corporeal substances") in terms of mind: "a corporeal substance [...] is one *per se*, and not a mere aggregate of many substances, for there is a great difference between an animal, for example, and a flock"; it "is either a soul or something analogous to a soul, and always naturally activates some organic body, which, taken separately, indeed, set apart or removed from soul, is not one substance but an aggregate of many, in a word, a machine of nature." (Untitled text from May 1702, G IV, pp. 395–396; translation under the title "On Body and Force, Against the Cartesians" in [Leibniz 1989, pp. 252–253]. Thanks to Sarah Tropper for this reference.)

44 Here, Packham is anticipating Sheehan and Wahrman's description of a kind of joint constitution of the self-organization concept, in biology and in society.

8 Conclusion: Swarm Intelligence and Superorganisms

What type of organizational model is the beeswarm? Aside from the variety of political and/or mathematical usages of the beeswarm which were not central to our story, we have focused on how this image functions as a way of articulating and explicating the twin motifs of organic (or organismic) unity and biological individuality. While Diderot could be seen as a culmination, as the most complex treatment of the beeswarm (and certainly of its explicit encounter with materialism), it is also possible to extract different, equally viable models from the other vitalist treatments of the swarm. For instance, as we discussed, Ménuret's version gave a greater role to the properties of the parts (= lives = organs), in a sense closer to what we might term componential analysis. This should be emphasized, as it is not generally an intellectual attitude associated with 'vitalism.' For Ménuret, there is a sense in which, aside from the existence of higher-level properties like health (or sickness), "the parts remain what they were," to borrow a phrase from the early-twentieth-century emergentist Samuel Alexander.[45] While Buffon is not to be counted among the small number of 'beeswarm theorists' that we examine here, unlike Bordeu and Diderot in particular, he too recognizes the phenomenon of emergent order, understood in the sense that (not in Buffon's words) rules of interaction between bees are, inseparably, what makes an individual bee a bee, and what makes the hive a hive [Epstein 1999, p. 55]. Much later, at the end of the nineteenth century, D'Arcy Wentworth Thompson dedicated a whole chapter of his sui generis work *On Growth and Form* to the construction of the regular cells in the beehive (the hexagonal structure has now been shown to be optimal: [Hales 2001]). To him, physical forces condition the shape of organisms and the geometry of the hive. The "beautiful regularity of the bee's architecture" is not due to apian ingenuity or instinct, or to the "geometrical forethought of the bees" but rather to "some automatic play of the physical forces" [Thompson 1992, pp. 132, 138].

Yet, however, 'emergent(ist)' the order of the beeswarm may be – and here, the difference between swarm and hive seems relevant – ontologically, it is entirely material, without 'spooky' features. Not only is the order not 'top-down,' but there are no spiritual or otherwise immaterial properties involved (unlike, say, what was always claimed about the 'entelechies' of later neo-vitalism). The unity of the animal described in the above texts is not the property or the 'doing' of a central self or controller:

45 Alexander discusses how "physiological complexes of a sufficient complexity carry mind or consciousness," yet "in the complex which thus acquires a new quality the parts retain their proper character and are not altered. The physiological elements remain physiological [...] The water in our bodies remains water still [...] *the parts remain what they were*" [Alexander 1927, p. 370 (emphasis S.K/C.W.)].

> There is no central, or 'top-down,' control over individual behavior in agent-based models. Of course, there will generally be feedback from macrostructures to microstructures, as where newborn agents are conditioned by social norms or institutions that have taken shape endogenously through earlier agent interactions. In this sense, micro and macro will typically co-evolve. But as a matter of model specification, no central controllers or other higher authorities are posited *ab initio*. [Epstein 1999, p. 42]

It is in this sense that complex structures emerge from the interaction of simpler agents, to the delight of eighteenth-century vitalists, as well as embodied roboticists, behavioral economists, and other researchers nowadays.

And here, our efforts to keep the political resonance of the swarm concept separate from its more 'naturalistic' dimensions can seem futile if we reflect on the way that swarms in some strands of contemporary thought present a "political paradox between 'control and emergence, sovereignty and multiplicity'"; "swarms organize the multiple into a relational whole – and one in which the collective is exactly defined by 'relationality'" [Parikka 2010, p. 47].[46] Here, Diderot's choice of the swarm over the hive, following the vitalist example, is telling: the swarm is "a community without a leader, where agency is distributed among equal, indistinguishable parts rather than located with a queen that can easily be singled out by the observer" [Wallmann 2017, p. 145]. In fact, in an expression of dynamism and processuality that Diderot doubtless would have endorsed, the term "swarm" is used to describe both the community of bees and the process by which it becomes divided, much like cell division in any living organism (growth and reproduction) extends the more informal body of the bees (genus) in space but also reproduces/duplicates the original swarm.

In a 1943 essay on Bergson, Georges Canguilhem refers to a distinction derived from Heinrich Rickert's 'philosophy of life,' which is both familiar to us and subtly different: between 'aristocratic' and 'democratic' tendencies in biology, where the latter are based on the principle of vital economy, and are deterministic (life is reducible to a mechanical and material phenomenon, and tends only to its own preservation); in contrast, the aristocratic tendency (Canguilhem mentions Friedrich Nietzsche) understands life as self-overcoming, as an instrument of hierarchical creation [Canguilhem 2007]. The swarm as discussed here indeed tends to be a more 'republican,' 'cooperative,' and 'emergent' affair. Now, our concern, unlike that of Giacomo Domenico Maraldi or Réaumur in earlier centuries, is not the role of the queen or the drone, but, rather, what sort of organizational whole the beeswarm is (or is meant to model). What Sheehan and Wahrman say of Diderot is actually true more broadly of numerous theorists of biological (or at least embodied) individuality of the period: "What in Bernard Mandeville had embodied the paradoxes of complex social systems, in Diderot embodied those of natural ones" [Sheehan and Wahrman 2015, ch. 4].

46 Parikka discussing the ideas of Eugene Thacker.

For Bordeu and his vitalist peers, the point was that bees in a swarm are like little lives composing a greater life; as Joshua Epstein puts it in a non-vitalist contemporary context, "the bee's interaction rules are what make it a bee – and not a lump. When [...] you get these rules right – when you get 'the individual bee' right – you get the hive, too" [Epstein 1999, p. 55]. To his earlier remark on the rules of interaction that govern bees and their hive, Epstein adds that in operational terms, "bee" might be defined as "that *x* that, when put together with other *x*'s, *makes* the hive (the 'emergent entity')" [ibid.]. What about the individuality of this larger collective? In what sense is a swarm an actual individual? This is what is referred to nowadays as the 'superorganism' concept, a term originally introduced in the early twentieth century by William Morton Wheeler, as a means of understanding how social insect collectives hold together. For Wheeler, a colony of social insects (although this could be extrapolated to other groups of animals, as is the case in current work on collective behavior) is properly identified as an organism, and not merely an analog of one: "The most general organismal character of the ant-colony is its individuality. Like the cell or the person, it behaves as a unitary whole, maintaining its identity in space, resisting dissolution and, as a general rule, any fusion with other colonies of the same or alien species" [Wheeler 1911, p. 310].[47] Building on this approach, Thomas Seeley describes the honey-bee colonies he studies as "superorganisms" because of their high degree of cohesiveness, made possible because each bee is free to move about the nest and exchange information with other members of the hive (through signals, such as the bee-dance, and cues) [Seeley 1989].[48] Other researchers speak of the "tightness of bee colonies" [Haber 2013, p. 197], and note that "animals often organize into groups that outperform the individuals that comprise them" [Feinerman 2018, p. 55]. In sum, bee (or ant, or termite) colonies are termed "superorganisms" because they exhibit many organism-like traits.

But extending the focus on the beeswarm as model into the present also yields a slightly different result: while the approach we discussed above focuses on unity and interconnectedness of the small 'lives' as forming a larger individual life, another focuses on how tiny agents in interaction yield (or indeed, *are*), as Cavendish argued, "one Minde." Thus, one of the most prominent researchers into the neurobiology of bees has written about the beehive as a "thinking machine" [Seeley and

47 To this one can add ontogenetic and phylogenetic development, and that the colony displays the Weismannian division of germ plasm and soma [Mitchell 1995, p. 238]. It is probably no coincidence that Wheeler was also actively interested in holistic/organismic ideas at a more abstract level, including in his commitment to "emergent evolution" [Parikka 2010, p. 51].

48 This fits with Turner's celebrated work on termite mounds as part of the "extended physiology" of the termites. There is debate, however, as to whether these colonies constitute real biological individuals or not: for instance, superorganisms do not reproduce (Haber). But beekeeper breeding plans seem indeed to focus on colony-level traits, not genetic traits of individual queens.

Levien 1987];[49] equally influential, the researcher in social insect physiology J. Scott Turner even credits Bordeu but straight away adds a notion of *intelligence*:

> Emergent intelligence traces its roots to the venerable observation that intelligence seems to crop up in unusual contexts. The most common of these unusual contexts is the social insect colony, and its emergent intelligence was articulated as early as the end of the eighteenth century by Theophile de Bordeu, who noted the seemingly intelligent and coordinated behavior of the bee swarm. [Turner 2016]

For Turner, Bordeu's question is less about organismic unity, as we described above, and more about how a swarm acts intelligently, even though there was no brain or other evident "specifier" to create it. This is not a brand-new idea: "Swarming was early on described as a peculiar group behavior that was of interest to entomologists and researchers of social insects. [...] for them, it represented a weird kind of organization that seemed to reside between instinct and intelligence" [Parikka 2010, p. 48]. This focus on the collective pattern of behavior of the bees as a kind of 'mind,' 'intelligence,' 'group cognition,' 'extended cognition,' and so forth (one thinks of Edwin Hutchins' influential study of life on a navy ship as a kind of giant brain in which the different individuals are like individual neurons [Hutchins 1995]) tends to take an increasingly abstract form, in which biological agents and their interactions become *models*, algorithms, and other objects of computation.

Bordeu's question (we might say the 'Bordeu-Diderot question') has remained alive until now, albeit in a less material form, given that the contemporary term for emergent intelligence is *swarm* intelligence, that is, the solving of cognitive problems by a group of individuals – typically, social insects – who pool their knowledge and process it through social interactions. Theories of swarm cognition or swarm intelligence emphasize that the individual bee (or ant, or termite) has vastly inferior cognitive capacities to the swarm as a whole; the latter has a richer cognitive repertoire because it inhabits a world full of more diverse cues and stimuli. Some researchers speak of 'nonconscious cognition' here, as collective action can take place in and through chemical signaling and other non-semantic modes of communication. Further, algorithms are now developed based on the intelligent behavior of bees (or ants), with names such as 'bee swarm optimization' or 'ant colony optimization' algorithms: these population-based techniques are used to find ideal ways of managing networks and other distributed systems, that is, "optimization research" [Bonabeau, Dorigo, and Theraulaz 1999, p. 7].[50]

If the beeswarm stands for life (organism) in the first case, and mind (intelligence, cognition) in the second, both share a fundamentally *distributed* character –

49 Recall that at least one interpreter of the beeswarm in Diderot, Tunstall, read this image as pointing to a theory of extended and embodied *mind*.

50 The term 'swarm intelligence' was first used by Beni and Wang in a paper of that title published in 1989.

in the latter case, as distributed cognition, that is, individual bees taken together as a collective are understood as forming a kind of brain, like Hutchins' navy ship. Bees (or ants in other discussions of swarm intelligence) are like individual neurons. As Georg Theiner and Tim O'Connor put it helpfully,

> ants exhibit a 'neuron-like' behavior insofar as inactive ants have a low propensity to become spontaneously active, but can become excited by other ants with whom they come into contact [...]. Conversely, ants are prone to lapse back into inactivity if their activation is not sufficiently reinforced, and even exhibit a short refractory period (similar to neurons) before they can be reactivated – a mechanism which keeps the swarm from getting permanently 'locked' into an excitatory state. [Theiner and O'Connor 2010, p. 90]

Now, this 'relational' quality of individual bees who only achieve their true potential in a collective unit can be characterized in a variety of ways. Not only does it allow of various sociopolitical, biological, metaphysical or cognitivist appropriations, as we have seen above; it can also be spelled out according to different theoretical vocabularies. In some hands, the beeswarm can also turn out to be Deleuzian, like Monsieur Jourdain, who wrote prose without knowing it. Thus Jussi Parikka argues that individual bees performing the 'bee dance' (as described in Karl von Frisch's pioneering work on bee communication, which showed how the orientation of the bee, its energy output, and the direction of the dance all communicate precise information about food sources) are not "representational entities" but "machinological becomings": they should be understood

> in terms of their capabilities of perceiving and grasping the environmental fluctuations as part of their organizational structures [...] where the intelligence of the interaction is not located in any one bee, or even a collective of bees as a stable unit, but in the 'in-between' space of becoming: bees relating to the mattering milieu, which becomes articulated as a continuum to the social behavior of the insect community. This community is not based on representational content, then, but on distributed organization of the society of nonhuman actors.[51]
>
> [Parikka 2010, p. 129]

Note that the 'algorithmic' approach is different, less speculative. If we compare Bordeu's beeswarm to that of swarm intelligence, of honeybee colony algorithms, one striking difference is that the latter is a *model* – in that sense closer to the mathematical obsession of Johann Samuel König, Réaumur, and D'Arcy Thompson. Granted, it is not the same mathematics at issue, but our point is simply to contrast the vitalist usage of the beeswarm in explicitly *material* terms – indeed as a form of organization, but one with concrete biochemical properties – with the more *formal(istic)* interest in the mathematics of hives and cells, or morphogenesis. In that sense, we would underline the fact that the eighteenth-century vitalist and materialist emphasis is on a type

51 Discussion in: [Hayles 2017, pp. 188–189].

of material *agencement* (obviously not "brute matter," given the emphasis on the concept of structure in the vitalist texts[52]) whereas models are dematerialized.[53]

The beeswarm is a core metaphor and even model for organicism as a theory, with its emphasis on organs as 'lives,' not just mere parts, and indeed it is one of the standard metaphors of organism. However, it is in perpetual danger of fragmentation, as is also captured by the recurrent fears of the swarm as an image of reductive materialism and/or chaos. But the beeswarm, *Janus bifrons*, is inseparably an image of multiplicity and fragmentation, and an image of order and equilibrium, albeit less than the hive, with its appeal to those pondering the mathematics of built structures, or the algorithms of swarm intelligence. As a metaphor of identity itself, it also crisscrosses the boundaries of the social and the biological, as the core intuitions regarding unity or atomization reveal.

Bibliography

Alexander S. Space, Time and Deity: The Gifford Lectures at Glasgow, 1916–1918. Vol. 2. London, Macmillan, 1927.

Allen DS. Burning The Fable of the Bees: Cultural Poetics and the Incendiary Authority of Nature. In: Daston L, Vidal F, eds. The Moral Authority of Nature. Chicago, University of Chicago Press, 2003, 74–99.

Andersen K, Bos HJM. Pure Mathematics. In: Park K, Daston L, eds. The Cambridge History of Science. Vol. 3. Early Modern Science. Cambridge, Cambridge University Press, 2006, 696–707.

Arbuthnot J. The Congress of Bees: Or, Political Remarks on the Bees Swarming at St. James's. London, A. Moore, 1728.

Aristotle. History of Animals. Vol. 1, 1. Trans. Peck AL. Cambridge MA, Harvard University Press, 1965.

Aristotle. Politics. In: Barnes J, ed. The Complete Works of Aristotle. Vol. 2. Princeton, Princeton University Press, 1984.

Baine Campbell M. Busy Bees: Utopia, Dystopia and the Very Small. Journal of Medieval and Early Modern Studies 2006, 36 (3), 539–560.

Beni G, Wang J. Swarm Intelligence. In: Proceedings of the Seventh Annual Meeting of the Robotics Society of Japan. Tokyo, 1989, 425–448.

52 In his fascinating programmatic article in the *Encyclopédie* on the notion of "animal economy," the vitalist physician Ménuret de Chambaud defines the latter term as "*l'ordre, le méchanisme, l'ensemble* des fonctions & des mouvemens qui entretiennent la vie des animaux" [Chambaud 1765b, 362a].

53 A recent revival of the bee trope in terms of "knowledge-gathering," "honeycombs of knowledge," speaks of the "revolt of the bees" but in a rather non-revolted way [Levy and Squire 2005] and, indeed, also more dematerialized. Our analysis is different, as the essays collected in the aforementioned volume take beehives and paper hives (e.g., archives and libraries) as utopias of memory storage, also related rather impressionistically to contemporary usage of beehive ideas in art and architecture.

Bentley R. The Folly and Unreasonableness of Atheism. London, Printed by J.H. for H. Mortlock, 1693.

Bergier NS. Examen du matèrialisme ou rèfutation du système de la nature. Vol. 1. Paris, Humblot, 1771.

Bergson H. Creative Evolution. Trans. A. Mitchell. New York, Henry Holt, 1911.

Bloch O. Sur les premières apparitions du mot 'matèrialiste'. Raison présente 1978, 47, 3–16, DOI: 10.3406/raipr.1978.1945.

Bloch O. Matière à histoires. Paris, Vrin, 1997.

Blum C. On the Uses of Bees. In: Yolton JW, Brown LE, eds. Studies in Eighteenth-Century Culture. Vol. 18. East Lansing, Michigan, Colleagues Press, 1988, 3–5.

Bonabeau E, Dorigo M, Theraulaz G. Swarm Intelligence: From Natural to Artificial Systems. Oxford, Oxford University Press, 1999.

Bonnet C. Palingènèsie philosophique ou Idèes sur l'ètat passè et sur l'ètat futur des ètres vivans. Vol. 1. Geneva, Claude Philibert, Barthelemi Chirol, 1769.

Bonnet C. Considérations sur les corps organisés. Paris, Fayard, 1985 [1762].

Bordeu TD. OEuvres complétes, précédées d'une notice sur sa vie et ses ouvrages par Monsieur le chevalier de Richerand. Vol. 2 . Paris, Caille et Ravier, 1818.

Bouchard F, Huneman P, eds. From Groups to Individuals: Evolution and Emerging Individuality. Cambridge, MIT Press, 2013.

Boyle R. The History of Fluidity. In: Birch T, ed. The Works of Robert Boyle. Vol. 1. London, J, Rivington F et al., 1772, reprint, Hildesheim, Georg Olms, 1965, 378–400.

Brewer B. The Sterility of Abundance: Marcela and Grisostomo in the Golden Age. In: Boyd S, O'Reilly T, eds. Artifice of Invention in the Spanish Golden Age. Abingdon, Routledge and the Modern Humanities Research Association, 2014, 33–44.

Brown GI. The Evolution of the Term "Mixed Mathematics." Journal of the History of Ideas 1991, 52 (1), 81–102.

Buffon GLLD. Histoire Naturelle. Vol. 4. Paris, Imprimerie royale, 1753.

Butler C. The Feminine Monarchie, or the Historie of Bees. Oxford, J. Barnes, 1609.

Campana J. The Bee and the Sovereign? Political Entomology and the Problem of Scale. Shakespeare Studies 2013, 41, 94–113.

Campana J. The Bee and the Sovereign (II): Segments, Swarms, and the Early-Modern Multitude. In: Cefalu P, Kuchar G, Reynolds B, eds. The Return of Theory in Early-Modern Studies. Vol. 2. Basingstoke, Palgrave Macmillan, 2014, 59–79.

Canguilhem G. Aspects du Vitalisme. In: Canguilhem G: La Connaissance de la Vie. Revised. Paris, Vrin, 1965, 83–100.

Canguilhem G. Commentaire au troisième chapitre de *L'Évolution créatrice*. In: Worms F, ed. Annales bergsoniennes. Vol. 3. Bergson et la Science. Paris, PUF, 2007, 113–160.

Cascardi AJ. Cervantes, Literature, and the Discourse of Politics. Toronto, Toronto University Press, 2012.

Cavendish M. Natures Pictures Drawn by Fancies Pencil to the Life. London, J. Martin and J. Allestrye, 1656.

Cavendish M. Observations upon Experimental Philosophy. O'Neill E, Ed. Cambridge, Cambridge University Press, 2001 [1666].

Cavendish M. Sociable Letters. Fitzmaurice J, Ed. Abington, Routledge, 2012.

Cayrou G. Le français classique. Lexique de la langue du dix-septième siècle. Paris, Didier, 1948.

Cervantes De Saavedra M. The History and Adventures of the Renowned Don Quixote. Trans. Smollett T. Vol. 2. London, A. Millar, 1755.

Citton Y. L'Envers de la Liberté. L'invention d'un imaginaire spinoziste dans la France des Lumières. Paris, Éditions Amsterdam, 2006.

Clarke S. A Letter to Mr. Dodwell. London, Knapton, 1731.

Coleridge ST. Hints Towards the Formation of a More Comprehensive Theory Of Life. Watson SB, Ed. London, John Churchill, 1848.

Daubenton LJM. Histoire naturelle. In: Daubenton LJM. Séances des Ecoles Normales, recueillies par des sténographes, et revues par les professeurs, nouvelle édition. Vol. 1. Paris, Imprimerie du cercle social, 1800, 92–97.

Daubenton LJM. Essaim. In: Diderot D, D'Alembert JR, eds. Encyclopédie ou Dictionnaire raisonné des arts et des métiers Vol. 5. Paris, Briasson, David, Le Breton & Durand, 1757, reprint Stuttgart, Frommann, 1966, 994a–995b.

Day J. The Parliament of Bees, with their proper Characters... Being an Allegoricall Description of the Actions of Good and Bad men in these our Daies. London, William Lee, 1641.

Demeter T, Schliesser E. The Uses and Abuses of Mathematics in Early Modern Philosophy: Introduction. Synthese 2019, 196, 3461–3464, DOI: 10.1007/s11229-017-1670-y.

Diderot D. Œuvres complètes. Vol. 9. Pensées sur l'Interprétation de la Nature. Varloot J, Ed. Paris, Hermann, 1975–.

Diderot D. Œuvres complètes. Vol. 17. Le Rêve de D'Alembert. Varloot J, Ed. Paris, Hermann, 1975–.

Dieckmann H. Théophile de Bordeu und Diderots *Rêve de D'Alembert*. Romanische Forschungen 1938, 52 (1), 55–122.

Dinsdale J. The Modern Art of Breeding Bees, a Poem. London, J. Davidson, 1740.

Drouin JM. L'image des sociétés d'insectes en France à l'époque de la Révolution. Revue de Synthése 1992, 113 (3/4), 333–345.

Duflo C. Diderot et Ménuret de Chambaud. Recherches sur Diderot et sur l'Encyclopédie 2003, 34, 5, DOI: 10.4000/rde.157.

Encyclopédie méthodique ou par ordre de matiéres. Vol. 8. Paris, Veuve Agasse, 1808.

Epstein J. Agent-Based Computational Models and Generative Social Science. Complexity 1999, 4 (5), 41–60, DOI: 10.1002/(SICI)1099-0526(199905/06)4:5<41::AID-CPLX9>3.0.CO;2-F.

Fairfax Withington A. Republican Bees: The Political Economy of the Beehive in Eighteenth-Century America. In: Yolton JW, Brown LE, eds. Studies in Eighteenth-Century Culture. Vol. 18. East Lansing, Michigan, Colleagues Press, 1988, 39–77.

Feinerman O. Leadership during Cooperative Transport by the Longhorn Crazy Ant. In: Gissis SB, Lamm E, Shavit A, eds. Landscapes of Collectivity in the Life Sciences. Cambridge MA, MIT Press, 2018, 55–65.

Feuerbach L. Darstellung, Entwicklung und Kritik der Leibnitz'schen Philosophie. Ansbach, C. Brügel, 1837.

Fouquet H. Sensibilité, Sentiment (Médecine). In: Encyclopédie. Vol. XV. Paris, Briasson, 1765, 38–52.

Fouquet H. Discours sur la Clinique. Montpellier, Izar & Ricard, an XI, 1802.

Fraser HM. Beekeeping in Antiquity. London, University of London Press, 1931.

Freeman JA. Milton and the Martial Muse. Princeton, Princeton University Press, 1980.

Gambarotto A. Vital Forces, Teleology and Organisation: Philosophy of Nature and the Rise of Biology in Germany. Dordrecht, Springer, 2017.

Gaukroger S. The Collapse of Mechanism and the Rise of Sensibility: Science and the Shaping of Modernity, 1680–1760. Oxford, Oxford University Press, 2010.

Gaukroger S. 'The Enlightenment Revolt Against Rationalism': Critical Notice of Peter Hanns Reill, *Vitalizing Nature in the Enlightenment*. Australian Review of Public Affairs 2005, 26. (Accessed June 21, 2021, at http://www.australianreview.net/digest/2005/11/gaukroger.html.)

Godfrey-Smith P. Other Minds: The Octopus, the Sea, and the Deep Origins of Consciousness. London, Harper Collins, 2016.

Goethe JW. Die Schriften zur Naturwissenschaft. Leopoldina-Ausgabe. 26 vols. Est. K. Lothar Wolf and Wilhelm Troll. Kuhn D, Engelhardt WV, Müller I, Eds, Weimar, Herman Böhlaus Nachfolger, 1947–2011.
Goldstein A. 'Sweet Science': Romantic Materialism and the New Sciences of Life. PhD thesis, Department of Comparative Literature, University of California, Berkeley, 2011.
Haber M. Colonies Are Individuals: Revisiting the Superorganism Revival. In: Bouchard F, Huneman P, eds. From Groups to Individuals: Evolution and Emerging Individuality. Cambridge, MIT Press, Vienna Series in Theoretical Biology, 2013, 195–217.
Hales TC. The Honeycomb Conjecture. Discrete Computational Geometry 2001, 25, 1–22.
Hancock J. Arguments to Prove the Being of God with Objections against it Answered. In: A Defence of Natural and Revealed Religion: Being a Collection of the Sermons Preached at the Lecture Founded by the Honourable Robert Boyle, Esq. Vol. 2. London, 1739, 197–258.
Haraway D. When We Have Never Been Human, What Is to Be Done? (Interview with Donna Haraway by Nicholas Gane). Theory, Culture & Society 2006, 23 (7–8), 135–158, DOI: 10.1177/0263276406069228.
Hartlib S. The Reformed Commonwealth of Bees. London, G. Calvert, 1655.
Harvey W. Anatomical Exercitations Concerning the Generation of Living Creatures. Trans. M. Llewellyn. London, Octaviani Pulleyn, 1653.
Hayles KN. The Cognitive Nonconscious and the New Materialisms. In: Ellenzweig S, Zammito J, eds. The New Politics of Materialism. History, Philosophy, Science. London, Routledge, 2017, 181–199.
Hill T. The Arte of Gardening, [...]. London, E. Allde, 1608.
Hobbes T. Leviathan. Vol. 2. Malcolm N, Ed. Oxford, Clarendon, 2012.
Hoquet T. Buffon's Natural History: The Catalogue of Facts and the Temptation of System. In: Rey AL, Bodenmann S, eds. What Does it Mean to be an Empiricist? Empiricisms in Eighteenth Century Science. Dordrecht, Springer, 2018, 139–167.
Hutchins E. Cognition in the Wild. Cambridge, MIT Press, 1995.
Jacobs NA. John Milton's Beehive, from Polemic to Epic. Studies in Philosophy 2015, 112 (4), 798–816.
Johach E. Der Bienenstaat: Geschichte eines politisch-moralischen Exempels. In: Von Der HA, Vogl J, eds. Politische Zoologie. Zurich, Diaphanes, 2007, 219–233.
Kalff S. Are Cranes Republicans? A Short Chapter in Political Ornithology. In: Enenkel KAE, Smith PJ, eds. Zoology in Early Modern Culture. Leiden, Brill, 2014, 437–459.
Le Menthéour R. De la ruche au polype: figures de l'organisation sociale. Dix-huitiéme siécle 2009, 41, 204–221, DOI: 10.3917/dhs.041.0204.
Leibniz GW. Die philosophischen Schriften. Vol. 7 . Gerhardt CI, Ed. Berlin, Weidmann, 1875–1890, reprint Hildesheim, New York, G. Olms, 1978.
Leibniz GW. Philosophical Essays. Trans. Ariew R, Garber D. Indianapolis, Hackett, 1989.
Levy A, Squire T, eds. The Revolt of the Bees: Essays by Anthony Grafton, Thomas Keenan, and Aaron Levy. Philadelphia, Slought Books, University of Pennsylvania Library, 2005.
Mancosu P. Philosophy of Mathematics and Mathematical Practice in the Seventeenth Century. Oxford, Oxford University Press, 1996.
Mandeville B. The Fable of the Bees, or Private Vices, Public Benefits. With a Commentary Critical, Historical and Explanatory by Kaye FB. Vol. 2. Oxford, Clarendon Press, 1924 [1714, 1729.
Mandeville B. A Treatise of the Hypochondriack and Hysterick Diseases, in Three Dialogues. Kleiman-Lafon S, Ed. Dordrecht, Springer, 2017, 1st ed. 1711, revised ed. 1730.
Maraldi GF. Observations sur les abeilles. Histoire de l'Académie royale des sciences. 1712. Paris, Imprimerie Royale, 1731, 297–331.
Marx K. Economic and Philosophic Manuscripts of 1844. Moscow, Progress Publishers, 1959 [1932].
Maupertuis PLMD. Système de la nature. Essai sur la formation des Corps organisés. In: Maupertuis PLMD. Œuvres. Vol. 2. Lyon, Bruyset, 1756, 139–168.

McLaughlin P. Naming Biology. Journal of the History of Biology 2002, 35, 1–4.
Ménuret De Chambaud JJ. Observation (Gram. Physiq. Méd.). In: Encyclopédie ou Dictionnaire raisonné des sciences, des arts et des métiers. Vol. 11. Paris, Briasson, 1765a, 313–321.
Ménuret De Chambaud JJ. Œconomie Animale (Médecine). In: Encyclopédie ou Dictionnaire raisonné des sciences, des arts et des métiers. Vol. 11. Paris, Briasson, 1765b, 360–366.
Merrick J. Royal Bees: The Gender Politics of the Beehive in Early-Modern Europe. In: Yolton JW, Brown LE, eds. Studies in Eighteenth-Century Culture. Vol. 18. East Lansing, Michigan, Colleagues Press, 1988, 7–37.
Mitchell SD. The Superorganism Metaphor: Then and Now. In: Maasen EMS, Weingart P, eds. Biology as Society, Society as Biology: Metaphors. Amsterdam, Kluwer, 1995, 231–248.
Moffett [Mouffet] T. The Theater of Insects: Or, Lesser Living Creatures. In: Topsell E, ed. The History of Four-Footed Beasts and Serpents. London, E. Cotes, 1658, 889–1130.
Nyhart LK. The Political Organism. Carl Vogt on Animals and States in the 1840s and '50s. Historical Studies in the Natural Sciences 2017, 47 (5), 602–628, DOI: 10.1525/hsns.2017.47.5.602.
O'Neill E. Margaret Cavendish, Stoic Antecedent Causes, and Early Modern Occasional Causes. Revue Philosophique de la France et de l'Étranger 2013, 138 (3), 311–326.
Packham C. Eighteenth-Century Vitalism: Bodies, Culture, Politics. Basingstoke, Palgrave Macmillan, 2012.
Parikka J. Insect Media: An Archaeology of Animals and Technology. Minneapolis, University of Minnesota Press, 2010.
Pépin F. Vitalisme, Chimie et Philosophie Autour de l'*Encyclopédie* et de Diderot. In: Nouvel P, ed. Repenser le Vitalisme. Paris, PUF, 2011, 131–143.
Petkov S. The Fitness Landscape Metaphor: Dead but Not Gone. Philosophia Scientiæ 2015, 19 (1), 159–174, DOI: 10.4000/philosophiascientiae.1050.
Preston C. Bee. London, Reaktion Press, 2005.
Prete FR. Can the Female Rule The Hive? The Controversy over Honey-Bee Gender Roles in British Beekeeping Texts of the Sixteenth-Eighteenth Centuries. Journal of the History of Biology 1991, 24 (1), 113–144.
Purchas S. A Theatre of Politicall Flying-Insects wherein Especially the Nature, the Worth, the Work, the Wonder, and the Manner of Right-Ordering of the Bee, is Discovered and Described. London, Thomas Parkhurst, 1657.
Quiviger F. Honey from Heaven, Aspects of the Topos of the Bees in Renaissance Artistic Literature. In: Pfisterer U, Seidel M, eds. Visuelle Topoi, Erfindung und Tradiertes Wissen in den Künsten der Italienischen Renaissance. Munich, Berlin, Deutscher Kunstverlag, 2003, 317–321.
Réaumur RAFD. Suite de l'histoire des mouches à deux ailes, & l'histoire de plusieurs mouches à quatre ailes, à savoir, des mouches à scies, des cigales, & des abeilles. In: mémoires pour servir à l'histoire des insectes. Vol. 5. Paris, Imprimerie Royale, 1740.
Rey R. L'Âme, le corps et le vivant. In: Grmek M, ed. Histoire de la Pensée Médicale en Occident. Vol. 2. De la Renaissance aux Lumières. Paris, Seuil, 1997, 117–156.
Rey R. Naissance et développement du vitalisme en France de la deuxième moitié du 18e siècle à la fin du Premier Empire. Oxford, Voltaire Foundation, 2000, Abridged version of PhD thesis, 1987.
Roux S. Forms of Mathematization (14th-seventeenth Centuries). Early Science and Medicine 2010, 15, 319–337.
Roux S. Pour une étude des formes de la mathèmatisation. In: Chabotand H, Roux S, eds. La Mathématisation comme Problème. Paris, Éditions des archives contemporaines, 2011, 3–38.
Rusden M. A Full Discovery of Bees. Treating of the Nature, Government, Generation & Preservation of the Bee. London, H. Million, 1685.
Schlanger J. Les métaphores de l'organisme. Paris, Vrin, 1971.

Schmidgen W. Exquisite Mixture. The Virtues of Impurity in Early Modern England. Philadelphia, University of Pennsylvania Press, 2013.

Seeley TD. The Honey Bee Colony as a Superorganism. American Scientist 1989, 77, 546–553.

Seeley TD, Levien RA. A Colony of Mind: The Beehive as Thinking Machine. The Sciences 1987, 27, 38–43.

Sheehan J, Wahrman D. Invisible Hands: Self-Organization and the Eighteenth Century. Chicago, University of Chicago Press, 2015.

Smith JEH. Divine Machines: Leibniz and the Sciences of Life. Princeton, Princeton University Press, 2011.

Swift J. The Works of the Rev. Jonathan Swift. Vol. 2. London, John Nichols, 1801.

Theiner G, O'Connor T. The Emergence of Group Cognition. In: Corradini A, O'Connor T, eds. Emergence in Science and Philosophy. New York, Routledge, 2010, 78–117.

Thompson DW. On Growth and Form. Abridged edition. Cambridge, Cambridge University Press, 1992.

Thorley J. Melisselogia: Or, the Female Monarchy. London, 1744.

Tunstall K. The Early Modern Embodied Mind and the Entomological Imaginary. In: McMurran MH, Conway A, eds. Mind, Body, Motion, Matter: Eighteenth-Century British and French Literary Perspectives. Toronto, University of Toronto Press, 2016, 202–229.

Turner JS. Homeostasis is the Key to the Intelligent Building. Intelligent Buildings International 2016, 8 (2), DOI: 10.1080/17508975.2015.1042958.

Wallmann E. The Political Economy of Eighteenth-Century Insects: Natural History and Political Economy in France 1700–1789. PhD thesis, Warwick University, School of Modern Languages, 2017.

Warder J. The True Amazons: Or the Monarchy of Bees. London, John Pemberton, 1712.

Wenger A. Le Médecin et le Philosophe, Théophile de Bordeu selon Diderot. Paris, Hermann, 2012.

Wheeler WM. The Ant Colony as an Organism. Journal of Morphology 1911, 22 (2), 307–325, DOI: 10.1002/jmor.1050220206.

Williams EA. A Cultural History of Medical Vitalism in Enlightenment Montpellier. Burlington, Ashgate, 2003.

Willis T. Pharmaceutice Rationalis, or an Exercitation of the Operations of Medicines in Human Bodies, 'Of Opiate Medicines on Causing Sleep. London, T. Dring, 1679.

Willis T. Two Discourses Concerning the Soul of Brutes, which is that of the Vital and Sensitive of Man. Trans. Pordage S. London, T. Dring, 1683.

Wolfe CT. Materialism. In: Garrett AV, ed. The Routledge Companion to Eighteenth-Century Philosophy. London, Routledge, 2014, 90–117.

Wolfe CT. Materialism. A Historico-Philosophical Introduction. Dordrecht, Springer, 2016.

Wolfe CT. Varieties of Vital Materialism. In: Ellenzweig S, Zammito J, eds. The New Politics of Materialism: History, Philosophy, Science. London, Routledge, 2017a, 44–65.

Wolfe CT. Vital Anti-Mathematicism and the Ontology of the Emerging life Sciences: From Mandeville to Diderot. Synthese 2017b, 196, 3633–3654, DOI: 10.1007/s11229-017-1350-y.

Wolfe CT. La philosophie de la biologie avant la biologie: une histoire du vitalisme. Paris, Garnier, 2019.

Wolfe CT, Terada M. The Animal Economy as Object and Program in Montpellier Vitalism. Science in Context 2008, 21 (4), 537–579, DOI: 10.1017/S0269889708001956.

Woolfson J. The Renaissance of Bees. Renaissance Studies 2010, 24 (2), 281–300.

Worlidge J. Apiarium, or a Discourse of Bees. London, T. Dring, 1676.

Sylvie Kleiman-Lafon is a senior lecturer in English literature at the Université Paris 8-Vincennes à Saint-Denis. She is primarily interested in the use of literary forms in early-modern scientific discourse, and by the history of translation in the eighteenth century. She has edited Bernard Mandeville's *Treatise of the Hypochondriack and Hysterick Diseases* (Springer, 2017), and translated it into French (Ellug, 2012). She has co-edited (with Rebecca Anne Barr and Sophie Vasset) a volume on digestion as metaphor –Bellies, Bowels and Entrails in the eighteenth century (MUP, 2018), and another (with Micheline Louis-Courvoisier) on animal spirits –Les Esprits animaux 16e-21e siècles: littérature, histoire, philosophie (Épistémocritique, 2018). She is currently writing a monograph on Antoine Lasalle, translator of Francis Bacon's works into French at the end of the eighteenth century. Most papers available at https://univ-paris8.academia.edu/SylvieKleimanLafon.

Charles T. Wolfe is a Professor in the Department of Philosophy, Université de Toulouse-2 Jean-Jaurés. He works primarily in history and philosophy of the early modern life sciences, with a particular interest in materialism and vitalism. He is the author of: Materialism: A Historico-Philosophical Introduction (2016), La philosophie de la biologie: une histoire du vitalisme (2019) and Lire le matérialisme (2020), and has edited or coedited volumes on monsters, brains, empiricism, biology, mechanism and vitalism, including currently (w. D. Jalobeanu) the Encyclopedia of Early Modern Philosophy and the Sciences and (w. J. Symons) The History and Philosophy of Materialism. He is co-editor of the book series 'History, Philosophy and Theory of the Life Sciences' (Springer). Papers available at https://univ-tlse2.academia.edu/CharlesWolfe.

Horst Bredekamp

Leibniz's Concept of *Agens* in Matter, Space, and Image

1 The Metaphysical Force of Nature

Gottfried Wilhelm Leibniz rejected the separation of *res cogitans* and *res extensa* that his esteemed rival René Descartes had declared the basis of all philosophy as an assumption that was incompatible with nature. His *New System* written in the mid-1690s on the relation between soul and body and the interactions of all substances represents the most concise exposition of his counterargument on dualistic modes of thought. According to this argument, "the efficient cause of physical actions arises from metaphysics."[1] Leibniz's considerations on the relation between movement and impetus go straight to the heart of the discussion about a general principle of intrinsic activities that proponents of a 'new materialism' associate with the concept of 'agency.' With all due appreciation for these endeavors, what is troubling here is that the historical derivation of this 'materialism' remains largely tied to a history of ideas. And more seriously still: with Leibniz, an author who like almost no other researcher and philosopher before or since understood the question of a moving substance in matter itself as a key to understanding the world as such has with few exceptions been largely ignored [Ellenzweig and Zammito 2017; Laerke 2017].

The question of the perhaps insoluble problem of the interpenetration of matter and spirit was Leibniz's lifelong theme. In a letter written a year before his death to his trusted correspondent Nicolas Remond, his fundamental conviction was summarized in the formulistic proposition that "everything in nature happens mechanically and at the same time metaphysically but [...] the source of mechanics is in metaphysics" [Leibniz 1969d, p. 655].[2] As opposed to the transcendental understanding of this term, this "metaphysics" for Leibniz lies in the autonomous activity of nature, which he identifies

1 "je trouve que la cause efficiente des actions physiques est du resort de la metaphysique" [Leibniz 1978 (1880a), p. 472]; see: [ibid., p. 108]. Unless stated otherwise, all the translations from French and Latin were made by Benjamin Carter.

2 "que tout se fait mechaniquement et metaphysiquement en même temps dans les phenomenes de la nature, mais que la source de la Mechanique est dans la Metaphysique" [Leibniz 1978 (1887), p. 607]; see: [Leibniz 1989a, p. 322; German trans.: p. 323].

Acknowledgment: The author would like to thank Benjamin Carter for translating this text from German.

The author acknowledges the support of the Cluster of Excellence 'Matters of Activity. Image Space Material' funded by the Deutsche Forschungsgemeinschaft (DFG, German Research Foundation) under Germany's Excellence Strategy – EXC 2025 – 390648296.

Open Access. © 2021 Horst Bredekamp, published by De Gruyter. BY-NC-ND This work is licensed under the Creative Commons Attribution-NonCommercial-NoDerivatives 4.0 International License.
https://doi.org/10.1515/9783110562064-014

with the central terms 'form' and 'force.' In his text *De ipsa natura sive de vi insita actionibusque Creaturarum*, published in 1698 – one of the first to use the term 'monad' – he remarks: "there is a certain efficacy residing in things, a form or force such as we usually designate by the name nature" [Leibniz 1969b, p. 501].[3] Things contain *formam vel vim*; with this pairing of form and force, he establishes a general principle.

In one of his texts aimed against Descartes, "Nullum quidem librum [...]," written in May 1702, Leibniz defines this combination as a latently unstable force, since within a permanently changing universe it has a framework that due to its mobility can be thought as infinite: "Moreover, this [...] elastic force, inherent in every body, shows that there is internal motion in every body as well as a primitive and (so to speak) infinite force, although in collision itself it is limited [*determinetur*] by derivative force as circumstances demand" [Leibniz 1989d, p. 255].[4]

Leibniz's examples are the arch, a taut string, and compressed air:

> For, just as in an arch, each part sustains the full force of that which puts weight on it, and just as, in a taut string, each part sustains the full force of that which tightens it, and just as each portion of compressed air has as much force as the weight of the air pressing down, so too each and every corpuscle is aroused into action by the combined force of the entire surrounding mass, and awaits only an occasion for exercising its power, as the example of gunpowder shows.[5] [Ibid., pp. 255–256]

Things tremble quasi-elastically between rigidity and explosion.

In the same text, one also finds the definition of *conatus* as a universal "drive," a "tendency to action" [ibid., p. 252],[6] which is expressed as an inner urge to move.[7] Nature as a whole is *conatus*, desiring movement. This is not directed toward a given telos that should be presupposed and attained; it is itself already the perfection that this has as its goal. Internal impetus is the nature of completion [Myrdal 2012, pp. 95–96].[8]

---

3 "jam concedendum est, quandam inditam esse rebus efficaciam, formam vel vim, qualis naturae nomine a nobis accipi solet" [Leibniz 1978 (1880c), p. 507].

4 "Ipsa autem vis Elastica omni copori insita ostendit, omni etiam corpori motum intestinum inesse et vim primitivam (ut ita dicam) infinitam, licet in ipso concursu, circumstantiis exigentibus vi derivativa determinetur" [Leibniz 1978 (1880a), p. 399].

5 "ut enim in fornice incumbentis aut in chorda tensa trahentis totam vim quaevis pars sustinet et quaevis portio aeris comprssi tantam vim habet quantam aeris incumbentis pondus, ita quodvis corpusculum totius massa ambientis vi conspirante ad agendum solicitatur nec nisi occasionem exercendae potentiae expectat, ut pulveris pyrii exemplo patet" [Leibniz 1978 (1880a), p. 399].

6 "Vis activa [...] involvit conatum seu tendentiam ad actionem" [Leibniz 1978 (1880a), p. 395].

7 "non tantum corpus praesenti sui motus momento inest in loco sibi commensurato, sed etiam conatum habet seu nisum mutandi locum, ita ut status sequens ex praesenti, per se, naturae vi consequatur" [Leibniz 1978 (1880c), p. 513]. On the derivation of this idea from the winter of 1670–1671: [Mercer 2001, pp. 273–299]; see: [Myrdal 2012, p. 35].

8 On the definition of *conatus* in the *Monadologie*, cf. [Bredekamp 2008, p. 111]. In this sense, *conatus* has an all-encompassing power, and thus possesses a property that Leibniz with a view to Spinoza paradoxically negates [Myrdal 2012, p. 99]. On Spinoza's concept of *conatus*, see: [Carriero 2011, pp. 69–92].

Conatus is linked with the concept of appetition as the idea of an all-pervasive *agens*. As Leibniz remarks in the *Monadologie*, "Souls act according to the laws of final causes through their appetitions, ends, and means" [Leibniz 1969c, p. 651, par. 79].[9] This concept initially takes up the Aristotelian 'final causes' as a teleological explanation for actions of all kind [Aristotle 1995, pp. 31–32, 194b23–195a3], but objects in principle to the idea of an autarkic, physically determined world, since this systematically raises the question of the cause, that is, of the 'why?' [Mercer 2001, p. 77]. Appetition, however, acts even before the determination of ends and the means to attain these ends. It has no concrete trigger, but is the principle of desire and longing as such. This basic assumption, which is as far as conceivably possible from Descartes' worldview, is related to matter, space, and image – which will be addressed in the following sections.

2 Leibniz's Active Matter

Aristotle also ascribed an internal motion to nature, a drive. The *agens* drives, for instance, the growth of seeds in their transformation into grain [Aristotle 1995, pp. 191–192, 1049b]. However, Leibniz's concept of *agens* does not mean the striving after a defined goal, but motion itself. Rather than as a means to an external end, it is defined as a continuous impulse that overcomes inertia. In this sense, it is an autonomous, original activity inscribed in the nature of a thing. Instead of Aristotle's original individualities, which carry the urge to completion – the *first entelechies* [ibid., pp. 191–195; Myrdal 2012, p. 40] – within themselves, Leibniz in the *Journal des Savants* (June 1695) postulates "primitive forces," which should not be understood as purposive executive bodies or extensions of possibilities. Rather, as "original *activity*," they carry their determination of essence within themselves: "Aristotle calls them *first entelechies*. I call them, more intelligibly perhaps, *primitive forces*, which contain not only the *actuality* or the *completion* of possibility but an original *activity* as well" [Leibniz 1969a, p. 454].[10] Whereas for Aristotle, the active form of the first, *primitive*, force represents a process toward perfection, for Leibniz this is a limitation, since it restricts the potentiality of all merely possible developments [Myrdal 2012, pp. 41, 47–48, 86, 91].

The conviction that a continuously renewing force must be given not as an external addition or subsequent restocking but as immediate nature motivated Leibniz's

9 "Les âmes agissent selon les loix des causes finales par appetitions, fins et moyens" [Leibniz 1998, pp. 56–57, par. 79].

10 "Aristote les apelle *entelechies premieres*, je les apelle peutestre plus intelligibilment *forces primitives*, qui ne contiennent pas seulement l'*acte* ou le complement de la possibilité, mais encor une *activité* originale" [Leibniz 1978 (1880b), p. 479]; see: [Rozemond 2009, p. 298].

attempt to describe and establish a universally operative activity. In a letter from 1698, he remarks that each creature possesses "the force of action that, according to my conception, constitutes the nature of substance. So much so that this value bestowed by God is indeed the energy or force that is given to things."[11] This remark contains the eminent element of Leibniz's considerations. His definition of a force linked with bodies – an *agens*, an appetition, a *conatus* – extends to the *rebus*, the "things"; to these too this *movens* is given, and in this way the sphere of matter is conceived in an elementary way.

3 Leibniz's Active Spaces

The almost unsurpassable radicality of Leibniz's theory of perception lies in the fact that it pushes the interaction between external impulse and internal, autonomous activity to a point at which responses in the form of active resonances are formed even when external impulses are no longer consciously perceived. This has repercussions for the conception of active spaces.

One of Leibniz's examples, which he names in his *Nouveaux Essais sur L'Entendement Humain*, completed in 1704, is the constant motion of water in the ebb and flow of waves. Although after a certain time someone who is exposed to this will no longer hear it due to the constant repetition, the rhythmic rise and fall of the sound conveys the laws of ebb and flow and with them the movements of the moon and, further, the rules of cosmic consistency, namely "those impressions which are made on us by the bodies around us and which involve the infinite; that connection that each being has with all the rest of the universe" [Leibniz 1996, p. 55].[12]

Like his antipode Descartes, Leibniz believed in a cosmos suffused with ether, a fine, transparent dust that was thought to bear the forces that keep the planets and stars in their orbits and positions (Fig. 1) [Kemp 2000, p. 38]. For Leibniz, as for many of his contemporaries, Newton's gravitational force was the expression of an unverifiable action at a distance: an occult anathema. Who is right here can only be decided once it has been determined what the 95% of the universe is made up of that has been given the name *dark matter* or *dark energy*. What is decisive is that the interconnectedness of the universe affords the 'small perceptions' (*petites perceptions*) extensive knowledge even before the reflection apparatus of the concepts can be started up. This marks the most far-reaching theory of preconceptual knowledge ever formulated.

---

11 "cui etiam agendi vis inest, quae ut ego arbiter substantiae naturam constituit, adeo ut valor ille a Deo tributus revera sit vigor seu vis indita rebus" [Leibniz 1863, p. 239]; see: [Myrdal 2012, p. 85].
12 "ces impressions que des corps environnans font sur nous, qui enveloppent l'infini, cette liaison que chaque estre a avec tout le reste de l'univers" [Leibniz 1985, p. XXIV, German trans.: p. XXV].

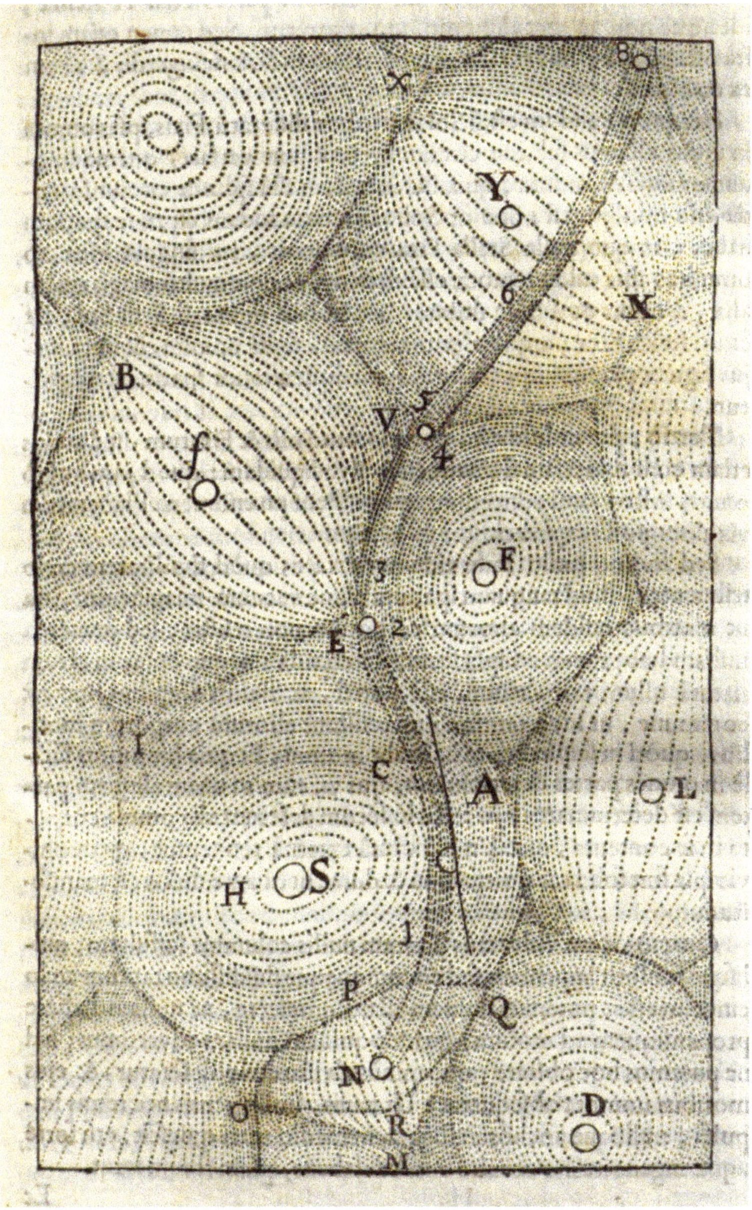

Fig. 1: René Descartes, ether-suffused cosmos, 1644. From: [Descartes 1664, p. 55]. Staatliche Bibliothek Regensburg, 999/Philos.430, p. 55, urn:nbn:de:bvb:12-bsb11109300-6.

The experience of the correspondence between the body and the shifting string spaces gives rise to small perceptions of their interaction. In this connection, the small perceptions are linked with the inner activity in the experience of space, whether in open nature or architecture, in landscapes or buildings. This interaction is of decisive importance for Leibniz's definitions of body and space. While, as indivisible units of creation, monads are spirit beings, they necessarily have a receptively agile body. For Hubertus Busche, this explains the "psychophysical expressionism" of all material substances.[13] The monads are shimmering perceptual instances that create their own means of disclosing, disseminating, and communicating the knowledge to be developed in them.[14] My consideration that the apparently autarkic, windowless monads necessarily dispose of windows [Bredekamp 2008] has been developed by Wolfram Hogrebe in his observation that it is not the case that monads do not *have* windows; rather, they *are* windows [Hogrebe 2013, p. 13]. It is the appetition that gives the monads a drive, a desire, which is responsible for all sensory motions and for the incessant motion in the universe [Leibniz 1969c, p. 644, par. 15; Leibniz 1998, par. 15, pp. 16–17; Bredekamp 2012, p. 89].

On February 25, 1716, Leibniz sent his third letter to Samuel Clark, a colleague and confidant of Newton's. This letter clearly reveals Leibniz's bitterness over the accusations of Newton's supporters that it was Newton and not Leibniz who founded infinitesimal calculus. In his counterattack, Leibniz accuses Newton of the assumption of a fixed, stable space that is a "real absolute being." According to Leibniz, this would lead to "great difficulties" [Leibniz 1969e, p. 682].[15] To this he opposes a relationality of space and time: "As for my own opinion, I have said more than once that I hold space to be something merely relative, as time is: that I hold it to be an order of co-existence as time is an order of successions" [ibid.].[16] With the conviction that time and space interact in a reciprocal relationalism, Leibniz's conviction of an all-pervasive *agens* deepened.

This is of decisive importance for Leibniz's definition of the perception of space. If without motion neither space nor time is able to exist, inertia has to be overcome in order to bring both phenomena into existence. The inertia-negating "original *activity*" quoted above appeared as a disturbance. Without deviation, which disturbs the stability of the geometric pattern, there can be no motion. Force is tendency in a double sense: as interest and appetition – in the sense of a *tendency-toward* and a *deviation-from*.[17] Internal force as appetition and *conatus* is synonymous as inclination: as deviating difference.

13 See for "psychophysischen Expressionismus," [Busche 1997, pp. 525–529; p. 59, fig. 1]. An early and subsequently forgotten argumentation was developed by Walter Feilchenfeld [Feilchenfeld 1923].

14 For seminal studies, see: [Busche 1997; Pape 1997].

15 "Ces Messieurs soutiennent donc, que *l'Espace* est un être reel absolue; mais cela les mene à de grande difficultés [...]" [Leibniz 1989b, p. 370, German trans.: p. 371].

16 "Pour moy, j'ay marqué plus d'une fois, que je tenois *l'Espace* pour quelque chose de purement relatif, comme *le Temps*; pour un ordre des Coexistences, comme le temps est un ordre de successions" [Leibniz 1989b, p. 370, German trans.: p. 371]; see: [Bredekamp 2008, p. 112].

17 "aliud est multoque plus continet rem non esse indifferentem sed vim habere et velut inclinationem ad statum retinendum atque adeo resistere mutanti" [Leibniz 1989c, p. 126, German trans.: p. 127]; see: [Myrdal 2012, p. 103].

4 Leibniz's Active Images

This principle can also finally be applied to 'active images,' which are central to the concept of 'image acts.' They have a long tradition that goes back to the ancient rhetorical teaching of *imagines agentes*, which while first appearing as mental images acquire an objective status through their power to act – and in this way also frame materially formed images [(Anonymous). Rhetorica ad Herrenium 1994, p. 176].

The status of images as a lively and thus active instance was developed by Leibniz in an inimitable way in his interpretation of the brain. In his *Nouveaux Essais sur L'Entendement Humain*, he imagines a room in which a screen has been stretched [Leibniz 1996, p. 144; Leibniz 1985, pp. 180–181]. An adequate visualization of this mental image is not yet known. Elements of it can be found in Johann Jakob Scheuchzer's illustration *The Silver Cord*, in which the brain and spinal cord appear on a giant cloth (Fig. 2).[18] The Swiss natural scientist was an esteemed friend of Leibniz's. In contrast to Scheuchzer's presentation of the brain *on* a screen, however, Leibniz's brain *is* a screen: a painting covered with an infinite number of motifs and above all folded in on itself: "there is a screen in this dark room to receive the images, and [...] it is not uniform but is diversified by folds representing items of innate knowledge" [Leibniz 1996, p. 144].[19] With this image, the brain becomes a microcosmic representation of the universe as a sphere folded in on itself, which forces all substances, without leaps, into ever new eddies and gyrations and new enfoldings.

An incomparable imagining of this motif has been created by the artist Giovanni Paolo Schor in his depiction of a Roman parade bed in which heaven and earth are united in a single, densely folded mountain of cloth (Fig. 3) [Karsten 2003, fig. 50; Walker 2007, pp. 154–155]. As if he had such an image before him Leibniz describes the universe as a continuous body that is "not divided but [...] like a tunic folded in various ways."[20]

Leibniz would have undoubtedly found confirmation for this idea in recent developments in astronomy in which the universe emerges as a vast structure in a permanent process of self-enfolding (Fig. 4). From the same cloth, so to speak, is Leibniz's image of the microcosmic brain as an infinitely folded, intricately painted canvas or screen in the dark chamber of the skull. To the extent that this "screen or membrane, being under tension, has a kind of elasticity or active force" [Leibniz 1996, p. 144],[21] it develops the inner *conatus*, which is realized as "a kind of elasticity or active force."

---

18 See: [Scheuchzer 1731, tab. DXCIII]; reproduced in [Felfe 2003, p. 52, fig. 52].

19 "dans la chambre obscure il y eut une toile pour recevoir les especes, qui ne fut pas unie, mais diversifiée par des plis, représentant les connoissances innées" [Leibniz 1985, p. 180, German trans.: p. 181] (translation modified).

20 "Totum universum est unum corpus continuum. Neque dividitur, sed instar cerae transfiguratur, instar tunicae varie plicatur" [Leibniz 1999, p. 1687, Z. 1–2].

21 "cette toile ou membrane estant tendue, eût une manière de ressort ou force d'agir" [Leibniz 1985, p. 180, German trans.: p. 181].

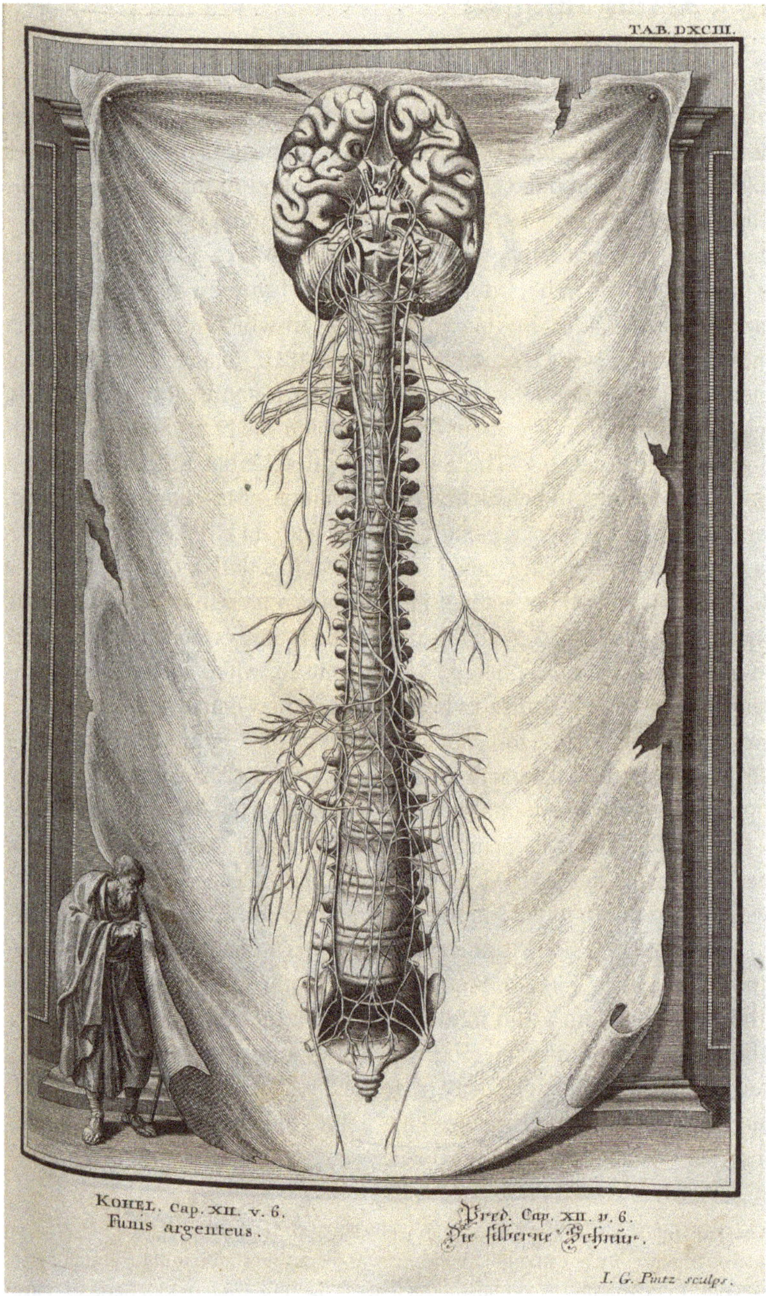

Fig. 2: Johann Georg Pintz after Johann Melchior Füßli, *The Silver Cord*, 1735, copper engraving. From: [Scheuchzer 1731, tab. DXCIII]. ETH-Bibliothek Zürich, Rar 5864, DOI: 10.3931/e-rara-10140. Public Domain Mark.

Fig. 3: Giovanni Paolo Schor, design of a bed for Maria Mancini, 1663, copper engraving. Photo © Dietmar Katz, bpk, Kunstbibliothek, Staatliche Museen zu Berlin.

It thereby "acts (or reacts) in ways which are adapted both to past folds and to new ones coming from the impressions of the images" [Leibniz 1996, p. 144].[22] In this interplay of old and new images, which causes the screen to vibrate and unfold what is embedded there, lies for Leibniz the dynamics of the images' agency.

But that is not all. To the extent that the screen undergoes constant vibrations and oscillations similar to the sound-producing vibration of a taut string, "vibrations or oscillations" arise "like those we see when a cord under tension is plucked" in such a way that it "gives off something of a musical sound" [Leibniz 1996, p. 145].[23] In this passage, the rivalry between the senses and the arts becomes a shared endeavor [Van Gastel, Hadjinicolaou, and Rath 2014]. The central proposition of this image – that "the screen which represents our brain must be active and elastic" [Leibniz 1996, p. 145][24] – corresponds to the principle of the universal validity of *agens*.

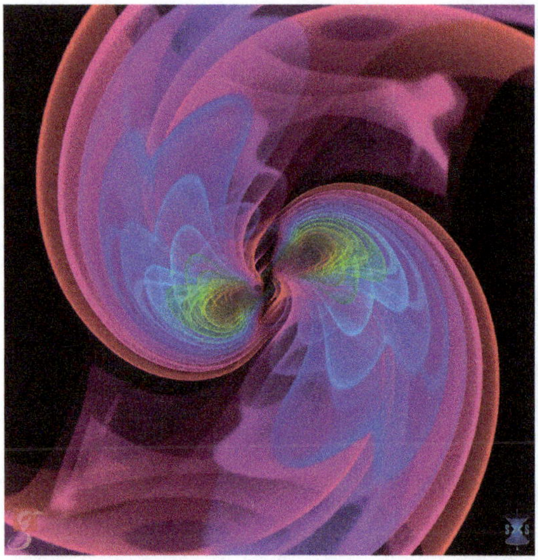

Fig. 4: Simulation of the gravitational waves of a black hole. © Serguei Ossokine, Alessandra Buonanno (Max Planck Institute for Gravitational Physics), Simulating eXtreme Spacetimes project, D. Steinhauser (airborne hydromapping).

22 "une action ou réaction accommodée tant aux plis passés qu'aux nouveaux venus des impressions des especes" [Leibniz 1985, p. 180, German trans.: p. 181] (English translation modified).

23 "vibrations ou oscillations, telles qu'on voit dans une corde tendue quand on la touche, de sorte qu'elle rendroit une manière de son musical" [Leibniz 1985, p. 180, German trans.: p. 181].

24 "la toile qui represente nostre cerveau soit active et elastique" [Leibniz 1985, p. 180, German trans.: p. 181] (English translation modified).

5 The *Vis Agendi*

Perfection is to be identified with force, the *vis agendi* [Myrdal 2012, p. 85], which represents a transferal of the *imagines agentes* to the whole of creation. For Leibniz, what gives things their internal force is God. Of course in our world seemingly purged of religious thought we flinch momentarily at the mention of a 'God' who is supposed to have given things their inherent force. This term retains its validity, however, if it is used as a sign for all phenomena that can be described but not explained: from the question of the cause of gravity to the problem of what preceded the big bang or the riddle of what causes the acceleration of the universe's expansion. One solution has been suggested by Baruch Spinoza with his formula *Deus, seu Natura*, which equates God with nature [Spinoza 2008, praefatio, pp. 382–383, Z. 24; Wollgast 1999, p. 15 and passim; Jung 2005, p. 131].[25] But even if this is rejected one must still reckon, on the level of phenomena, with all that Leibniz has proposed: a universally operative activity given to all creations that presses for connectivity and interaction.

The art lies in taking account of a general activity that – from the gravity supporting the universe to the cohesion of the monads – acts in the microcosmic domain without lapsing into mystification. A sharpened phenomenology could help to avoid this short circuit. Here, nature should not be thought solely as a passive-female potential for activation that is either subjected to violence or made the object of a peace settlement. Rather, it should be ascribed an active status.

This applies, in particular, to Leibniz and his comprehensive proposal of a force operating in each thing and creature, an acting vector, a cosmic *conatus*, and a microcosmically operating appetition. Leibniz, who in the context of his own time postulated an autonomously active *vis* collectively operating in matter, space, and artifacts, and an all-pervasive *vigor*, has provided us with a standard against which to sharpen the concepts being developed today.

Bibliography

[Anonymous]. Rhetorica ad Herennium. Ed. and trans. Nüßlein T. München, Zürich, Artemis & Winkler, 1994.

Aristoteles. Physik: Vorlesung über die Natur. German trans. Zekl HG. In: Aristoteles: Philosophische Schriften in sechs Bänden. Vol. 6. Hamburg, Meiner, 1995.

Aristoteles. Metaphysik. German trans. Bonit H and Seidl H. In: Aristoteles: Philosophische Schriften in sechs Bänden. Vol. 5. Hamburg, Meiner, 1995.

Bredekamp H. Die Fenster der Monade: Gottfried Wilhelm Leibniz' Theater der Natur und Kunst. 2nd edn. Berlin, Akademie Verlag, 2008.

25 The quote is "Deus, seu Natura," and not as commonly quoted the equivalent 'Deus sive Natura.'

Bredekamp H. Leibniz und die Revolution der Gartenkunst: Herrenhausen, Versailles und die Philosophie der Blätter. Berlin, Klaus Wagenbach Verlag, 2012.
Busche H. Leibniz' Weg ins perspektivische Universum: Eine Harmonie im Zeitalter der Berechnung. Hamburg, Meiner, 1997.
Carriero J. Conatus and Perfection in Spinoza. Midwest Studies in Philosophy, 2011 (XXXV), 69–92.
Descartes R. Principia Philosophiae. 4th edn. Amsterdam, Elzevir, 1644.
Ellenzweig S, Zammito JH, eds. The New Politics of Materialism: History, Philosophy, Science. London, New York, Routledge, 2017.
Feilchenfeld W. Leibniz und Henry More: Ein Beitrag zur Entwicklungsgeschichte der Monadologie. Kant-Studien 1923, 28, 323–334.
Felfe R. Naturgeschichte als kunstvolle Synthese: Physikotheologie und Bildpraxis bei Johann Jakob Scheuchzer. Berlin, Akademie Verlag, 2003.
Gastel JV, Hadjinicolaou Y, Rath M, eds. Paragone als Mitstreit. Berlin, Akademie Verlag, 2014.
Hogrebe W. Der implizite Mensch. Berlin, Oldenburg Verlag, 2013.
Jung M. Spinoza: Gott ist Natur – Natur ist Gott. Lahnstein, EMU-Verlag, 2005.
Karsten A. Künstler und Kardinäle: Vom Mäzenatentum römischer Kardinalnepoten im 17. Jahrhundert. Cologne, Weimar, Vienna, Böhlau, 2003.
Kemp M. Visualizations: The Nature Book of Art and Science. Berkeley CA, University of California Press, 2000.
Laerke M. Materialism, Constructivism, and Political Scepticism: Leibniz, Hobbes, and the Erudite Libertines. In: Ellenzweig S, Zammito JH, eds. The New Politics of Materialism: History, Philosophy, Science. London, New York, Routledge, 2017, 250–268.
Leibniz GW. Leibnizens mathematische Schriften. Vol. 7. Ed. Gerhardt Cl. Berlin, Halle, Schmidt, 1863.
Leibniz GW. A New System of the Nature and the Communication of Substances, as well as the Union between the Soul and the Body, 1695. In: Ed. and trans. Loemker LE. Leibniz GW. Philosophical Papers and Letters. 2nd edn. Dordrecht, Boston MA, London, Kluwer Academic, 1969a, 453–459.
Leibniz GW. On Nature Itself, or on the Inherent Force and Actions of Created Things, 1698. In: Ed. and trans. Loemker LE. Leibniz GW. Philosophical Papers and Letters. 2nd edn. Dordrecht, Boston MA, London, Kluwer Academic, 1969b, 498–508.
Leibniz GW. The Monadology, 1714. In: Ed. and trans. Loemker LE. Leibniz GW. Philosophical Papers and Letters. 2nd edn. Dordrecht, Boston MA, London, Kluwer Academic, 1969c, 643–653.
Leibniz GW. Letters to Nicolas Remond, 1714–15. In: Ed. and trans. Loemker LE. Leibniz GW. Philosophical Papers and Letters. 2nd edn. Dordrecht, Boston MA, London, Kluwer Academic, 1969d, 654–660.
Leibniz GW. The Controversy between Leibniz and Clarke, 1715–16. In: Ed. and trans. Loemker LE. Leibniz GW. Philosophical Papers and Letters. 2nd edn. Dordrecht, R. Reidel, 1969e, 675–721.
Leibniz GW. Nullum quidem librum [...]. In: Leibniz GW. Die philosophischen Schriften. Vol. 4. Ed. Gerhardt Cl. Berlin, Weidmann, 1880a, reprint Hildesheim. New York, Georg Olms Verlag, 1978, 393–400.
Leibniz GW. Système nouveau de la nature et de la communication des substances, aussi bien que de l'union qu'il y a entre l'âme et le corps: Erster Entwurf. In: Leibniz GW. Die philosophischen Schriften. Vol. 4. Ed. Gerhardt Cl. Berlin, Weidmann, 1880b, reprint Hildesheim. New York, Georg Olms Verlag, 1978, 471–477.
Leibniz GW. De ipsa natura sive de vi insita actionibusque creaturarum, pro dynamicis suis confirmandis illustrandisque. In: Leibniz GW. Die philosophischen Schriften. Vol. 4. Gerhardt Cl. Berlin, Weidmann, 1880c, reprint Hildesheim. New York, Georg Olms Verlag, 1978, 504–516.
Leibniz GW. Leibniz an Nicolas Remond, 10. 1.1714. In: Leibniz GW. Die philosophischen Schriften. Vol. 3. Gerhardt Cl. Berlin, Weidmann, 1887. reprint Hildesheim. New York, Georg Olms Verlag, 1978, 605–608.

Leibniz GW. Philosophische Schriften. Vol. 3,1–2. Nouveaux Essais sur l'entendement humain. Neue Abhandlungen über den menschlichen Verstand. Ed. and trans. von Engelhardt W and Holz HH. 2nd edn. Darmstadt, Wissenschaftliche Buchgesellschaft, 1985.
Leibniz GW. 1. Brief an Remond (10.1.1714). In: Leibniz GW. Philosophische Schriften. Vol. 5,2. Briefe von besonderem philosophischem Interesse. Ed. and trans. Wiater W. Darmstadt, Wissenschaftliche Buchgesellschaft, 1989a, 318–325.
Leibniz GW. Leibniz' drittes Schreiben [an Samuel Clark] (25.2.1716). In: Leibniz GW. Philosophische Schriften. Vol. 5,2. Briefe von besonderem philosophischem Interesse. Ed. and trans. Wiater W. Darmstadt, Wissenschaftliche Buchgesellschaft, 1989b, 370–371.
Leibniz GW. 1. Brief an B. de Volder (24.3./3.4.1698). In: Leibniz GW. Philosophische Schriften. Vol. 5,2. Briefe von besonderem philosophischem Interesse. Ed. and trans. Wiater W. Darmstadt, Wissenschaftliche Buchgesellschaft, 1989c, 122–133.
Leibniz GW. On Body and Force, Against the Cartesians (1702). In: Leibniz GW: Philosophical Essays. Ed. and trans. Ariew R and Garber D. Indianapolis, Hackett, 1989d, 250–256.
Leibniz GW. New Essays on Human Understanding. Trans. and ed. Peter Remnant and Jonathan Bennett. Cambridge, 1996.
Leibniz GW. Monadologie. Ed. and trans. Hecht H. Stuttgart, Reclam, 1998.
Leibniz GW. Sämtliche Schriften und Briefe. Col. VI, vol. 4,B. Ed. Berlin-Brandenburgische Akademie der Wissenschaften und Akademie der Wissenschaften Göttingen. Berlin, Akademie Verlag, 1999.
Mercer C. Leibniz's Metaphysics: Its Origins and Development. Cambridge, Cambridge University Press, 2001.
Myrdal P. Leibniz on Agency and Force. Doctoral Thesis, Uppsala University, 2012.
Pape H. Die Unsichtbarkeit der Welt: Eine visuelle kritik neuzeitlicher Ontologie. Frankfurt am Main, Suhrkamp, 1997.
Rozemond M. Leibniz on Final Causation. In: Newlands S, Jorgensen LM, eds. Metaphysics and the Good: Themes from the Philosophy of Robert Merrihew Adams. Oxford, New York, Oxford University Press, 2009, 272–291.
Scheuchzer JJ. Physica Sacra. Vol. 4. Augsburg, Ulm, Johann Andreas Pfeffel, 1731.
Scheuchzer JJ. Kupfer-Bibel, in welcher die Physica Sacra oder geheiligte Naturwissenschafft derer in der Heiligen Schrifft vorkommenden natürlichen Sachen deutlich erklärt und bewährt. Vol. 4. Augsburg, Ulm, Christian Ulrich Wagner, 1735, DOI: 10.3931/e-rara-10140.
Spinoza BD. Ethica. Pars quarta: De servitute humana seu de affectuum viribus. In: Blumenstock K, ed. Baruch de Spinoza: Opera – Werke. Latin/German. Vol. 2. Darmstadt, Wissenschaftliche Buchgesellschaft, 2008, 380–562.
Walker S. "Gantz irregulier": Ein barocker Betstuhl aus Rom. Barockberichte: Informationsblätter des Salzburger Barockmuseums zur bildenden Kunst des 17. und 18. Jahrhunderts 2007, 48–49, 154–161.
Wollgast S. Deus sive Natura: Zum Pantheismusstreit in der europäischen Philosophie und Religionsgeschichte. Sitzungsberichte der Leibniz-Sozietät 1999, 27 (8), 5–40.

Horst Bredekamp, born 29 April 1947 in Kiel, Germany, studied art history, archaeology, philosophy and sociology in Kiel, Munich, Berlin and Marburg, where he received his PhD in 1974. After an internship at the Liebieghaus in Frankfurt am Main, Germany (1974–1976) he became assistent professor at the university of Hamburg, where he tenured as Professor of Art History in 1982. Since 1993 he has tenured as Professor of Art History at the Humboldt University of Berlin. In addition 2003–2012 he has been a permanent fellow of the Institute for Advanced Study in Berlin (Wissenschaftskolleg zu Berlin). From 2015–2018 he was one of the three founding directors of the Humboldt Forum, Berlin.

Stephan Kammer

Rhetoric's Active Matters

1 Preliminary Remarks

In the following pages, I attempt, in the field of the historical theory of language, and thus also of poetics – and that means, in a long-term perspective, above all in the field of rhetoric – to trace a structural analogy to conceptions of active matter developed in the fields of physics, chemistry, biology, and the materials sciences. This attempt is risky for at least two reasons. First, is it not possible that the author has simply not or poorly understood the concerns and implications of the debates arising within these disciplines in which he is not a specialist? In that case, rather than a structural analogy, what was intended as part of a parallel undertaking would at best give the impression of a distorted and deceptive similarity – this similarity would then rest solely on the misunderstanding or incomprehension of what produced this anamorphic illusion. Second, and perhaps less seriously, might not the sought-for structural analogy be the mere effect of an illegitimate transference?[1] Even if the author has been able to grasp a few of the methodological concerns of active matter research, this still would not guarantee the transferability of the parallel findings. Indeed, as might be objected, even a more or less legitimate similarity should not automatically be accorded the dignity of an epistemic argument. Such transference problems are illustrated with one of the most frequently used examples of the constitution of metaphor as an abbreviated comparison: while Achilles may well have been as courageous as a lion, as far as we know he did not thereby acquire a hide, main, and quadrupedalism, let alone a lion's musculature and metabolism [Kohn et al. 2011] as a result of the 'as' of the comparison being omitted and replaced by a metaphor.[2]

As the previous example suggests, with these risks (as with their explication, which has assumed here the securitized form of a *dubitatio*, that is, the figure of an ostentatious doubt) we are already moving within the contentious history of

1 I will discretely pass over the aggravating circumstance that, while in my academic training and profession I have specialized in modern German literature, in the following I am forced to trespass into the territories of classical philology. Consequently, it cannot be ruled out that this too could lead to transference problems.

2 A fictional *experimentum crucis* of such tropic transference problems can be found in: [Lewitscharoff 2011].

Acknowledgment: The author would like to thank Benjamin Carter for translating the text.

The author acknowledges the support of the Cluster of Excellence 'Matters of Activity. Image Space Material' funded by the Deutsche Forschungsgemeinschaft (DFG, German Research Foundation) under Germany's Excellence Strategy – EXC 2025 – 390648296.

Open Access. © 2021 Stephan Kammer, published by De Gruyter. BY-NC-ND This work is licensed under the Creative Commons Attribution-NonCommercial-NoDerivatives 4.0 International License.
https://doi.org/10.1515/9783110562064-015

the elementary language-related *technē*[3] known as rhetoric, and thus, if standard epistemological definitions of this *technē* can be trusted, ultimately outside any claim to certainty. In this sense, the two risks mentioned at the beginning perhaps only prepare the ground for an additional rhetorical gesture: the disclosure of the *doxa*, and hence of the two or three things I *think* I know about active matter. First and most obviously, the classical distinction between form and matter, which ascribes to the former the prior and literally defining role in the ontological framework, and understands the latter as a secondary, passive-receptive category. Already in its choice of epithet, however, active matter has renounced its allegiance to this well-established understanding of matter with its many affiliated semantizations. Operating under the premise of self-movement and self-organization, matter here is conceptualized "as a nonrandom assemblage of particles whose self-movement, just as the resulting total motion" follow common laws [Friedman and Krauthausen 2017, p. 167]. Second, these laws are not imposed on it by a superordinate principle named spirit or form, or used in the context of human and/or mechanical processing; they are articulated dynamically and emergently as principles of motion and organization, and for this reason often involve an extraordinary degree of complexity with regard to their observation, description, and modeling. Third, and as a consequence of this, the traditional philosophical site of the question of matter should not be permitted to shape the discursive field in which the question of active matter is negotiated. At issue here is rather a phenomenal, often directly application-related material complex, which is primarily the object of pragmatics. Even when it is a matter of questions of observation and description, one will still tend to seek agreement concerning the operativity of this matter rather than its ontological status. The activity here gives rise, namely, to formal and structural effects, which need to be inquired into. Fourth and finally, what appears to have been called into question is a no less time-honored fundamental conceptual difference to the phenomena, discourses, and practices around active matter – that is, the culture-founding distinction between natural object and artifact. The table of contents of a publication on active matter lists a series of hybrids that for ontological fundamentalists just as for strict guardians of the border between nature and culture or between nature and technics are likely to appear equally monstrous: programmable bacteria,

3 In the following, I employ the Greek loanword *technē* (from ἡ τέχνη; the *Greek-English Lexicon* [1940] by Liddell and Scott provides the following basic meanings: "art, skill, cunning of hand"; "way, manner, *or* means whereby a thing is gained"; "a set of rules, system *or* method of making *or* doing") that can play an opposition role to nature in the present-day differential pairs nature/culture and nature/technics. *Technē* links the system claim that we associate with the concept of culture with operativity or operationalizability, which is at the basis of the *dispositif* of technics. Also and particularly for rhetoric, this double figuration is fundamental. Hence, from this Greek perspective, the German term *Kulturtechnik*, which brings together both areas of focus, and is often used in such contexts, would have to be seen as a pleonasm. In *technē*, just as in the Latin term derived from this, *ars*, the semantics of 'technics' and 'art,' which in the modern period have become increasingly divergent, are integrated to an equal extent and without substantial distinction.

heat-active textiles, self-folding polymer sheets, communicatively interacting materials, and so on [Tibbits 2017].

For the literary historian, all this has a structural resemblance to the history of key questions relating to the theory of language as they have been developed in the field of rhetoric from antiquity to the modern period. While with the abandonment of the paradigms and claims of 'old rhetoric' in the eighteenth century[4] a considerable mistrust of this kind and site of linguistic questioning of language began to establish itself, might it not be possible, with a sideways glance at concepts of active matter, to provide a different reading of the history of this challenge of/through rhetoric? It would be both presumptuous and superfluous to speak of a vindication of rhetoric; however, from the perspective of the discussions outlined above, the lines of conflict of an enduring debate nevertheless appear in a new light. Is language considered by itself a passive, neutral medium that transports and advances the concerns of the beautiful, the true, and the good – as well as of course, if we are not careful, those of language-shapers with less pure intentions? Or is it rather on the basis of the structure and use of language that the ordering framework that makes it possible to distinguish the beautiful from the ugly, the true from the false, and the good from the bad first emerges? In relation to what is the proper use of language to be measured? To the truth of its communications or to the knowledge of its rules and their implementation? If the experienced speaker (who to this end must undergo a rigorous training and supervision) is responsible for the effects of persuasion and the production of conviction, is the user therefore responsible for the use of this *technē*? Or is it the power of speech itself that affects the listener almost like a psychoactive substance? Passivity or self-activity, ontology or pragmatics, instrumentality or inherent logic – since its beginnings, the history of rhetoric has been inscribed along the threshold of these opposing poles. Accordingly, it is hopefully a little more than professional bias that allows the literary historian to surmise the said structural analogies in the discussions around active matter. In short: (1) rhetoric understands language not as a material that is formed, but as a forming principle that acquires a regulatory and performative function; (2) the emergence of rhetorical order is reflected in the fact that the *ars rhetorica* is articulated both as an analytical and as a poietic systemic praxis – which means that, according to its premises, the rhetorical artifact can be both produced and judged; (3) that rhetoric is a pragmatic discipline cannot be doubted even by its severest enemies – indeed, one is in the habit of reproaching it for precisely this in order to deny it an independent epistemological status; (4) it is only in relation to the *physis*/*poiesis* distinction that a comparable claim to hybridity for rhetoric and its products seems unlikely – nevertheless, it will be shown that the problems raised by this distinction also pertain to the present field of inquiry.

4 For a pertinent study, see: [Bender and Wellbery 1990].

On the legitimacy of the suppositions outlined above, even the one who makes them can hardly decide. It must be left to the experts in their respective fields. Will they recognize and accept an object of study that shows the traits and characteristics set out in the following as an object of their own research interests and questions? Will they recognize in the problems and issues raised structural parallels to their own? In view of this uncertainty, I will simply attempt to show what a consideration of particular moments in the history of rhetoric inspired by the initial findings of the discussions around active matter believes to have concretely found, without overstretching this inspiration. To this end, I have selected a number of scenes in which the negotiation of rhetoric's status becomes especially visible, as well as debates that have had a particularly lasting influence. I will begin (in Section 2) in the fifth century BCE, where following a prologue on the theater I will turn (in Section 3) with Gorgias to a threshold and founding figure of this history, and reconstruct the establishment of a specific power of speech. Section 4 focuses, if only in exemplary highlights, on the status of the relation between nature and technics in rhetoric's founding (Gorgias) and systematization (Aristotle). Lastly, in Section 5, I will consider the return or reactualization of these inaugural debates at the beginning of the twentieth century in structuralist linguistics.

If I promise to turn a spotlight on the history of rhetoric, it is important to bear in mind that 'rhetoric' here is not to be understood as the name for the slightly disreputable soft skill of self-marketing, or the even more disreputable technique for the manipulation of others, as is now often the case far from the historical contexts of the discipline. According to its status within an order of knowledge that has remained stable for over two millennia, in which world knowledge is more or less synonymous with the knowledge of (the proper use of) language, rhetoric is considered as a kind of relay, one that both mediates between the mastery of language (grammar) and the rules of argumentation (dialectics) and reflects on their relation. A mediating foundational science that is also linked with an explicit praxis claim – does this not show at least a promising elective affinity to the present-day interdiscipline of active-matter research?

2 Prologue on the Theater: Aristophanes' *Aἱ νεφέλαι*

In the first scene of Aristophanes' comedy *Clouds* (423 BCE),[5] we are introduced to an Attic landowner who after a troubled and sleepless night is up at dawn examining his ledger. This is the elderly Strepsiades, who because of his son's passion for

5 All ancient texts are quoted from the bilingual editions of the Loeb Classical Library to the extent that they are available there. In this instance, see: [Aristophanes 1988]. In the following, the quotations will be indicated with the verse number. Also consulted was [Aristophanes 1968].

horses, and despite his not inconsiderable assets, has become burdened by debts, and now fears being forced "right off my property" (ἐκ τῶν ἐμῶν, v. 33). At the origin of this economic – and that means financial and familial – misery is, as we learn, what seems to have been a slight mésalliance. The prosperous farmer found his bride in the urban aristocracy – and the latter's concern for the outward signs of her social standing has clearly been passed on to their son. As the despairing father suspects, this was already in evidence at the time of their son's naming. The mother had insisted on the name having a knightly semantics – and she got her way. The boy was called Pheidippides, because the mother "was for adding *hippos* to the name" (ἵππον προσετίθει πρὸς τοὔνομα, v. 63). What to do? Strepsiades has learned of "a Thinkery of sage souls" (ψυχῶν σοφῶν [...] φροντιστήριον, v. 94), where he would like to send his son to study. He has heard that its members are able to "argue convincingly" (λέγοντες ἀναπείθουσιν, v. 96) about the most unbelievable things, and whoever goes there to study – in exchange for a fee – is able to win every debate (λέγοντα νικᾶν, v. 99) he participates in. What Strepsiades expects from such a training, he openly declares to his son:

> I'm told they have both Arguments there, the Better, whatever that may be, and the Worse. And one of these Arguments, the Worse, I'm told, can plead the unjust side of a case and win. So, if you learn this Unjust Argument for me, then I wouldn't have to pay anyone even a penny of these debts that I now owe on your account.[6]

Pheidippides rejects his father's offer, but not because of moral scruples, but for fear of social humiliation within his peer group. In his view, the residents of the Thinkery have a far too unknightly constitution, being "pasty-faced" and "unshod" (v. 103). And so it is the old man himself who knocks at the door of the Thinkery. Strepsiades' confidence in his ability to fully master the arts he has heard spoken of is rightly low; and so it comes as no surprise when his teacher Socrates – this, we learn, is the name of the school's head – soon despairs of the rustic habits and, in particular, the intellectual failings of his new client, and turns him out. However, by putting to use what he believes to be the skills he would have acquired from the school, Strepsiades is nonetheless ultimately able to rid himself of his creditors. While it is certainly boorish antics rather than subtle sophisms that form the basis of his arguments, the distraction and confusion into which this casts his creditors allows Strepsiades to believe, at least for a time, that he has achieved his purpose.

---

6 εἶναι παρ' αὐτοῖς φασὶν ἄμφω τὼ λόγω,
τὸν κρείττον', ὅστις ἐστί, καὶ τὸν ἥττονα.
τούτοιν τὸν ἕτερον τοῖν λόγοιν, τὸν ἥττονα,
νικᾶν λέγοντά φασι τἀδικώτερα.
ἢν οὖν μάθῃς μοι τὸν ἄδικον τοῦτον λόγον,
ἃ νῦν ὀφείλω διὰ σέ, τούτων τῶν χρεῶν
οὐκ ἂν ἀποδοίην οὐδ' ἂν ὀβολὸν οὐδενί (v. 112–118).

Here is not the place to consider the details of the plot or the comedy's controversial reception history. A brief observation will have to suffice. The main characteristics of Socrates and his school depicted in the comedy [Gelzer 1956] correspond exactly with the accusations that led, a quarter of a century after the play's first performance, to the trial of the historical Socrates: the corruption of youth, who, goaded on by him, revolted against traditional values and customs of Athenian society; and impiety (ἀσέβεια), which is, to say, the rejection of the official Athenian gods and the adoption of new, self-chosen deities. When at the end of the play, Strepsiades – having just been given a beating by his son Pheidippides – sets the Socratic Thinkery alight, he does so, as he says, "most of all for wronging the gods" (μάλιστα δ'εἰδὼς τοὺς θεοὺς ὡς ἠδίκουν, v. 1509). In his *Apology*, Plato lets Socrates refer explicitly to *Clouds* when he responds to the accusation brought against him by remarking, "you yourselves saw these things in Aristophanes' comedy" (19c).

Kenneth J. Dover was certainly not wrong to point out in this connection "that the Athenians did not necessarily do what Ar[istophanes] told them to do" [Dover 1968, p. lvi], and thus to reject the assumption that a comedy that at the time of the trial was hardly fresh in the minds of the Athenians might have contributed even in part to Socrates' conviction. With respect to the accusation of impiety, however, the first scene of the comedy does give wide scope to Socrates' self-authorization to appoint more suitable gods than those of the Athenian canon. The manner in which this is justified explains my decision to begin my history with this prologue on the theater. *Chaos*, *Tongue*, and the titular *Clouds* – according to the Aristophanean Socrates, it is these deities and no others that are suited to his method of schooling, and therefore the gods that should be honored at the Thinkery, and to which he initiates Strepsiades (τὸ Χάος τουτὶ καὶ τὰς Νεφέλας καὶ τὴν Γλῶτταν, τρία ταυτί, v. 424). *Chaos* would have been known to the Athenians from Hesiod's *Theogony* as the gaping, empty abyss from which the world first arose – and thus as part of a well-established cosmological metaphor of emergence.[7] The *Tongue* as a personification of the organ of speech should hardly surprise as a god of choice for a school of rhetoric. But what of the titular *Clouds*? In the first scene, the Aristophanean Socrates outlines the area of responsibility of these specially selected deities as follows:

> they're heavenly Clouds, great goddesses for idle gentlemen, who provide [παρέχουσιν] us with judgment and dialectic and intelligence, fantasy and circumlocution and verbal thrust and parry.[8]

---

7 Whether the understanding of the term as a disordered confusion that has been popular up to the present had already established itself in the outgoing fifth century has not been settled. Given the depiction in the comedy of the conditions and events in the Thinkery, however, this does not seem improbable.

8 ἀλλ' οὐράνιαι Νεφέλαι, μεγάλαι θεαὶ ἀνδράσιν ἀργοῖς,
αἵπερ γνώμην καὶ διάλεξιν καὶ νοῦν ἡμῖν παρέχουσιν
καὶ τερατείαν καὶ περίλεξιν καὶ κροῦσιν καὶ κατάληψιν (v. 316–318).

In the Greek wording of Socrates' praise of the *Clouds*, these do not appear exclusively as mere agents or 'providers' of epistemic and practical competencies. παρέχειν – particularly when natural contexts are in play – means, very generally, 'to bring forth,' that is, to bring into appearance and to make available. To the eager Strepsiades, to whom the divine immortality of what he had previously considered mere atmospheric matter, "mist and dew and smoke" (v. 339), refuses to fully manifest itself, the *Clouds*' song appears despite these reservations to have already placed him in a certain disposition:

> So that's why my soul has taken flight at the sound of their voice, and now seeks to split hairs, prattle narrowly about smoke, and meet argument with counterargument, puncturing a point with a pointlet.[9]

According to Socrates, the *Clouds*, which in the form of the chorus also appear performatively on the stage, "feed" a whole host of eulogists, principally sophists (v. 332). Furthermore, their appearance allows the comedy by means of a metaleptic joke to treat another important aspect. Strepsiades wants to know why the *Clouds* that appear on the stage, unlike those in the sky, "look like mortal women" (v. 341–342), and learns that they are equipped with an unlimited potential for metamorphosis, and are capable of coming into being in any conceivable form (γίγνονται πάνθ' ὅτι βούλονται, v. 348).

When, in her considerations on the digital techniques used in animated films, Esther Leslie notes that clouds as a "source of any imaginable form," rather than being "a metaphor of something else" [Leslie 2017, p. 230] are capable here "of acting like metaphors" themselves [ibid., p. 234], she only names a present-day technological and aesthetic realization of a principle of activity for which the Aristophanean Socrates had proposed the deification of the clouds:

> But with all these forms of cloud, the cloud, every cloud, painted clouds, cinematic clouds, clouds of fantasy and clouds of particles, there is no fixed form or essence. The cloud, these clouds are made of an assemblage of materials, assorted ideas, beliefs and practices. The cloud is itself and something else. It is the cloud of all history and of none at all. The cloud is the shape of something really in the world, as much as it is the sign of an effervescent affective actuality. [Ibid., p. 234]

In Aristophanes' comedy, the fifth century BCE had already realized its own staging of this cloud-like potential for self-activity and self-animation: rhetorically shaped speech, λόγος (*logos*).

---

9 ταῦτ' ἄρ' ἀκούσασ' αὐτῶν τὸ φθέγμ' ἡ ψυχή μου πεπότηται
καὶ λεπτολογεῖν ἤδη ζητεῖ καὶ περὶ καπνοῦ στενολεσχεῖν
καὶ γνωμιδίῳ γνώμην νύξασ' ἑτέρῳ λόγῳ ἀντιλογῆσαι (v. 319–321).

3 *Dynamis* of Speech

If Aristophanes' *Clouds* mocks sophist rhetorical didactics under the name of Socrates, what strikes us as odd from today's point of view is that the Platonic Socrates was himself one of rhetoric's severest critics, and another protagonist for such an attack seems much more plausible: the Sicilian orator and teacher of rhetoric Gorgias.[10] In the history of rhetoric, Gorgias is considered the first representative of the discipline whose work and teaching has been at least partly preserved and documented; his fellow citizens Tisias and Corax, who are considered the 'inventors' of rhetoric, have more of a legendary status. Gorgias was a native of Leontini, a city close to present-day Syracuse in Sicily. While there is some disagreement about the exact dates of his birth and death (the two surviving testimonies differ by approximately 15 years), all sources agree that Gorgias reached an age of well over 100, which is astonishing not only for Greek antiquity. He was born either around 500 BCE (according to one source), or more plausibly in 484 BCE (according to the other), and died either around 390 or in 376. In either case, this means he would have lived through a large part of the fifth century BCE that was so important for the cultural history of ancient Greece. However, Gorgias did not live permanently in the geographical center of these political, scientific, and artistic developments. He is said to have arrived in Athens for the first time only in 427 – and thus already at an advanced age – as part of a diplomatic mission on behalf of Leontini. Nevertheless, his life in the provinces did no harm to his supra-regional fame as a rhetorician. Later, Cicero would report that he was so highly honored in Greece that a statue erected for him in Delphi was not merely gilded, but – uniquely – made of solid gold (*non inaurata statua sed aurea statueretur*) [Cicero 1942, p. 100, sec. 129].

Besides his lucrative teaching activities, Gorgias also appears to have been politically active – that is suggested at least by the said diplomatic mission to Athens, during which he successfully petitioned for military support for his home polis of Leontini against the neighboring city of Syracuse. Moreover, he was not only a practicing teacher of rhetoric; according to a number of sources, he also authored a system for teaching rhetoric, a so-called *technē*. If true, this would have been one of the first rhetorical textbooks. However, nothing of this has survived, not even fragments, as with almost all of Gorgias' other writings. With the exception of two speeches, namely, none of Gorgias' writings have been preserved in their entirety; and even the surviving versions of the two speeches are paraphrases made long after Gorgias' death. These circumstances are further exacerbated by the fact that the only extensive contemporary depiction of Gorgias – it is also the most famous and influential – is found in a highly partisan account, or at least not one that could fall in favor of the rhetoric teacher from Leontini. This is Plato's dialogue *Gorgias*, whose partiality is already shown by the fact that, of all

10 On the affinity between Aristophanes and Socrates, see [Euben 1997, pp. 109–138].

those participating in the debate, and despite his notorious eloquence, Socrates' titular antagonist has, both quantitatively and qualitatively, the least to say. It can probably be stated without exaggeration that the Platonic Gorgias is ultimately a strategic straw man in a war of words for the intellectual high ground.

What has been said of Gorgias can also be said of the anything but homogenous group of sophists. In this respect, the divine patronage of the Aristophanean *Clouds* does not appear to have been much help. Even the contemporary discussion about the achievements and merits of sophistics turned out to be totally to their disadvantage. And just as with Gorgias, the same is true of the posthumous treatment of their writings. Of the numerous treatises that are known through their titles, none have been fully preserved. Only fragments of various lengths have come down to us – and of these only a fraction can claim the status of verbatim copies. Accordingly, most of these fragments are merely paraphrases of what this or that sophist is supposed to have said, and thus representations of representations, which in view of the negative assessment of the Sophist's (theoretical) practices must be considered with caution. To put it bluntly, we know about the deeds, words, and thoughts of the sophists through accounts provided by their adversaries, and chiefly via the writings of their fiercest enemy: Plato. The French sophistics expert, Barbara Cassin, summarizes the picture of sophistics that has arisen from this transmission situation as follows:

> If we consider the Platonic dialogues as a whole, we can indeed discern the figure of sophistics, which will henceforth belong to the tradition. It is devalued on all grounds – ontologically, because the sophist is not concerned with being, but seeks refuge in non-being and what is accidental; logically, because he is not in pursuit of truth or dialectical rigor, but merely opinion, seeming coherence, persuasion, and victory in the oratorical joust; ethically, pedagogically, and politically: his goal is not wisdom and virtue, for the individual or for the city, but rather personal power and gain; the sophist is even devalued on literary grounds, since the figures of speech he makes use of, his style, are merely the bulges of an encyclopedic vacuity. If one makes use of the standard of being and truth in order to judge the teaching of the sophist, it must be condemned as pseudo-philosophy: a philosophy of appearances and a mere appearance of philosophy. [Cassin 2000, p. 106]

But how did the sophists acquire such a poor reputation? That is not easy to say, for the sophists in no way formed a philosophical school with a unified doctrine. Quite the opposite in fact. Strictly speaking, their different, even opposing philosophical and political views have only one common denominator: they were for the most part professional and well-paid teachers primarily of rhetoric, which they understood as the key instrument of a practical education. A few of them, however – but chiefly Gorgias – went so far in their estimation of rhetoric as to view artful and properly wrought speech as a "psychagogic charm."[11] And it is this conception of *logos* that makes the Gorgian understanding of rhetoric of interest to my inquiry.

---

11 "psychagogisches Zaubermittel" [Meister 2010, p. 154]. Unless stated otherwise, all the translations from German quotes were made by Benjamin Carter.

For considered closely, this understanding is not limited to the manipulative instrumentality suggested by the above designation.

The central passage on the efficacy of speech is found in Gorgias' famous *Encomium of Helen*:

> The power of speech has the same relation with the arrangement of the soul as the arrangement of drugs has with the nature of bodies. For just as some drugs draw some fluids out of the body, and others other ones, and some stop an illness and others stop life, in the same way some speeches (*logoi*) cause pain, others pleasure, others fear, others dispose listeners to courage, others drug and bewitch the soul by some evil persuasion.[12]

This definition of the efficacy of speech cannot be explained with a representational understanding of language. Rather, it points to the performative function of speech – or, to use Gorgias' own term, its *dynamis*. The latter has been described "as a specific, *non-physical force* [...], albeit one whose dynamics approximate an external, bodily force."[13] However, there are also other passages in Gorgias in which one finds sufficient indications that he profoundly mistrusted the representational function of language. Thus, as we read in the third thesis of his treatise *On Nonbeing or On Nature*, which has only survived in later paraphrases: if there were an entity, and if this entity were to be an object of knowledge, this knowledge could not be communicated.[14] Sextus Empiricus, to whom we also owe a commentary on this unpreserved treatise, laconically notes that, in view of these Gorgian aporias, "the criterion of truth is swept away" – for where something neither is, nor can be known, nor communicated to someone else, naturally no such criterion can exist [Sextus Empiricus 1935, pp. 45, I, 87]. While the relativistic, even nihilistic coloring characterizing the paraphrases of his treatise is surely due in large part to Gorgias himself through his choice of theme and line of argument, it can only be fully maintained if one ignores or overlooks the positive

12 τὸν αὐτὸν δὲ λόγον ἔχει ἥ τε τοῦ λόγου δύναμις πρὸς τὴν τῆς ψυχῆς τάξιν ἥ τε τῶν φαρμάκων τάξις πρὸς τὴν τῶν σωμάτων φύσιν. ὥσπερ γὰρ τῶν φαρμάκων ἄλλους ἄλλα χυμοὺς ἐκ τοῦ σώματος ἐξάγει, καὶ τὰ μὲν νόσου τὰ δὲ βίου παύει, οὕτω καὶ τῶν λόγων οἱ μὲν ἐλύπησαν, οἱ δὲ ἔτερψαν, οἱ δὲ ἐφόβησαν, οἱ δὲ εἰς θάρσος κατέστησαν τοὺς ἀκούοντας, οἱ δὲ πειθοῖ τινι κακῇ τὴν ψυχὴν ἐφαρμάκευσαν καὶ ἐξεγοήτευσαν [Gorgias 2016b, pp. 178, 180, D26,14].

13 "als einen spezifischen *nicht-physischen Zwang* [...], der in seiner Dynamik jedoch einem äußerlichen, körperlichen Zwang nahekommt" [Franz 1999, p. 122]. See also the succinct presentation in [Strowick 2009, pp. 54–64].

14 "But even if they [= things] are knowable, how could someone, he asks, indicate them to someone else? For what one sees, how, he asks, could one say this by a speech (*logos*)? Or how could that thing become clear to someone who hears, but does not see it? For just as sight does not know sounds, so too hearing does not hear colors, but sounds: and someone who speaks utters a speech, but not a color or a thing" (εἰ δὲ καὶ γνωστά, πῶς ἄν τις, φησί, δηλώσειεν ἄλλῳ; ὃ γὰρ εἶδε, πῶς ἄν τις, φησί, τοῦτο εἴποι λόγῳ; ἢ πῶς ἂν ἐκεῖνο δῆλον ἀκούσαντι γίγνοιτο, μὴ ἰδόντι; ὥσπερ γὰρ οὐδὲ ἡ ὄψις τοὺς φθόγγους γιγνώσκει, οὕτως οὐδὲ ἡ ἀκοὴ τὰ χρώματα ἀκούει, ἀλλὰ φθόγγους· καὶ λέγει ὁ λέγων, ἀλλ' οὐ χρῶμα οὐδὲ πρᾶγμα) [Gorgias 2016c, pp. 226–227, D26,21].

and productive implications for speech that he also derives from his argument. The two surviving sources of the treatise do this in a more or less radical way. In the first, anonymous paraphrase at least (D26), there is hardly a trace of this productive aspect. In Sextus Empiricus, a few, possibly decisive, hints are to be found, even if they are not pursued systematically – the suggestion, for example, that "it is not the speech that presents the external thing, but the external thing that indicates the speech" [Gorgias 2016c, p. 241].[15]

Despite the problematic nature of the sources, there is little doubt that Gorgias did not want the functional analogy between drugs (*pharmaka*) and speeches (*logoi*) referred to above to be understood simply in the mode of a comparison or allegory, but as a real equivalence. In this sense, speech becomes an active substance in its own right, thus fulfilling the first two criteria of our activity search profile: speech itself and its *dynamis* take the place of the concrete and the real. As a result, this functional position does not, as in a semiotic understanding of language, remain subject to what the speech refers to or what it represents. As Thomas Buchheim notes:

> Speech for him should under no circumstances be divided into sound form as the sensual aspect, and meaning as the notional content. [...] In the sought for space of development of speech, the sound character and signifying function of language are grasped together in a specific play of intensity and configuration.[16]

For the pragmatics and operativity of rhetorical efficacy (this corresponds to the third of our activity criteria), ancient rhetoric had its own category, and this is also already found developed in Gorgias' writings and fragments. The equivalence between drugs and speech mentioned above is at the basis of a distinct function: persuasion or πειθώ. As Wolfram Groddeck summarizes the related discussion in his foundational work on rhetoric: "in Gorgias, [...] πειθώ appears to have an original meaning: it means the active efficacy of speech, a force that posits reality through and in language."[17] The resulting power of speech is treated in large parts of Gorgias' *Encomium of Helen*. Here, Gorgias places speech in an equivalent operative set with violence and eros. Thus, according to his remarks on the fateful event, it is not improbable that Helen traveled to Troy as the result of the rhetorical persuasiveness of a speech – that is, it was perhaps by being "persuaded by words" (λόγοις πεισθεῖσα) that Helen followed Paris [Gorgias 2016b, pp. 170–171, D24,6]. This hypothesis is found literally at the center of Gorgias' speech, and it is given by far the most extensive

15 οὐχ ὁ λόγος τοῦ ἐκτὸς παραστατικός ἐστιν, ἀλλὰ τὸ ἐκτὸς τοῦ λόγου μηνυτικὸν γίνεται [Gorgias 2016c, p. 240].

16 "Was ihm Rede ist, darf auf keinen Fall zerlegt werden in Lautgestalt einerseits als sinnlichen Aspekt und Bedeutung andererseits als gedanklichen Gehalt. [...] In dem gesuchten Entfaltungsraum der Rede sind Lautcharaktere und Bedeutungsfunktion der Sprache amalgamiert in einem eigenen Spiel von Intensität und Konfiguration begriffen" [Buchheim 2012, pp. xii–xiii].

17 "Bei Gorgias scheint [...] πειθώ eine ursprüngliche Bedeutung zu haben; es meint das aktive Bewirken der Rede, eine Kraft, die Wirklichkeit durch und in Sprache setzt" [Groddeck 1995, p. 29].

discussion: it is treated in 7 of the 21 sections of the speech – and these are sections 8–14. In the eighth section, and thus the opening of the middle third of the speech, Gorgias narrows the related argument down:

> But if it was speech that persuaded and deceived her soul, it is not difficult to make a defense with regard to this too and to secure acquittal from the accusation in the following way. Speech is a great potentate that by means of an extremely tiny and entirely invisible body performs the most divine deeds.[18] [Gorgias 2016b, p. 175, D24,8]

Gorgias derives this efficacy first of all from poetry, which he defines as the first special form of speech: metrical speech, or in his own words λόγον ἔχοντα μέτρον, as he writes in the following ninth section. By means of this, the affects enter into the hearer – fearful shuddering, tearful pity, and mournful desire. Furthermore, poetry, according to Gorgias, enables one to sympathize with the affects of others. Shortly afterward, already in Aristotle, this becomes the central idea in the theory of tragedy. My first intention here, however, was to postulate that Gorgias took this detour via the special forms of speech in order to generalize its efficacy. That is to say that the efficacy of poetry should serve to illustrate the efficacy of speech as a whole. One finds evidence of this in the metaphor of the "entry" of the affects of speech into the hearer – the verb used by Gorgias, εἰσέρχομαι, also designates the entry of the chorus or the actor onto the stage. Just as the chorus and actors enter the stage as agents, so too do the affect-words enter as agents into the hearers of a speech. One should not then overlook the point that for Gorgias this faculty of language is at the basis of social behavior – it carries out what we would now designate with the term empathy. Following on from this, the tenth section invokes a division of speech effects that goes beyond the creation of affect. Here it is matter of physical effects: the summoning of pleasure, the dispelling of pain, and so on. To illustrate these speech effects, Gorgias selects incantation as a special form of speech: "For the power of an incantation, when it is conjoined with the opinion of the soul, beguiles it, persuades it, and transforms it by sorcery" [Gorgias 2016b, p. 175, D24,10].[19] Hence, magic would be an additional and, according to common opinion, relatively dubious practice, whose efficacy is analogous to that of speech (perhaps only the extension of rhetorical *dynamis* to include things and/or teleaction). This middle part of Gorgias' *Encomium of Helen* relating to the power and efficacy of speech concludes in the fourteenth section the abovementioned definition of speech as a *pharmakon*.

Gorgias' rhetoric promises to make the power of speech illustrated in the example operative. An echo of this promise is still found in the Platonic Gorgias, when in

18 εἰ δὲ λόγος ὁ πείσας καὶ τὴν ψυχὴν ἀπατήσας, οὐδὲ πρὸς τοῦτο χαλεπὸν ἀπολογήσασθαι καὶ τὴν αἰτίαν ἀπολύσασθαι ὧδε· λόγος δυνάστης μέγας ἐστίν, ὃς σμικροτάτῳ σώματι καὶ ἀφανεστάτῳ θειότατα ἔργα ἀποτελεῖ [Gorgias 2016b, pp. 174, D24,8].

19 συγγινομένη γὰρ τῇ δόξῃ τῆς ψυχῆς ἡ δύναμις τῆς ἐπῳδῆς ἔθελξε καὶ ἔπεισε καὶ μετέστησεν αὐτὴν γοητείᾳ [Gorgias 2016b, pp. 174, D24,10].

answer to the question about rhetoric's effectiveness he remarks that it is "the greatest good, and a cause [...] of freedom to mankind at large" [Plato 1925, pp. 279, 452d].[20] According to Gorgias, *τὸ πείθειν* is the faculty that distinguishes the active matters of rhetoric from all other social forms of knowledge and praxis; for him, this gathers all the latter's powers under rhetoric's command. That might sound boastful if like Socrates one considers the articulations of speech merely as a cosmetic supplement to the content, and understands speech itself merely as the representation of a (mental) state of affairs instead of as a *pharmakon* that brings something into effect. In short, only when one presupposes the Platonic reprimand whereby rhetoric is accused of having ended up in the "role of the however perfect perfumer and decorator of a significant content"[21] does rhetoric begin to lose its quality as a matter of activity.

4 Nature/*Technē*

The fourth analytical finding that I have incorporated into my search profile on the active matters of rhetoric appears to have slipped out of focus in the previous section. In the case of rhetorical system formation, a hybridity between nature and *technē* seems to be a rather improbable requirement – for both the human actors and the products. As we know, the sophists extended to a maximum the limits of what can be learned and practiced, and thus of what can be achieved through pedagogical discipline – indeed, in the opinion of its critics, overextended this.[22] For the private and itinerant teachers mindful of their own profit that made up the majority of this richly heterogeneous group, this was offered for practical reasons alone. According to a criticism by Isocrates, however, they exaggerated – precisely with regard to the teaching of the *rhetorikē technē* – the learnability of the art of speech

20 μέγιστον ἀγαθὸν καὶ αἴτιον ἅμα μὲν ἐλευθερίας αὐτοῖς τοῖς ἀνθρώποις [Plato 1925, pp. 278, 452d].

21 "Rolle des wie auch immer perfekten Parfumeurs und Dekorateurs eines bedeutenden Inhalts" [Buchheim 2012, p. xiv].

22 It is probably no accident that the Platonic critique of the sophists concurs structurally in an essential feature with the critique of writing in *Phaidros*: the sophists/writing want(s) to make all knowledge available to everybody without distinction, but without disposing of this knowledge themselves/itself. "Writing, Phaedrus, has this strange quality, and is very like painting; for the creatures of painting stand like living beings, but if one asks them a question, they preserve a solemn silence. And so it is with written words; you might think they spoke as if they had intelligence, but if you question them, wishing to know about their sayings, they always say only one and the same thing. And every word, when once it is written, is bandied about, alike among those who understand and those who have no interest in it, and it knows not to whom to speak or not to speak; when ill-treated or unjustly reviled it always needs its father to help it; for it has no power to protect or help itself" [Plato 1914, pp. 565–567, 275d–e].

and its efficacy to such an extent that they no longer paid any attention to the experience (ταῖς ἐμπειρίαις) and the natural abilities of their students (τῇ φύσει τῇ τοῦ μαθητοῦ). Instead, they attempted "to transmit the science of discourse as simply as they would teach the letters of the alphabet" [Isocrates 1929, p. 169].[23] Thus, it is not only due to its systemic position as a *technē* that rhetoric appears to stand fully and unreservedly on the side of the 'artificial.'

In the history of rhetoric, however, this attribution is accompanied by a far less unambiguous attitude – indeed, one that questions the judgment of the artificiality or unnaturalness of this *technē* shared by partisans and enemies of rhetoric alike on the basis of this simple distinction itself. In the following, I can only outline in an exemplary way two articulations of this attitude, which can be encountered historically from the founding to the codification of rhetoric.

a) That Gorgias appears to have suggested an independent organ of perception and reception for speech, and in this way to have provided a physical basis for the *technē* that he had largely cofounded, has already been mentioned in the previous section. The passages in question are found scattered in the fragmentary sources, and are in much need of commentary. In the following, I will orient my discussion to elucidations provided by Buchheim. In the context of the third thesis of the treatise *On Nonbeing or On Nature* according to which even something known cannot be communicated, we find the question: what happens when we speak? The apparently tautological answer is: the speaker speaks (λέγει ὁ λέγων).

At least according to one of the surviving paraphrases of Gorgias' text, however, this does not, as one might expect, amount to a distinction between senses and perceptions, in the way that, for instance, one might see things, or hear speech (remember, a representational understanding of language is not a concern of the sophists). "For on principle," we read further – although in a passage that has come down to us in a fragmentary form – "'someone who speaks' does not say ‹a sound› or a color, but a speech."[24] Thus, alongside sound and color, speech would be an object of perception in its own right, and therefore situated outside the distinction between natural and artificial. If, according to Buchheim, "all of man's practical possibilities" are concentrated in the Gorgian *logos*, and Gorgias strives to turn this into "a kind of universal effective organ of man," then this vanishing point of his theoretical and didactic efforts appears to be attributable to both a "physical" and a "cultural" anthropology of man.[25] Accordingly, Gorgias asserted in man something like an independent organ or faculty for the reception of speech, and thus "a sense specifically for speech" that, parallel to the extension of color and sound on the part of the perceptual materials referred to

23 ἀλλά φασιν ὁμοίως τὴν τῶν λόγων ἐπιστήμην ὥσπερ τὴν τῶν γραμμάτων παραδώσειν [Isocrates 1929, p. 168].

24 ἀρχὴν γὰρ οὐ ‹ψόφον› λέγει ‹ὁ λέ›γων οὐδὲ χρῶμα, ἀλλὰ λόγον [Gorgias 2016c, pp. 226–227].

25 "alle praktischen Möglichkeiten des Menschen"; "einer Art universellem Wirkungsorgan des Menschen" [Buchheim 2012, pp. ix–x].

above, would join the eye and the ear.[26] In the *Encomium of Helen*, this sense is designated with the enigmatic catachresis "the eyes of opinion" (τῆς δόξης ὄμμα[τα]) [Gorgias 2016b, pp. 178–179, D24,12]. The paraphrase, which here uses the later term *doxa* vouched for by Aristotle, obscures the nuance that the passage refers to a specific but undetermined faculty of reception. As Buchheim has noted, the *logos* uses the *doxa* "as a link to man's soul":

> *Peitho* seen from the *logos* is the same as the *doxa* seen from the soul; together they form a link, a bridge over which the powers of the logos flow into the soul. ψυχή-δόξα-πειθώ-λόγος: on such building blocks, the logos establishes its power as μέγας δυνάστης.[27]

Whether and how this receptive organ can be tuned for the reception of the *logos*, whether and how it *will* be able to be tuned for this, this dual function-related anthropological question is then pursued by Gorgias from the perspective of the efficacy of speech outlined in the preceding section.

b) Aristotle's *Rhetoric* constitutes in many respects a counter-project to sophist rhetoric of the Gorgian type and its 'pharmaceutical' understanding of efficacy, as well as to the paradoxical, since self-contradictory, Platonic judgment of rhetoric as a non-art found in *Gorgias* (which I will pass over here, as it has nothing to contribute to my inquiry). That Aristotle appears at all as a protagonist of this paper may cause surprise, as he is generally considered the theoretical founder of a 'passive' understanding of matter, as well as the philosophical godfather of its transfer between metaphysics and physics. The result of positing and transfer alike is manifested in an abstract concept of matter whereby this is understood as the bearer, or at best medium, of form processes – in short, "as possibility (*dynamis*) of being formed."[28] Michael Franz understands the organizational core of Aristotelian ontology as follows: "All becoming (*genesis*) takes place as a transition from *dynamis* to *energeia* to the extent that material becomes form: through nature (*physis*) or through art (*technē*)."[29] Thus, Aristotle thinks the relation between nature and art treated in this section as a strictly parallel one after all [Bartels 1965].

Is such a parallelism to be encountered in Aristotle's understanding of rhetoric too? At first glance, the attempt at objectification and the systematizing intention of

---

26 "daß es ebenso, wie es einen Sinn für Farben und einen für Laute, auch einen Sinn gibt spezifisch für die Rede" [ibid., p. xii].

27 "geradezu als ein Bindeglied zur Seele des Menschen"; "Dasselbe, was die Peitho vom Logos aus gesehen ist, ist die Doxa von der Seele aus; zusammen bilden sie eine Klammer, eine Brücke, über welche die Wirkkräfte des Logos in die Seele hinüberfließen. ψυχή – δόξα – πειθώ – λόγος, auf solchen Bausteinen gründet der Logos seine Macht als μέγας δυνάστης" [Buchheim 1986, pp. 21–22].

28 "als Möglichkeit (dýnamis) des Geformtwerdens" [Mainzer 1996, p. 15]. This short summary is naturally under complex. Cf. [Cohen 2012, pp. 205–226; Happ 1971; Gasser 2015], in particular the detailed overview of earlier research literature in [Happ 1971, pp. 2–49].

29 "Alles Werden (*génesis*) vollzieht sich als Übergang von *dýnamis zu enérgeia*, indem aus Stoff Form wird: durch Natur (*phýsis*) oder durch Kunst (*téchnē*)" [Franz 1999, p. 225].

his treatise on rhetoric leave little room for hybridization. Already the first sentences of *Rhetoric*, for instance, make the decisive claim that this *technē* – which is designated as "a counterpart of Dialectic" (ἀντίστροφος τῇ διαλεκτικῇ) [Aristotle 1926, pp. 2–3, 1354a][30] – should be understood as an elementary, foundational discipline. As such, it deals with "matters that are in a manner within the cognizance of all men and not confined to any special science."[31] So when at the beginning of the third book, in the discussion on elocution, the voice as its fundamental element is dismissed quickly and reluctantly with the remark that, as a matter of "natural talent," it is rather alien to the domain of a *technē* [Aristotle 1926, pp. 349, 1404a], this appears to confirm the rigid division between nature and art/technics, and a decision to side with the latter. However, even if at this point Aristotle declares everything connected with the use of the "natural" voice to be irrelevant, and hence unsuitable for his *technē*, characteristic of the systemic pragmatics of his rhetoric is the fact that while strictly speaking he does not consider its cultivation to be right, he considers it useful, even necessary.[32] Thus, while from a systematic point of view the modulation of affect by means of the voice is not one of the legitimate categories of rhetorical persuasion, to dispense with it would be senseless. This pragmatic relativization of the systemic limits to what lies outside the proper domain of *technē* can on closer inspection be seen everywhere. That means that nature, strictly speaking, has long become a playing chip of Aristotelian rhetoric itself. Therefore, in speech as a form of prose, Aristotle recommends, for instance, dissimulating the tricks of the style (*lexis*) used and also discussed in the third book of *Rhetoric*. As he points out in the introductory remarks of the third book: "those who practise this artifice must conceal it and avoid the appearance of speaking artificially instead of naturally [μὴ δοκεῖν λέγειν πεπλασμένως ἀλλὰ πεφυκότως]; for that which is natural persuades, but the artificial does not," and he extends this advice in what follows to include lexical and grammatical specifications [Aristotle 1926, pp. 352–353, 1404b]. If one wants to generalize this maxim in all its paradox, that means that the *technē* of rhetoric, in order not to jeopardize its goal (the production of conviction), must conceal its artistry in the products made according to its principles. Obviously, however, the 'natural' clarity and objectivity of speech are not left unaffected by this maneuver; on the contrary, this is itself the most perfect product of the *technē*

30 The Greek formulation takes a clear stance against the provocative Socratic joke about rhetoric being "the counterpart to cookery in the soul" (ἀντίστροφon ὀψοποιίας ἐν ψυχῇ) [Plato 1925, pp. 320–321, 465d].

31 περὶ τοιούτων τινῶν εἰσὶν ἃ κοινὰ τρόπον τινὰ ἁπάντων ἐστὶ γνωρίζειν καὶ οὐδεμιᾶς ἐπιστήμης ἀφωρισμένης [Aristotle 1926, pp. 2–3, 1354a].

32 On this strategy, Christof Rapp notes: "That is a simple schema to expand the area of rhetoric, starting from objective methods found to be good and right, with extra-objective methods." "Das ist ein einfaches Schema, um den Bereich der Rhetorik ausgehend von den für gut und richtig befundenen, sachbezogenen Methoden um außersachliche Methoden zu erweitern" [Aristoteles 2002, p. 814].

rhētorikē. As a result, in the case of rhetoric, a strategic complication of this relation appears in the functional position of the abovementioned ontological parallelism of *physis* and *technē*.[33]

5 Return/Repetitions of Rhetoric?

"Gorgias transferred the style of poetry to political speeches, for he did not think that the orator is similar to ordinary people," Dionysius of Halicarnassus is said to have remarked;[34] as I have just shown, Aristotle strongly advised against demonstrations of technical competence – at least its ostentatious display, to which Gorgias of course owes something of his legendary reputation. One can hardly avoid perceiving in these tactical suggestions a basic structural pattern that will be (re-)encountered many centuries after rhetoric's founding in entirely different theoretical and historical settings. In the critical, if not openly anti-rhetorical, understanding of language of the modern period, this pattern will lead first of all to the distinction between a 'natural,' unformed, basic, or zero level of speech, possibly even a language of things themselves, and speech's 'artificial' rhetorical reshaping.[35] Exemplary for this distinction, which is generally linked with a discrediting of rhetoric as a *technē*, is René Descartes' theory of the power of reasoning (*raisonnement*) found in the first part of his *Discourse on Method*:

> Those who have the strongest power of reasoning, and who most skillfully arrange their thoughts in order to render them clear and intelligible, have the best power of persuasion even if they can but speak the language of Lower Brittany and have never learned Rhetoric.[36]
>
> [Descartes 1973, p. 85]

The assumption that rhetorical *dispositio* (arrangement) and *elocutio* (style) should be understood as an overcoding of both conceptual clarity and "plain speech" are among the topoi of a critique whose objections "rendered the traditional practice

---

33 Here it should be added that, to the irritation of many modern readers, despite the said parallelism, in ontology too the suspicion occasionally arises that the Aristotelian understanding of nature is excessively prefigured by the practice of art. Thus, for example, "generally recurring, the process of becoming is illustrated mainly by use of examples borrowed from areas of human activity and production." ("allgemein wiederkehrend wird der Vorgang des Werdens überhaupt durch Beispiele verdeutlicht, welche dem Gebiete menschlichen Wirkens und Hervorbringens entlehnt sind" [Hertling 1871, p. 62]).

34 Γοργίας μὲν τὴν ποιητικὴν ἑρμηνείαν μετήνεγκεν εἰς λόγους πολιτικούς, οὐκ ἀξιῶν ὅμοιον τὸν ῥήτορα τοῖς ἰδιώταις εἶναι [Gorgias 2016a, pp. 164–165].

35 For a profound and critical examination, see: [Schüttpelz 1996].

36 "Ceux qui ont le raisonnement le plus fort, & qui digerent le mieux leurs pensées, affin de les rendre claires & intelligibles, peuuent tousiours le mieux persuader ce qu'ils proposent, encore qu'ils ne parlassent que bas Breton, & qu'ils n'eussent iamais apris de Rhetorique" [Descartes 1902, p. 7].

and doctrine of rhetoric obsolete" [Bender and Wellbery 1990, p. 22].[37] Resting on the same assumption was the conviction that all rhetorical tricks were dispensable for the speech forms of truth and sincerity; indeed, that rhetorical reshaping would even sabotage the reliability and integrity of such speech forms. This last critical stage changes the 'quantitative' idea that the supplementary surplus of rhetorical shaping would cover over the 'plainness' of zero-level speech into a 'qualitative' phantasm of distortion. In this now largely anthropologically and socially grasped threat scenario of deception, lies, and deceit, modern anti-rhetoric is able to draw on the ontologically and epistemologically grounded Platonic judgment of its enemy.[38]

Second and more crucially, however, the modern(ist) refounding of the rhetorical – John Bender and David E. Wellbery have gathered this under the term 'rhetoricality' – may also be due in large part to this basic pattern. This has shown itself to be productive in two ways: first, by becoming the occasion for polemical rejection; second, as a result of its displacement into the structural architecture of language itself.

On the first point: When, in the winter semester of 1872/1873 at the University of Basel, Friedrich Nietzsche read "The Rhetoric of the Greeks and Romans" before two intrepid listeners, he added to his expositions oriented to the customary *partes orationis* a paragraph on a foundational topic: namely, on the relation between language and rhetoric, which he formulated with maximum concision as "language is rhetoric," thereby annulling precisely the idea of the distinction between original/'natural' and rhetorical/'artificial' language and speech events [Nietzsche 1995, p. 425].[39] As we know, Nietzsche owes his thesis not to the tradition or the founding history of rhetoric itself, but to the reading of a now-forgotten philosopher of language, Gustav Gerber [Meijers 1988; Meijers and Stingelin 1988]. Nietzsche had come across the first, just-published volume of Gerber's *Sprache als Kunst* (*Language as Art*) at the university library in the autumn of 1872, and thus during the immediate preparations for his lecture. Here, one could find passages that might indeed make one think of a

37 Alongside the mentioned ideal of transparency and objectivity, for Bender and Wellbery four additional factors contribute decisively to this delegitimation: the transition from the literary-aesthetic system to the paradigm of subjective expression; the liberal idea of a political discourse understood as "communal exchange"; the replacement of a predominantly oratorical model of communication (forum, pulpit, court, etc.) with printing and the press; the nation state as a political model of governance that replaces ideas of the polis and the court [Bender and Wellbery 1990, p. 22].

38 And it does this even though such an ethical framing of the Platonic critique was entirely alien. One should not forget that Plato's *Politeia* adhered of all things to the social steering function of rhetorical performing arts. Even in the ideal polis, deception remains an indispensible pedagogical instrument (see in particular the distinction between τὸ τῷ ὄντι ψεῦδος and τὸ ἐν τοῖς λόγοις ψεῦδος, 382a–383c). If it issues from a regime of truth, the Platonic Socrates generally has little to oppose to the pragmatics of deception [Latour 1999, pp. 216–265 (chaps. 7 and 8)].

39 The full sentence is: "*language is rhetoric*, because it desires to convey only a *doxa* [opinion], not an *epistēmē* [knowledge]" [Nietzsche 1989, p. 23] ("die Sprache ist Rhetorik, denn sie will nur eine δόξα, keine ἐπιστήμη übertragen" [Nietzsche 1995, p. 425]).

reactualization of Gorgian rhetoric within the theory of language: "All words are sound images and, in relation to their meaning, are, in themselves and from the beginning, tropes," Gerber pointed out, and in this way took up again the thesis of discontinuity found in the treatise *On Nonbeing or On Nature.*[40] Only, he used this insight relatively specifically in order to correct modernist, or naturalist, notions of meaning: "Just as the origin of the word was an artificial one, it also changes its meaning essentially only through artistic intuition. There are no 'proper words' – i.e. prose – in language."[41]

Nietzsche's radicalization, however, no longer confines itself to the repudiation of the linguistic common sense of the time but expands and converts rhetoric into a "methodological instrument of a linguistically grounded epistemology."[42] How exactly this is to be understood is shown by the paper "On Truth and Lying in an Extra-Moral Sense." Leaving aside for the moment all its anthropological ramifications, the text's extended 'fable' can simply be read as the story of a canceling of rhetoric's material activity in favor of a stable conceptual architecture:

> As the Romans and Etruscans carved up the sky into rigid mathematical sectors and assigned a god to each delimited space as in a temple, so every nation has such a mathematically divided conceptual sky above it and understands by the demand for truth that each conceptual god must be sought only in *his* own sphere. In this respect man can probably be admired as a mighty architectural genius who succeeds in building an infinitely complicated conceptual cathedral on foundations that move like flowing water; of course, in order to anchor itself to such a foundation, the building must be light as gossamer – delicate enough to be carried along by the wave, yet strong enough not to be blown apart by the wind. As an architectural genius, man excels the bee; for it builds out of wax which it collects from nature, while man builds out of the much more delicate material of the concepts, which he must fabricate out of his own self. In this respect he is quite admirable, but not because of his desire for truth, for pure knowledge of things.[43] [Nietzsche 1989, p. 251]

---

40 "Alle Wörter sind Lautbilder und sind in Bezug auf ihre Bedeutung an sich und von Anfang an Tropen" [Gerber 1871, p. 333 (emphasis spaced out in the original)].

41 "Wie der Ursprung des Wortes ein künstlicher war, so verändert es auch seine Bedeutung wesentlich nur durch künstlerische Intuition. 'Eigentliche Worte' d.h. Prosa giebt es in der Sprache nicht" [Gerber 1871, p. 333 (emphasis spaced out in the original)].

42 "methodische Instrument einer sprachkritisch fundierten Erkenntniskritik" [Kopperschmidt 1994, p. 42].

43 "Wie die Römer und Etrusker sich den Himmel durch starre mathematische Linien zerschnitten und in einen solchermaassen abgegrenzten Raum als in ein templum einen Gott bannten, so hat jedes Volk über sich einen solchen mathematisch zertheilten Begriffshimmel und versteht nun unter der Forderung der Wahrheit, dass jeder Begriffsgott nur in *seiner* Sphäre gesucht werde. Man darf hier den Menschen wohl bewundern als ein gewaltiges Baugenie, dem auf beweglichen Fundamenten und gleichsam auf fliessendem Wasser das Aufthürmen eines unendlich complicirten Begriffsdomes gelingt; freilich, um auf solchen Fundamenten Halt zu finden, muss es ein Bau, wie aus Spinnefäden sein, so zart, um von der Welle mit fortgetragen, so fest, um nicht von dem Winde auseinander geblasen zu werden. Als Baugenie erhebt sich solcher Maassen der Mensch weit über die Biene: diese baut aus Wachs, das sie aus der Natur zusammenholt, er aus dem weit zarteren Stoffe

In the very "drive to form metaphors," however, which Nietzsche's short text understands as a "fundamental desire in man" [ibid., p. 254],[44] one recognizes without difficulty the four activity criteria formulated at the beginning: the self-movement of "the mass of images that originally gushed forth as hot magma out of the primeval faculty of human fantasy" [ibid., p. 252];[45] the dynamic emergence of the primary "intuitive metaphor[s]" of which each "is individual and unique and therefore always eludes any commentary" [ibid., p. 250];[46] a twofold pragmatics – on the one hand, that of a primary behavior whose "most universal effect is deception" [ibid., p. 246], on the other, the stabilizing "peace agreement" [ibid., p. 247] of conceptual suspension;[47] a hybridity, finally, that already illuminates the coupling of drive and metaphor formation.

Nietzsche's considerations on rhetorical epistemology received attention only later and ultimately as the effect of a second, even more improbable reconfiguration of rhetoric. This second, let us call it structuralist, repetition of rhetoric did then indeed lead to a renewed engagement with the (founding) history of rhetoric. Thus, in his précis of "old rhetoric," Roland Barthes is able to distinguish between a "rhetoric of the syntagm" ("rhétorique du syntagme") and a rhetoric "of the feature" ("du trait"): "There are, by and large, in a complete art of rhetoric [...] two poles: a syntagmatic pole (the order of the parts of the discourse, *taxis* or *dispositio*) and a paradigmatic pole (the 'figures' of rhetoric, *lexis* or *elocutio*)" [Barthes 1988, pp. 17–18].[48]

Barthes lets this unfolding begin with Gorgias, who introduced "a code immanent to prose (though borrowed from poetry)" into the structuring and mapping of speech, and thereby – by means of "words of similar consonance, symmetrical sentences, antitheses reinforced by assonance, alliteration, metaphor" – opened up a "paradigmatic perspective" to the *logos* [ibid., p. 18].[49] One easily sees how the double pattern of rhetorical shaping is reactualized in this systematization, and to which

der Begriffe, die er erst aus sich fabriciren muss. Er ist hier sehr zu bewundern – aber nur nicht wegen seines Triebes zur Wahrheit, zum reinen Erkennen der Dinge" [Nietzsche 1973, p. 376].

44 "Jener Trieb zur Metapherbildung, jener Fundamentaltrieb des Menschen" [Nietzsche 1973, p. 381].

45 "einer ursprünglich in hitziger Flüssigkeit aus dem Urvermögen menschlicher Phantasie hervorströmenden Bildermasse" [Nietzsche 1973, p. 377].

46 "Während jede Anschauungsmetapher individuell und ohne ihres Gleichen ist und deshalb allem Rubriciren immer zu entfliehen weiß" [Nietzsche 1973, p. 376. Engl. trans. *Friedrich Nietzsche on Rhetoric and Language*, p. 250].

47 "Seine allgemeinste Wirkung ist Täuschung"; "Friedensschluss" [Nietzsche 1973, pp. 370, 371].

48 "Il y a en gros dans l'art rhétorique complet [...] deux pôles : un pôle syntagmatique : c'est l'ordre des partis du discours, la *taxis* ou *dispositio* ; et un pôle paradigmatique : ce sont les 'figures' de rhétorique, la *lexis* ou *elocutio*" [Barthes 1970, p. 176].

49 "un code immanent à la prose (bien qu'emprunté à la poésie)"; "mots de même consonance, symétrie des phrases, renforcements des antithèses par assonances, métaphores, allitérations"; "perspective paradigmatique" [Barthes 1970, p. 176].

linguistic approach it is due. Appearing in the place of the referentialist distinction between a 'zero-level' utterance and a rhetorical reshaping – and thus of a two-layer model – is a biaxial system of syntagm and paradigm. The theoretical source of this rereading is Roman Jakobson, who, as Barthes notes, "[o]ut of all rhetoric [...] retained only two figures, metaphor and metonymy, making them into an emblem of the two axes of language" [ibid., p. 46].[50] If I insist on the distinction between the layer model and the axis model, this is because Jakobson's communication model could itself easily result in the misunderstanding that both are logically equivalent. Does the relay not offer the 'poetic function' for this, owing to which the paradigmatic "rhetoric of the figure" can be converted into that of the syntagm? One would then rediscover the basic pattern outlined at the beginning of this section – only that it is now no longer named with the distinction between 'proper' and 'rhetorically shaped' speech, but is concretized in one between 'grammatical' and 'poetic' speech. However, such a change of name would be the result of an unduly reductive reading of Jakobson's modeling. Indeed, if we read the relevant sentence in his paper "Linguistics and Poetics" more attentively, another picture emerges: "The poetic function projects the principle of equivalence from the axis of selection into the axis of combination" [Jakobson 1981 (1960), p. 27].[51] Thus, it is by no means a question of tipping the paradigmatic axis of the figure as it were onto the syntagmatic, but rather of *projecting* the principle of equivalence from one into the other. The function, not the ontology, of speech forms is negotiable; in other words, the 'poetic function' measures the "operative degrees of freedom" of the relation between syntagm and paradigm;[52] it does not designate a status of speech, as in the two-layer model the figured speech in relation to 'plain speech' for instance. That such models are of little use analytically had, incidentally, already been emphasized by Jakobson himself in the 1930s when he described them sarcastically as "equations with two unknowns" [Jakobson 1981 (1933/34), p. 743].

It is precisely due to this functionality, too, that the improbability of a structuralist reconfiguration of rhetorical matters of activity is slightly reduced, but without it disappearing altogether – for obviously a theory of language and communication that wants to build on the foundation of linguistic invariants will hardly be able to be brought in line with rhetoric's situational and performative understanding of language. Even if Jakobson criticized Ferdinand de Saussure's strict suppression of the pragmatic dimension of language (*parole*) early on, he never doubted the precedence of structure (*langue*). And when he decrees that "the speaker, as a rule, is only a word-user, not a word-coiner" [Jakobson 1971 (1956), p. 242], then on a systematic level this idea and the underlying understanding of language diverges fundamentally

50 "De toute la rhétorique, Jakobson n'a retenu que deux figures, la métaphore et la métonymie, pour en faire l'emblème des deux axes du langage" [Barthes 1970, p. 195].
51 Original in italics. Cf. on this [Simon 2018].
52 "operative Freiheitsgrade" [Kammer and Krauthausen 2020, p. 40].

from the Gorgian "wor(l)d-coiner" as well as from his or her Nietzschean double. Should one therefore be concerned about the active matter search profile of my reading of rhetoric?

What cannot be disputed at any rate is the hybridity that plays a literally central role in Jakobson's structuralist understanding of language, and doubtless appears more prominently than in the models of "old rhetoric." "Language straddles nature and culture" Jakobson remarks. Linguistics is seen as a "link between science and the humanities."[53] Starting from this hybrid point (in the architecture of Jakobson's model, this can be designated positively: the phoneme [Jakobson 2002 (1939)][54]), the biaxial structure model is now suddenly set in motion as a dynamic machine of both structure formation and articulation. Language, according to Jakobson, is a structure in motion. The task of linguistics will be delineated in a correspondingly kinetic fashion: "Linguistics is concerned with language in all its aspects – language in operation, language in drift, language in the nascent state, and language in dissolution" [Jakobson 1971 [1956], p. 239]. But what is the material of this activity? The point of Jakobson's theory of language consists in the fact that these materials of a moving structure are mobilized as well. To describe this, he uses the biaxial model that he opposes to the stasis of Saussure's understanding of language. Thus, beginning with the irreducible starting point of the phoneme, all linguistic entities can become the object/agent of a dual operation themselves: "Speech implies a *selection* of certain linguistic entities and their *combination* into linguistic units of a higher degree of complexity" [ibid., p. 241]. The differential unity of contiguity and remoteness that determines the relationship mode of combination, and which Jakobson associates with the trope of metonymy, also delineates all aspects of language – from the communication structure between addresser and addressee to the syntactic, 'horizontal' linking of the sentence elements and the textual integration of the latter. The second differential unity of similarity and contrast – which Jakobson associates with the trope of metaphor – keeps the processuality of the dynamic language machine running by adding to the 'horizontal' axis of the code the 'vertical' axis of the context. The reason for Jakobson's choice of the aphasic as the test case of his theory of language is because the latter exemplifies the coupling of these two constitutive axes as it were *ex negativo*,

53 "Die Sprache sitzt rittlings zwischen Natur und Kultur"; "Bindeglied zwischen den Natur- und den Geisteswissenschaften" [Jakobson 1970, p. 33].

54 The phoneme is the founding element of a "science of form" ("Formwissenschaft" [Jakobson 2002 [1939], p. 281]) in which "the meaning-conferring act but by no means the meaning-fulfilling act is given" ("der bedeutungsverleihende Akt, keineswegs aber der bedeutungserfüllende Akt gegeben [ist]" [ibid., p. 292]). The biaxial model of language is also developed in the two Copenhagen lectures delivered in 1939 – conceived temporally and to break through the Saussurian dichotomy of *langue* and *parole*: "Just as the linguistic structure as a whole, the linguistic signs in particular are also two-dimensional" ("Wie das Sprachgebilde im Ganzen, so sind auch die Sprachzeichen im Besonderen zweidimensional" [ibid., p. 307]).

and is therefore able to expose their fundamentally physiologically/psychologically grounded emergence:

> Thus every level of linguistic units [i.e. phoneme-morpheme-word-sentence-utterance] presents a different relationship between code and context, and these differences are of great consequence for the various problems of linguistic structure and especially for the study of aphasia.
> [Jakobson 1971 (1955), p. 234]

Consequently, aphasics suffer from "similarity disorder" when the context-forming interconnection of the levels of language becomes impossible, and from "contiguity disorder" when they are incapable of the syntagmatic linking of linguistic elements [ibid., p. 238].[55]

Thus, it is no longer the gods – and no longer veiled speech, poetry, or metalanguage – that are enthroned over the speech events determined by this "bipolarity" [Jakobson 1971 (1956), p. 259]. While the Aristophanean Socrates wanted to consign the control over the powerful dynamics of speech to the divine triumvirate of *Chaos*, *Tongue*, and *Clouds*, the secular governing authorities of structure-opening sound events, the structuring dynamics of speech, and the structure-stabilizing two-dimensionality of paradigm and syntagm have superseded Aristophanes' unruly divinities. The touchstone of the *rhetorikē technē* was a parsimonious farmer; the touchstones of Jakobson's dynamic understanding of language are similarity- or contiguity-disturbed aphasics. In the meantime, rhetoric's active matters continue to play under this new patronage.

Bibliography

Aristophanes. Clouds. Dover KJ, Ed. Oxford, Oxford University Press, 1968.

Aristophanes. Clouds, Wasps, Peace. Ed. and trans. Henderson J. Loeb Classical Library 488. Cambridge MA, London, Harvard University Press, 1988.

Aristoteles. Werke in deutscher Übersetzung. Vol. 4,2. Rhetorik. Ed. Flashar H. Trans. and annotated Rapp C. Berlin, Akademie Verlag, 2002.

Aristotle. The "Art" of Rhetoric. Trans. Freese JH. Cambridge MA, London, 1926.

Bartels K. Der Begriff Techne bei Aristoteles. In: Flashar H, Gaiser K, eds. Synusia: Festgabe für Wolfgang Schadewaldt zum 15. März 1965. Pfullingen, Neske, 1965, 175–287.

Barthes R. L'ancienne rhétorique: Aide-mémoire. Communications 1970, 16, 172–223, DOI: 10.3406/comm.1970.1236.

Barthes R. The Old Rhetoric: An Aide-Mémoire. In: Barthes R. The Semiotic Challenge. Oxford, Basil Blackwell, 1988, 11–93.

Bender J, Wellbery DE. Rhetoricality: On the Modernist Return of Rhetoric. In: Bender J, Wellbery DE. The Ends of Rhetoric: History, Theory, Practice. Stanford, Stanford University Press, 1990, 3–39.

55 For a more detailed discussion, see Jakobson [1971 (1956), pp. 244–254].

Buchheim T. Die Sophistik als Avantgarde des Normalen Lebens. Hamburg, 1986.
Buchheim T. Einleitung. In: Gorgias: Reden, Fragmente und Testimonien. 2nd edn. Ed. and trans. Buchheim T. Hamburg, Meiner, 2012, vii–xxxiv.
Cassin B. Who's Afraid of the Sophists. Against Ethical Correctness. Hypatia 2000, 14 (4), 2000, 102–120, DOI: 10.1111/j.1527-2001.2000.tb00354.x.
Cicero. De Oratore, III. In: Cicero. On the Orator: Book 3. On Fate. Stoic Paradoxes. Divisions of Oratory. Trans. Rackham H. Loeb Classical Library 349. Cambridge, MA, London, Harvard University Press, 1942, 1–186.
Cohen SM. Alteration and Persistence: Form and Matter in the Physics and De Generatione et Corruptione. In: Shields C. The Oxford Handbook of Aristotle. Oxford, New York, 205–226.
Descartes R. Discours de la méthode pour bien conduire sa raison et chercher la vérité dans les sciences [1637]. In: Adam C, Tannery P, Ed. Descartes R. Œuvres de Descartes. Vol. 6. Paris, Léopold Cerf, 1902, 1–78.
Descartes R. The Philosophical Works of Descartes. Vol. 1. Trans. Haldane ES, Ross GRT. London, New York, Dover, Cambridge University Press 1973.
Dover KJ. Introduction. In: Dover KJ, Ed. Aristophanes. Clouds. Oxford, Oxford University Press, 1968, xvii–90.
Euben JP. Corrupting Youth: Political Education, Democratic Culture, and Political Theory. Princeton, Princeton University Press, 1997.
Franz M. Von Gorgias bis Lukrez: Antike Ästhetik und Poetik als vergleichende Zeichentheorie. Berlin, Akademie Verlag, 1999.
Friedman M, Krauthausen K. Inspired Mechanics: Active Matter as Machine and Structure. In: Doll N, Bredekamp H, Schäffner W, eds. +ultra: Knowledge and Gestaltung. Leipzig, E. A. Seemann, 2017, 171–176.
Gasser A. Form und Materie bei Aristoteles: Vorarbeiten zu einer Interpretation der Substanzbücher. Tübingen, Mohr Siebeck, 2015.
Gelzer T. Aristophanes und sein Sokrates. Museum Helveticum 1956, 13 (2), 65–93.
Gerber G. Die Sprache als Kunst. Vol. 1. Bromberg, Mittler, 1871.
Gorgias. D21. In: Early Greek Philosophy. Vol. 8: Sophists, Part 1. Ed. and trans. Laks A, Most GW. Loeb Classical Library 531. Cambridge, MA, London, Harvard University Press, 2016a, 164–165.
Gorgias. Encomium of Helen (D24). In: Early Greek Philosophy. Vol. 8: Sophists, Part 1. Ed. and trans. Laks A, Most GW. Loeb Classical Library 531. Cambridge, MA, London, Harvard University Press, 2016b, 166–185.
Gorgias. On Nonbeing or On Nature (D26). In: Early Greek Philosophy. Vol. 8: Sophists, Part 1. Ed. and trans. Laks A, Most GW. Loeb Classical Library 531. Cambridge, MA, London, Harvard University Press, 2016c, 218–241.
Groddeck W. Reden über Rhetorik: Zu einer Stilistik des Lesens. Basel, Frankfurt am Main, Klostermann, 1995.
Happ H. Hyle: Studien zum aristotelischen Materie-Begriff. Berlin, New York, De Gruyter, 1971.
Hertling GV. Materie und Form und die Definition der Seele bei Aristoteles: Ein kritischer Beitrag zur Geschichte der Philosophie. Bonn, E. Weber, 1871.
Isocrates. Against the Sophists 10. In: Isocrates, Vol. 2. Trans. Norlin G. Cambridge MA, London, W. Heinemann, New York, G.P. Putnam's Sons, 1929, 160–177.
Jakobson R. Sprache und Einbildungskraft (Ein Interview) [interview with Michel Tréguer and François Châtelet; French original in La Quinzaine littéraire, 15/31 May 1968; title attributed by the German translator Felix Philipp Ingold]. In: Jakobson R. Poesie und Sprachstruktur: Zwei Grundsatzerklärungen. Trans. Ingold FP. Zurich, Verlag der Arche, 1970, 29–42.
Jakobson R. Two Aspects of Language and Two Types of Aphasic Disturbances [1956]. In: Jakobson R. Selected Writings. Vol. 2. Word and Language. The Hague, Paris, Mouton, 1971, 239–259.

Jakobson R. Aphasia as a Linguistic Topic [1955]. In: Jakobson R. Selected Writings. Vol. 2. Word and Language. The Hague, Paris, Mouton, 1971, 229–238.

Jakobson R. Linguistics and Poetics [1960]. In: Rudy S, Ed. Jakobson R. Selected Writings. Vol. 3. Poetry of Grammar and Grammar of Poetry. The Hague, Paris, New York, Mouton, 1981, 18–51.

Jakobson R. What is Poetry [1933/34]?. In: Rudy S, Ed. Jakobson R. Selected Writings. Vol. 3. Poetry of Grammar and Grammar of Poetry. The Hague, Paris, New York, Mouton, 1981, 740–750.

Jakobson R. Zur Struktur des Phonems [1939]. In: Jakobson R. Selected Writings. Vol. 1. Phonological Studies. 3rd edn. Berlin, New York, De Gruyter Mouton, 2002, 280–310, DOI: 10.1515/9783110892499.280.

Kammer S, Krauthausen K. Für einen strukturalen Realismus. Einleitung. In: Kammer S, Krauthausen K, eds. Make it Real: Für einen Strukturalen Realismus. Zurich, Berlin, Diaphanes, 2020, 7–80.

Kohn TA, Burroughs R, Hartmann MJ, Noakes TD. Fiber Type and Metabolic Characteristics of Lion (Panthera Leo), Caracal (Caracal Caracal) and Human Skeletal Muscle. Comparative Biochemistry and Physiology Part A: Molecular & Integrative Physiology 2011, 159 (2), 125–133, DOI: 10.1016/j.cbpa.2011.02.006.

Kopperschmidt J. Nietzsches Entdeckung der Rhetorik oder Rhetorik im Dienste der Kritik der unreinen Vernunft. In: Kopperschmidt J, Schanze H, eds. Nietzsche oder “Die Sprache ist Rhetorik.” Munich, Fink, 1994, 39–62.

Latour B. Pandora's Hope: An Essay on the Reality of Science Studies. Cambridge, MA, London, Harvard University Press, 1999.

Leslie E. Cloud Animation. Animation: An Interdisciplinary Journal 2017, 12 (3), 230–243, DOI: 10.1177/1746847717740840.

Lewitscharoff S. Blumenberg: Roman. Berlin, Suhrkamp, 2011.

Liddell HG, Scott R, Jones HS, McKenzie R. A Greek-English Lexicon. Oxford, Clarendon Press, 1940.

Mainzer K. Materie: Von der Urmaterie zum Leben. Munich, Beck, 1996.

Meijers A. Gustav Gerber und Friedrich Nietzsche: Zum historischen Hintergrund der sprachphilosophischen Auffassungen des Frühen Nietzsche. Nietzsche-Studien 1988, 17, 369–390.

Meijers A, Stingelin M. Konkordanz zu den wörtlichen Abschriften und Übernahmen von Beispielen und Zitaten aus Gustav Gerber: Die Sprache als Kunst (Bromberg 1871) in Nietzsches Rhetorik-Vorlesung und in 'Über Wahrheit und Lüge im aussermoralischen Sinne'. Nietzsche-Studien 1988, 17, 350–368.

Meister K. Aller Dinge Maß ist der Mensch": Die Lehren der Sophisten. Munich, Fink, 2010.

Nietzsche F. Ueber Wahrheit und Lüge im aussermoralischen Sinne. In: Colli G, Montinari M, eds. Nietzsche F. Werke: Kritische Gesamtausgabe. Vol. 3,2. Nachgelassene Schriften 1870–1873. Berlin, New York, De Gruyter, 1973, 367–384.

Nietzsche F. On Truth and Lying in an Extra-Moral Sense. In: Nietzsche F. On Rhetoric and Language. Ed. and trans. with a critical introduction Gilman SL, Blair C, Parent DJ. New York, Oxford, Oxford University Press, 1989.

Nietzsche F. § 3: Verhältniß des Rhetorischen zur Sprache. [In: Geschichte der griechischen Beredsamkeit]. In: Bornmann F, Carpitella M, eds. Nietzsche F. Werke: Kritische Gesamtausgabe. Vol. 2,4. Vorlesungsaufzeichnungen (WS 1871/72–WS 1874/85). Berlin, New York, De Gruyter, 1995, 425–428.

Plato. Phaedrus. In: Plato. Eutyphro, Apology, Crito, Phaedo, Phaedrus. Trans. Fowler HN. Cambridge, MA, London, Harvard University Press, 1914, 412–579.

Plato. Gorgias. In: Plato. Lysis, Symposium, Gorgias. Trans. Lamb WRM. Loeb Classical Library 166. Cambridge, MA, London, Harvard University Press, 1925, 247–533, DOI: 10.4159/DLCL.plato_philosopher-gorgias.1925.

Schüttpelz E. Figuren der Rede: Zur Theorie der Rhetorischen Figur. Berlin, E. Schmidt, 1996.

Sextus Empiricus. Against the Logicians. Trans. Bury RG. Loeb Classical Library 291. Cambridge MA, London, Harvard University Press, 1935, DOI: 10.4159/DLCL.sextus_empiricus-against_logicians.1935.

Simon R. Was genau heißt: 'Projektion des Äquivalenzprinzips?' Roman Jakobsons Lehre vom Ähnlichen. In: Endres M, Herrmann L, eds. Strukturalismus, heute: Brüche, Spuren, Kontinuitäten. Stuttgart, J.B. Metzler, 2018, 121–138.

Strowick E. Sprechende Körper – Poetik der Ansteckung: Performativa in Literatur und Rhetorik. Munich, Fink, 2009.

Tibbits S, ed. Active Matter. Cambridge, MA, MIT Press, 2017.

Stephan Kammer is Professor of Modern German Literature at Ludwig-Maximilian University Munich. After studying German Language and Literature, Modern History, Sociology and Art History in Basel, he received his doctorate from the University of Basel with a thesis on Robert Walser (2000). He completed his habilitation at Goethe University Frankfurt with a thesis on philological and antiquarian knowledge in early 18th century (2011). His research interests are: German-language literatures from 17th to 21st centuries in an European context; history of media; theory and history of the artificial; history of philology and writing. Recent publications include: Überlieferung. Das philologisch-antiquarische Wissen im frühen 18. Jahrhundert (Berlin/Boston 2017; paperback: Berlin/Boston 2019); Make it Real. Für einen strukturalen Realismus (ed. with Karin Krauthausen, Zurich/Berlin 2020); Audiophilologie, in: Handbuch Literatur und Audiokultur, ed. by Natalie Binczek and Uwe Wirth, Berlin/Boston 2020, 177–191; Weltempfänger, in: Welt-Komposita. Ein Lexikon, ed. by Thomas Erthel and Robert Stockhammer, Paderborn 2020, 44–48.

Sonia Dheur, Sven J. Saupe

Proteins as Monads

> "As the wind of time blows into the sails of space, the unfolding of the universe nurtures the evolution of matter under the pressure of information."[1]

> "Placed on the opaque canvas, these folds, cords, or springs represent an innate form of knowl - edge, but when solicited by matter they move into action."[2]

1 Introduction

Active matter deals with elementary chemical and biological components and their collective properties-among others, macromolecular cell constituents capable of converting chemical energy into motion [Sanchez et al. 2012]. It is proposed that "the field of active matter is concerned with non-equilibrium systems in which the individual, constitutive units are themselves internally driven: machines made from machines" [Needleman and Dogic 2017, p. 1]. We would like to recall here that the formation of these elementary biological machines is – in itself – an internally driven process. If active matter indeed defines materials "able to change their shape in response to changes in environmental conditions and thus perform mechanical work on the nano-, micro- and macroscales" [Ionov 2014, p. 5015], then proteins in general, viewed as the Leibnizian machines that make up other organic machines [Leibniz 1991],[3] are active materials. Proteins, regardless of their biological function, have

1 [Lehn 2002, p. 4763].

2 [Deleuze 2006, p. 4]. "Ces plis, cordes ou ressorts constitués sur la toile opaque, représentent les connaissances innées mais qui passent à l'acte sous les sollicitations de la matière" [Deleuze 1988, p. 6].

3 Leibniz defines organic bodies as machines made of machines. "Thus the organic body of each living being is a kind of divine machine or natural automaton, which infinitely surpasses all artificial automata. For a machine made by the skill of man is not a machine in each of its parts. [...] But the machines of nature, namely, living bodies, are still machines in their smallest parts *ad infinitum*" [Leibniz 1898, p. 254]. ("Ainsi chaque corps organique d'un vivant est une espèce de machine divine, ou d'un automate naturel, qui surpasse infiniment tous les automates artificiels. Parce qu'une machine faite par l'art de l'homme, n'est pas machine dans chacune de ses parties [...]. Mais les machines de la nature, c'est-à-dire les corps vivants sont encore des machines dans leurs moindres parties, jusqu'à l'infini" [Leibniz 1991, pp. 161–162].) After Descartes, the use of the word "machine" to designate an organic body is common, but the word is used loosely with very different meanings and intentions and sometimes even without an explicit underlying philosophical thesis [Andrault 2011]. In the dualist view of Descartes, the organic body is presented as a machine in order to define it as separated and subordinated to the soul, While the Leibnizian perspective of machine made of machines implies a radically distinct philosophical meaning and incorporates the

Open Access. © 2021 Sonia Dheur, Sven J. Saupe, published by De Gruyter. (cc) BY-NC-ND This work is licensed under the Creative Commons Attribution-NonCommercial-NoDerivatives 4.0 International License.
https://doi.org/10.1515/9783110562064-016

the inherent property to set themselves in motion, to fold into their native structure. The notion of motion often embedded explicitly in the active matter concept does not take here the form of a self-propelled displacement of an object in three-dimensional space but rather corresponds to a change of shape, a folding, that nonetheless corresponds to a movement. This biological self-organization principle, here at the most basic level, is then followed by various levels of supramolecular self-organization of increasing complexity, leading to the Leibnizian view of a living machine made of machines. We will insist here on the relation of genetic information to protein structure, to describe proteins as instructed components or informed matter that attains a determined shape (and supramolecular-organization state) thanks to the information contained in their primary sequence and in response to their environment [Dill and MacCallum 2012]. Protein folding envisioned as the most basic form of biological development and morphogenesis simultaneously contains elements of preformism and epigenesis.[4] After a discussion of general aspects of protein folding and self-assembly and the interconnected role of the genetic and folding/assembly code in these processes, we will describe how proteins can display emergent properties through their collective behavior in the form of phase transition phenomena, and turn to the specific example of prion proteins to depict how in a further step of supramolecular self-organization biological matter can store and convey genetic information. Protein activity is classically framed into a structure-function paradigm which gradually gets replaced by a plasticity-function concept that takes into account how proteins react to their environment by changing shape. To think

notion of a machine animated from within. The materialistic monism of Julien Offray de La Mettrie, with *l'homme machine*, rejects that dualistic view of Descartes and the notion of a separate soul that would escape mechanism. His intention is to materialize the soul rather than to spiritualize matter. The use of the term "desiring machines" by Gilles Deleuze and Félix Guattari is closer to the Leibnizian sense as it confers a level of autonomy to the machines, the very wording (desiring machines) implies that machines have an internal agenda. For Deleuze and Guattari, the word also applies to social bodies and is not used in a theological context, in contrast to the Leibnizian perspective in which organic bodies are presented as divine machines.

4 These opposing concepts traditionally structure the debate on animal embryology and biological development in general [Van Speybroeck, De Waele, and Van De Vijver 2002]. Preformism envisions development as the growth and unfolding of preexisting structures already present in the egg, spore, or sperm and is traditionally associated to a theological dimension as the preexisting structures are presented as implanted by divine intervention. Epigenesis views development as the differentiation of new structures form homogenous matter lacking preexisting organization and philosophically emphasizes an autonomous nature of organic matter. Deleuze in *Le Pli* (*The Fold*) finds that preformism and epigenesis both conceive the organism as a fold ("An organism is defined by endogenous folds" [Deleuze 2006, p. 7] ("Un organisme se définit par des plis endogènes" [Deleuze 1988, p. 11])) and makes a direct reference to protein folding. The modern conception of protein folding is neither purely preformist as protein structures do not grow out of preexisting analogous structures nor purely epigenetic as the protein contains primary structure (the sequence of amino acids) and in that sense is not homogenous and undifferentiated.

of these basal biological processes, we follow the Deleuzian reading of Gottfried Wilhelm Leibniz [Deleuze 1988] and propose that protein could be envisioned as *monads* and recall that Gilles Deleuze and Félix Guattari have explicitly defined proteins as *machines désirantes* (*desiring machines*) [Deleuze and Guattari 1972]. Throughout, we will call upon idiosyncratic examples because biological phenomena, in spite of all categorization efforts, often tend to bend and break constructed theoretical frames [Bergson 2013 (1907)] and are maybe best approached in the style of the schizo-analysis promoted by Deleuze and Guattari [Deleuze and Guattari 1980]. We will therefore attempt to combine philosophical and biological perspectives to review certain aspects of protein biochemistry.

2 The Protein Folding Problem, Anfinsen's Principle

Most of the structural and chemical work in living organisms is carried out by proteins. Proteins are central biological agents. The rules of the complex stereochemical templating procedures establishing the relation of the DNA sequence of genes to the amino acid sequence of the corresponding proteins have been laid out in the 1970s and termed the genetic code.[5] But there is a second code, a folding code. After their mere material synthesis, to develop into functional entities, proteins have to fold into their *native* conformation. It is through this process of folding that the genetic information acquires biological meaning. The so-called Anfinsen principle states that the information specifying the fold is contained in the protein sequence, hence in the DNA sequence of the corresponding gene.[6] The experimental basis behind the Anfinsen principle is deceptively simple and rests on the fact that protein denaturation (chemically induced unfolding) is reversible in vitro. Take an active purified protein, for instance a nuclease (an enzyme degrading nucleic acids) from *Staphylococcus* (Fig. 1A). In the presence of

5 The book by Lily Kay has fully described the historical conditions of emergence of the code metaphor for the protein translation process [Kay 2000].

6 It is of interest here to cite directly from Anfinsen's Nobel lecture of 1972: "the three-dimensional structure of a native protein in its normal physiological milieu (solvent, pH, ionic strength, presence of other components such as metal ions or prosthetic groups, temperature, and other) is the one in which the Gibbs free energy of the whole system is lowest; that is, that the native conformation is determined by the totality of interatomic interactions and hence by the amino acid sequence, in a given environment. In terms of natural selection through the 'design' of macromolecules during evolution, this idea emphasized the fact that a protein molecule only makes stable, structural sense when it exists under conditions similar to those for which it was selected – the so-called physiological state. [...] we can predict, in advance, the three-dimensional, phenotypic consequences of a genetic message" [Anfinsen 1973, pp. 223, 229]. These citations emphasize the articulation of information to meaning that is embedded in the protein folding process and stress that the conversion of information to meaning does not occur at the step of the decoding of the triplet information into amino acid by way of the genetic code but during the active folding of the protein.

high concentrations of the chemical urea, the protein is *denatured* (i.e., loses activity and fold), but when the chemical is removed, it spontaneously re-acquires activity and native fold, the protein *renatures*. It follows that the molecule contains endogenously the information to set itself in motion to fold into its native structure.[7]

Following Anfinsen's principle, of the astronomical number of shapes a nascent protein chain can adopt, the native fold is the one of lowest energy. Energetically, protein folding is a downhill process occurring spontaneously with a change in free energy in the range of 5–10 kcal/mol. The formation of novel non-covalent chemical interactions in the native fold has a positive contribution to folding while the decrease in entropy of the protein chain negatively contributes to folding. One of the main drivers of folding is the so-called hydrophobic effect leading to a collapse of the protein chain that secures hydrophobic amino acid side chains from the energetically unfavorable interaction with water molecules. In the general case, folding roughly corresponds to the transition from an extended highly mobile state to a more compact globule-like shape, with extreme diversity and variation from protein to protein.[8] The protein folding process is often represented graphically as a protein transiting downhill in a funnel or landscape to a lower free energy state and from many conformations to a unique native fold (Fig. 2).

Ultimately, most forms of biological motion, as performed by biological nanomotors, are underpinned by conformational changes in proteins; in these cases, however, force generation is coupled with the expenditure of chemical energy (often ATP) [Guo et al. 2016]. To convey the general impression that proteins set themselves into

7 The semantics of protein folding embeds the dualistic preformist and epigenetic aspects of protein morphogenesis. The protein is synthesized as a *nascent* chain which then folds into its *native* state (in *native* conditions), can be *denatured* by heat or chemicals, and can further *renature* if conditions are native anew. This formulation implies paradoxically that although the fold is designated native (inborn) it is not present in the newborn (nascent) protein and its native (inborn) character is determined by interaction with the environment (the native or denaturing conditions). The unfolding process deprives the protein of its nature, but this lost part may be regained, the protein is equipped with the information to re-nature itself from within, given the right conditions.

8 In a paper founding the field of structural biology of proteins and describing for the first an time X-ray crystallography 3D protein structure, the authors note that "perhaps the most remarkable features of the molecule are its complexity and its lack of symmetry. The arrangement seems to be almost totally lacking in the kind of regularities which one instinctively anticipates, and it is more complicated than has been predicted by any theory of protein structure. Though the detailed principles of construction do not yet emerge, we may hope that they will do so at later stages of analysis" [Kendrew et al. 1958, p. 665] (Fig. 1B). These words are in sharp contrast with those spawned by the discovery of the structure of DNA. For instance, Max Delbrück stated "The whole business was like a child's toy that you could buy at the dime store, all built in this wonderful way that you could explain in *Life* magazine so that really a five-year-old can understand what's going on [...] This was the greatest surprise for everyone" [Judson and Delbruck 1996, p. 41]. While DNA is immediately intelligible, proteins are not. DNA is most often referred to in the singular – against rationality – and proteins in the plural.

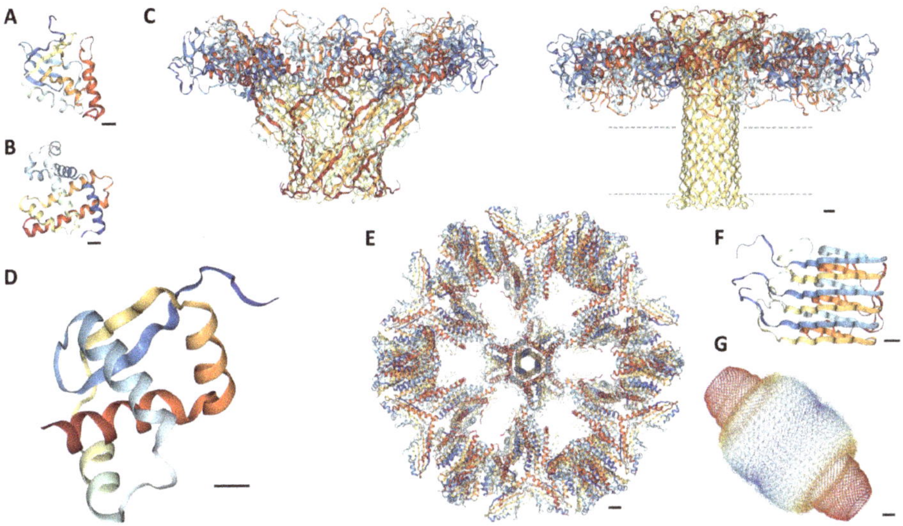

Fig. 1: Protein folds. The figure depicts the folds of different natural and designed proteins. For each protein, the name of the PDB file (protein database) containing the coordinates of all atoms of the protein and which was used to generate the protein image is given in parentheses: (A) Nuclease from *Staphylococcus aureus* (pdb, 2 sns) the protein used in the classical experiments of Christian Anfinsen on the general principles of protein folding. (B) Myoglobin, the first protein structure that was solved by X-ray crystallography, here the myoglobin of the sperm whale (*Physeter macrocephalus*) (pdb, 1mbn). (C) Two conformations of aerolysin, a pore-forming toxin produced by the human pathogen *Aeromonas hydrophila* capable of opening pores in cell membranes and causing cell death. The left image corresponds to the pre-pore state (pdb 5jzh), the right image to the pore state (pdb 5jzl) producing membrane damage. The dotted lines depict the position of the two membrane leaflets. (D) Protein MJ0366 of unknown function of the archaea *Methanocaldococcus jannaschii* (pdb 2 efv) as an example of a protein forming a knot, in this case a trefoil knot. (E) A designed protein complex forming a icosahedral cage, a polyhedron with 20 faces that has been produced by the group of David Baker (pdb 5im6). (F) The HET-s prion of the filamentous fungus *Podospora anserina* (pdb 2kj3), the first solved prion structure. (G) Major vault protein from *Rattus norvegicus* (pdb 4V60), this protein assembles into 78 subunit cage-like structure found inside cells and whose exact function remains unclear. Scale bar is 0.5 nm in (A)–(D) and (F), 1 nm in (E) and 5 nm in (G). All images were generated by the authors (S.D./S.S.) with Mol* at http://www.rcsb.org (Sehnal D, Bittrich S, Deshpande M, Svobodová R, Berka K, Bazgier V, Velankar S, Burley SK, Koča J, Rose AS. Mol* Viewer: modern web app for 3D visualization and analysis of large biomolecular structures. Nucleic Acids Res. Accessed May 6, 2021).

motion during folding (even without ATP expenditure) a few idiosyncratic examples can be of use. A class of bacterial proteins termed autotransporters uses the energy of folding of a specific protein domain (misnamed passenger domain) to transport virulence factors across the membrane. This transport is ATP-independent and energy for the transport is provided by the folding of the passenger domain [Renn et al. 2012]. So, in this specific example, folding is translated to displacement of the protein in space

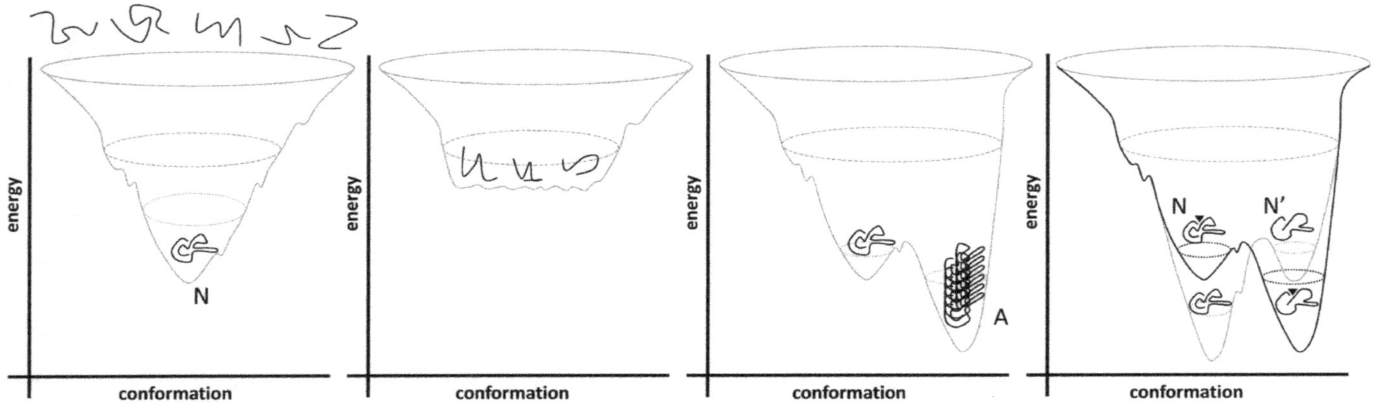

Fig. 2: Protein folding landscapes. The first panel on the left corresponds to the classical representation of folding as a protein folding funnel or protein folding landscape where the horizontal axis represents conformations and the vertical axis energy. The protein, represented as a solid curved line, exists as many different conformation displaying similar energies (at the top of the funnel). During the folding process, the protein acquires a unique conformation and a state of minimal free energy (at the bottom of the funnel), the native state marked N. This diagram is an illustration of protein folding process as described by Christian Anfinsen. The second panel represents the protein folding landscape in the case of IDPs. In that case, there is no single energy minimum corresponding to a native folded state, instead the protein exists in a plethora of interconverting conformations with comparable energy level, leading to this representation of a funnel with a flattened bottom. The third panel depicts the case of proteins capable of supramolecular assembly such as for instance proteins forming amyloid polymers. In this case, in addition to the energy minimum corresponding to the native state (as in the first panel) there is a second energy well of even lower energy corresponding to proteins assembled into higher order polymers (marked A). The assembled protein may or may not maintain the native conformation in the polymeric form. The last panel illustrates the process of allostery by which a protein can exist in two different conformations depicted by the different geometry of the solid curved line and marked N and N′ and the relative free energy of the two conformations varies depending on the physico-chemical environment, more specifically by the presence or absence of a ligand (represented by the black triangle). In the absence of the ligand (gray funnel), the N conformation is the one of lower energy and is adopted, while in the presence of the ligand (black funnel) the N′ conformation is favored. This process illustrates how sensing of the physico-chemical environment translates to a conformation change in the protein. Sensing of the ligand causes the protein to change from the N to the N′ conformation. © Graphic by the authors (S.D./S.S.).

and through a membrane compartment. One can also mention the case of bacterial pore-forming toxins such as aerolysin, in which the re-folding of a region of the protein is able to poke holes into the membrane of the host cell under attack [Bokori-Brown et al. 2016] (Fig. 1C).[9] Finally, to illustrate the complexity of the movements involved in the folding process, one might recall that about 1% of the proteins of known structure are not only folded but knotted; the functional roles of such knots remain poorly understood and cover a range of different topologies from twist knots that can be tied or untied with a single threading to more complex slipknots [Jackson, Sunna, and Micheletti 2017] (Fig. 1D).

3 So Many Shapes, the Levinthal Paradox

Anfinsen's principle comes with a paradox. The nascent or unfolded protein exists under astronomical numbers of interconverting conformers and gradually acquires a unique defined native fold with a decrease in entropy. Folding is thus a path from an extreme disordered to a unique ordered state. Each protein molecule experiences a transition from chaos to order, its own private cosmogony. Cy Levinthal pointed to the fact that the number of degrees of freedom in polypeptide chains is such that it would take longer than the age of the universe for a protein to sample all possible conformations until the native one is reached. Yet this process takes only microseconds to seconds depending on the protein. This so-called Levinthal paradox can be resolved by supposing that folding occurs through specific pathways that gradually lead the protein into the native fold, *Holzwege* are thus avoided. Currently, two opposing views on how exactly the Levinthal paradox is solved exist. The first hypothesis proposes that discrete subportions of the proteins of about 25 residues termed 'foldons' gradually take shape in a hierarchical order and that native folding occurs through assembly of such 'foldons' through a discrete, defined pathway (in a divide-and-conquer or local-to-global fashion) [Englander and Mayne 2017]. The opposing view supports the existence of a multitude of folding pathways and is described as the energy landscape theory in which the folding pathways are represented graphically as multiple possible displacements of the protein chain along a funnel-like landscape in which the *x*- and *y*-axes represent conformational space and the *z*-axis the decrease in free energy [Eaton and Wolynes 2017] (Fig. 2). In the first model, folding occurs in a defined pathway, in the second there are multiple possible folding pathways leading

9 Cryo-EM structure of lysenin pore elucidates membrane insertion by an aerolysin family protein (https://www.youtube.com/watch?v=ZOBtKZcePZo, accessed November 26, 2019).

to the native state. This is no minor disagreement [Baldwin 2017], so that, in a way, the protein folding problem still stands.[10]

The Levinthal paradox in fact extends to the next step of molecular self-organization, at the level of supramolecular assembly into higher order complexes. Some emphasize that by analogy to protein folding the formation of the cellular interactome, that is, the higher order assembly of all the cellular macromolecular complexes (machines made of machines), cannot be achieved by a simple combinatorial trial-and-error process but occurs following ordered pathways that are collectively encoded in genomes and guided by preexisting cellular structures. It follows that a cell can only be assembled from a preexisting cell [Tompa and Rose 2011]. This concept of preexisting structural and super-structural information that every cellular component is supplied with evokes a form of preformism distinct from the strictly gene-centered view; giving a new meaning to the *omnis cellula e cellula* statement enunciated by Rudolf Virchow in the formulation of the cell theory [Virchow 1855, p. 311]. This extension of the Levinthal paradox at the level of the interactome also represents a critique of the performances claimed by the field of synthetic biology. The so-called synthetic bacteria engineered by the group of Craig Venter do indeed contain a synthetic genome but were obtained by introducing this genome in a preexisting natural cell.

4 Chaperones, Protein Memory, and Silent Mutations, Some Refinements of Anfinsen's Principle

Like many 'laws' in biology, Anfinsen's principle suffers exceptions, yet it serves as a general background rule onto which particular cases can be contrasted. If folding is fundamentally viewed as an intrinsic, immanent process, in practice, many proteins are difficult (to say the least) to 'renature,' that is, to fold de novo in vitro. In the biological context, folding often relies on the assistance of so-called chaperone proteins that ensure that folding proceeds normally in the highly crowded cellular environment. Such chaperones can also be required for the supramolecular assembly steps as mentioned in the previous section. The chaperone metaphor was put forward to stress that such proteins function to "prevent incorrect [...] interactions" [Laskey et al. 1978, p. 420]. One typical example of such chaperone proteins is the bacterial GroEL/GroES complex which functions as an isolation chamber fitting a

10 Molecular dynamics simulation now allows modeling the folding pathway at least for small fast folding domains that fold in the nano- to microsecond range. See, for instance, simulation of millisecond protein folding: NTL9 (from Folding@home) (https://www.youtube.com/watch?v=gFcp2Xpd29I, accessed November 26, 2019).

single nascent protein chain that then folds shielded from the multiple interactions encountered in the cytoplasm and exits the chamber once folded, as if execution of the intrinsic development program of the protein chain requires isolation to be completed.

Another interesting quasi-violation of Anfinsen's principle is provided by the phenomenon of protein memory [Shinde, Liu, and Inouye 1997]. Certain proteases (proteins that degrade proteins) are expressed as precursors with a so-called propeptide. This propeptide binds to the enzyme and inhibits its activity but is then cleaved off by self-maturation (the protein cleaves itself to eliminate the propeptide and attains full activity). Without this propeptide, however, the protein cannot fold and mutations introduced in the propeptide region affect the folding state of the protease, so that two proteases with the exact same sequence but having interacted with distinct propeptides acquire slightly different folds and have different activities. This process has been termed protein memory because the protein fold is shaped by past interactions and not only by the present conditions.

It is also becoming increasingly apparent that proteins are dependent on information contained in the triplet codons that is distinct from the specification of the amino acid type. The genetic code is termed degenerated because different codons (so-called synonymous codons) can encode the same amino acid, and mutations that do not alter the amino acid sequences are therefore termed silent (and often correspond to those occurring on the third position of the codons). It turns out that in many cases silent mutations are not, and alter the meaning of the genetic message. This is because different codons (encoding the same amino acid) lead to different rates of translation and the rate of translation or pauses in the translation process can alter the folding pathway of nascent proteins. So fold and function can be modified, even though the sequence of the protein is unchanged. For instance, *Neurospora* mutants bearing only silent (synonymous) mutations in the *frq* biological clock gene lose circadian rhythmicity [Zhou et al. 2013]. So, not unlike synonyms in natural language, no two codons are actually perfectly synonymous. Through this additional layer of informational content the genetic code and the folding code are intimately interwoven. In contrast to the genetic code, the folding code, this code within the code, is not yet fully cracked.[11]

---

11 This latter code-script derived metaphor is not well suited for a heuristic description of the folding code, that is probably best understood as governed by a rule-code. It has been stated that the term "code-script" appearing in Erwin Schrödinger's *What is life?* of 1944 was a sort of prescient forerunner of the discovery of the genetic code. But Walsby and Hodge argue that Schrödinger's code is to be understood as a rule-code rather than a cipher-code and could therefore apply to other aspects of biological development such as protein folding rather than being restricted to the cipher-code-like character of the genetic code [Walsby and Hodge 2017].

5 Protein Design, from *Machines Désirantes* to *Machines Désirées*

In protein science, there is currently a marked shift of emphasis from causal explanation to empirical prediction, a trend common to different fields of knowledge.[12] The main aim is no longer to explain how proteins fold but to predict fold from sequence and, most of all, to devise sequences with desired folds and functions. Active design now overshadows passive (naturalistic) description of the folding process. Structure prediction methods have gradually acquired better accuracy thanks to refinements in the energy functions used to calculate the minimal energy states, progress in available computational power and increase in the number of reference experimentally obtained protein structures. Protein structure prediction is also fueled by the gigantic amount of protein sequence produced by genomic and metagenomic approaches and it is now possible to identify many novel protein folds in metagenomics data sets corresponding to microbial organisms that are otherwise unknown and unculturable [Ovchinnikov et al. 2017]. A critical improvement was also the use of contact prediction based on residue co-evolution. Evolutionary information is implemented in the prediction scheme by analyzing residue co-variation in a large number of homologs of the target protein from different species. Residue pairs which co-vary during evolution are predicted to be close in space and to form contacts in the folded protein. The collective information contained within the protein sequences of homologs (which are the molecular descendants of an ancestral protein) and resulting from evolutionary trajectories spanning often millions of years in hundreds of different species provide the structural map of the protein shape. Shape is enciphered in history and history reveals the present shape. Total ab initio structure prediction (without help of any type of evolutionary information) is still out of reach in most cases.[13]

12 Michael Scriven noted that however 'enlightening,' causal explanation is doomed to always remain partial and a never-ending endeavor and, following Jean-François Lyotard, it may be possible to trace back the deflation of the valence attributed to explanation to the wreck of the great metanarratives of modernity [Scriven 1959; Lyotard 1979].

13 From the point of view of the sociology of science, the protein fold prediction issue spawned several original initiatives. The CASP competition (for Critical Assessment of protein Structure Prediction) is a community-wide protein structure prediction experiment held every two years in which the different competing labs critically assess the performance of their prediction methods and software in a double-blind setting using a list of proteins whose actual experimentally determined structures are unknown to the organizers and the predictors [Kryshtafovych et al. 2018]. The protein folding field also led to different projects of citizen-science, like rosetta@home, in which more than a million participants contribute spare computing time on their personal computer or smartphone to the calculation intensive functioning of the Rosetta protein folding energy prediction algorithm. Likewise, the Foldit software (implemented also as foldit@home), is a citizen-science video game in which the participants solve heuristically protein folding problems [Cooper et al. 2010]. The latest

On the side of protein design, David Baker, head of the Institute of Protein Design at the University of Washington, is certainly the uncontested leader. His group could successfully design de novo proteins with a variety of folds, types of supramolecular organization and desired functions, including, in the most sophisticated case, a synthetic virus-like protein capsule capable of incorporating its own RNA genome [Butterfield et al. 2017] (Fig. 1E). The sequence space that can be explored in de novo protein design is proposed to be much larger than the actual sequence space explored by proteins from extant organisms (which is compared to an "archipelagoes in a vast sea of unexplored territory" [Huang, Boyken, and Baker 2016, p. 321]). In an unrestrained style, the advent of de novo protein design is described by this group as analogous to the transition from the Stone to Iron Age. A promotion video explains, "Early humans adapted objects they found in their surrounding environment to solve the challenge they faced, modern man, on the other hand, has learned to solve problems by building new objects from scratch," as the video fades from images of prehistoric hunters to the Golden Gate bridge and 3D printers. The catch phrase for the institute, "we aim to design a new world of proteins,"[14] clearly displays the change in scientific aims from naturalistic description and explanation of the natural world to de novo design. David Baker stated, "There's a lot of things that nature has come up with just by randomly bumbling around" and "as we understand more and more of the basic principles, we ought to be able to do far better" [Zimmer 2017, n.p.], or along the same line the webpage of the Institute reads: "Evolution has done an incredible job making these proteins with exquisite functions, but nature can only do so much. Modern day challenges demand new proteins with new functions on a much faster timescale than evolution is capable of."[15] In addition to fulfilling the desired task (digest gluten, protect from flu, bind an opioid), the designed proteins that have been experimentally obtained are indeed in general superior in performance to the natural ones in terms of stability ("do far better"); natural protein folds are in general marginally stable and easily denatured, while several designed proteins were found to be highly stable and to withstand heat and chemical denaturation. The protein designers insist on the fact that these novel objects are designed "from scratch" stressing the anti-genealogic move by which these engineers wish to emancipate from (natural) historical constraints.

Here biology has led itself into its own negation, its Hegelian *Aufhebung*. Natural processes are outdated, obsolete, outcompeted. Interest in organic machines (Deleuze and Guattari might say *machines désirantes* (desiring machines) [Deleuze and Guattari 1972]) is replaced by creation of desired machines (*machines désirées*). Natural proteins are informed, and this information derives directly from evolutionary processes, from

developments in this area correspond to the introduction of the Alphafold AI program based on Google's DeepMind.

14 *What is the Institute for Protein Design?* (https://www.youtube.com/watch?v=jBNCj2vqau0, accessed November 26, 2019.)

15 http://www.ipd.uw.edu/research/basic-areas/ (accessed April 4, 2018).

lived time. Proteins are active because they have acted before, their sequences have been shaped by time and action, and thus this information is encapsulated in the sequences of the corresponding genes. De novo designed proteins are also informed, but information is mind-derived and computer-generated. As such they can also be envisioned as an indirect derivation of evolutionary processes and their sequences also reflect the parameters of the energy functions that were used to build them, that is, human knowledge. The current blur in the nature/nurture divide invites us to consider these newly built second-order biological objects as part of the transition from biosphere to noosphere predicted by Vladimir Vernadsky (1945).

6 Intrinsic Disorder, Proteins at the Edge of Chaos

So far, we have treated the relation of sequence to fold and function within Anfinsen's principle that proteins have a minimum energy state and spontaneously fold to attain this shape, which in turn defines their function. But it is now known that in a number of cases proteins do not fold into a determined structural state but have a chaotic behavior and remain disordered [Habchi et al. 2014]. Instead of having a unique energy minimum they display multiple quasi-equivalent energetic states (Fig. 2). Initially, the existence of intrinsically disordered proteins (IDPs) was considered with much skepticism as this concept of natively unfolded proteins (as they are also called) contradicted the general paradigm of the structure to function relation that governed the understanding of structural biology. Also, their chaotic behavior rendered these objects recalcitrant to classical experimental structural description; this intractable character made it more comfortable to just disregard them for some time. It is now widely accepted that IDPs are common, and present in all forms of life. IDPs exist as a dynamic, permanently interconverting ensemble of structural states in which the protein chain is extended rather than collapsed into a globule-like state; the collection of the different states has been termed “a dancing protein cloud” by Vladimir Uversky (2016, p. 6681), a pioneer in the field. The order in ‘normal’ folded proteins is encoded in their sequences and naturally, so is also the disorder encoded in the sequences of these proteins (proteins at the edge of chaos as Uversky terms them). Disordered protein sequences display a bias in amino acid composition and are therefore also termed low complexity sequences. These are characterized by a depletion in order-promoting residues (for instance aromatic and hydrophobic residues) and enrichment in disorder-promoting residues (like glycine or polar residues). IDPs typically evolve faster than folded proteins as their sequences are less constrained; exact sequence matters less as long as a form of disorder is maintained. This structural plasticity transcends the classical one protein-one structure-one function concept and entails such proteins with functional polyvalence. The semantic catalogue associated with the description of this type of proteins includes terms as malleable machines, morphs, chameleon, dual-personality, fuzzy, moonlighting, rheomorphic, or promiscuous sequences, all intended to convey this impression of plasticity. For instance, such proteins can interact with many different cellular partners and in an induced folding mechanism acquire a partial fold when interacting with a given binding partner. In this case, the folding code of the protein is not entirely intrinsically contained within the protein sequence but partially externalized in the sequence and fold of its cognate binding partner. Some IDPs, however, perform their function while remaining fully in a disordered state and have been termed stochastic machines [Xue et al. 2013].[16]

As the biological importance of disorder in protein function continues to be further documented, the one protein–one structure–one function concept needs to be gradually replaced or complemented by a more complex one protein–many structures–many functions paradigm. There appears to be an interesting trend of "blue-collar/white-collar" tasks distribution in cellular activities, folded proteins ensuring primary metabolism and transport functions while IDPs are more often involved in signaling and regulatory functions (specifically because IDPs do not adopt a stable 3D structure, they do not perform enzymatic functions which are based on the precise stereochemical positioning of catalytic chemical groups in the active sites; in contrast, IDPs are particularly suited for molecular recognition tasks involving multiple binding interactions). In that sense, folded proteins act while non-folders preferentially interact [Tompa et al. 2015]. Also, apparently there is a positive correlation between organismal complexity and prevalence of IDPs, the more complex the organism the more frequent IDPs are in its proteome [Habchi et al. 2014]. One should measure how much unrest the introduction of these anti-platonic objects introduced in the paradigm of a firm relation of form to function in biological entities.

7 Protein Fluctuation and Function

We have contrasted here the behavior of 'normal' ordered proteins to IDPs. This dichotomy is largely unjustified, and in fact some, as perhaps most prominently the biological physicist Hans Frauenfelder, have explained that proteins move. Even "normal" folded proteins are intrinsically dynamic and change shape on timescales ranging from the picosecond to the second ("A protein does not exist in a unique conformation but can assume a very large number of somewhat different conformations

16 "The [protein] complex works by random movements of a 'stochastic machine' not by cooperative conformational changes. Unlike typical machines, the different parts of the device are loosely connected, with random movements bringing components together" [Xue 2013, p. 1588]. In that sense, the word "loosely" is used in the same sense as in mathematics and probability theory as relating to a random process (here random motions).

or conformational substates" [Frauenfelder, McMahon, and Fenimore 2003, p. 8616; Henzler-Wildman and Kern 2007]. Some deeply rooted platonistic tendency fueled by the beautiful – static – images of protein structures produced by structural biology leads to the repeated oblivion of this fact.[17] What is important here is not so much that proteins move, but that protein structural fluctuation is a condition of function. As Hans Frauenfelder puts it, "if a protein had just a single conformation, it could not function and would be dead like a stone" [cited in Kovac 2006, p. 564]. Structural fluctuation is critical for many aspects of protein function including signal transduction, enzymatic catalysis, ligand binding, and assembly of supramolecular complexes. In the end, the structure-function paradigm extends into a disorder-function concept.[18]

The concept of allostery is essential for the understanding of the relation of conformational change to function. The term "allostery" was coined by Jacques Monod [Monod, Changeux, and Jacob 1963] and denotes the ability of a protein to change its conformation in response to the binding of a ligand (Fig. 2). Initially framed in the context of multimeric enzymes whose activity is modulated by the binding of a ligand at a site distinct from the substrate binding site, this allostery concept has evolved considerably from a two-state to a multi-substate model and from multisubunit proteins to all types of proteins [Liu and Nussinov 2016]. This ability of proteins to change between substates in response to changes in their environment is considered so fundamental that it has been elevated to the level of a "second secret of life,"[19] second only to the genetic code [Fenton 2008, p. 420].[20]

---

17 Thanks to a series of technological innovations – such as solution X-ray scattering – structural biology becomes more and more apt to describe protein structural fluctuation [Palamini, Canciani, and Forneris 2016]. The discipline undergoes a kind of photography-to-cinematography transition. Novel graphic methods to display structural substates are being developed [Melvin and Salsbury 2016] and will perhaps play a heuristic role in the increased awareness of protein plasticity.

18 Aside from internal movement associated with transition from one substate to another, recent experiments enabled by development of specific spectroscopic methods have revealed that movement (in the form of displacement of the object in space) is also associated to enzymatic catalysis per se. Enzymes were found to be propelled, to leap by ballistic-like movements in a length scale of 100 nm when performing their catalytic function. Such leaps are more frequent at high substrate concentration, so that enzymes collectively distribute in a gradient inverse to the concentration gradient of the substrate, in a process that was designated antichemotaxis [Jee et al. 2018].

19 Agnès Ullmann provides a recollection of the discovery of the allostery concept: "By the end of 1961, one evening quite late Jacques Monod walked into my laboratory looking rather tired and worried. Monod stood silently at my bench and after a few long minutes he said: '*I think I have discovered the second secret of life*.' I was quite alarmed by this unexpected revelation and asked him if he needed a glass of whisky. After the second or maybe the third glass, he explained the discovery, which he had already given a name: 'allostery' " [Ullmann 2011, p. 1029].

20 In a following section, we will question this order and suggest that the possibility exists that allosteric behavior has a temporal primacy and that – if at all – it is the genetic code that might be qualified as 'second secret of life.'

To describe the existence of structural substates in proteins, for instance determined by allostery, the energy landscape model is used (we have already mentioned this model in the context of protein folding) [Henzler-Wildman and Kern 2007]. While the energy landscape is generally represented as a funnel-shaped landscape with a single minimum, protein landscapes are better understood as rugged landscapes with a number of local minima that can be explored. These local energy minima are separated by barriers that can be overcome thanks to heat and only become effective at very low temperatures (which in a way 'freeze' protein motion); higher energy barriers are overcome by the binding of a ligand or various changes in the solvent condition or physical conditions. Frauenfelder states poetically that "to perform functions, proteins must dance in the energy landscape" [Frauenfelder 2017, p. 1]. It should be noted that there is no fundamental contradiction between Anfinsen's principle and the existence of protein conformation substates as Anfinsen considers energy "of the whole system," protein and solvent, stressing that protein folds are fundamentally co-constructions of the protein sequence and the environment.

8 Protein Coalescence, Flocks of Proteins

Gradually over the last few years, a remarkable new view of the collective behavior of proteins in living cells has emerged. While proteins in general act in a soluble monomeric state or as multimeric supramolecular complexes of defined composition, they also display a collective behavior involving coordinated self-associations leading to phase transitions [Shin and Brangwynne 2017; Kaganovich 2017]. In response to environmental changes, a large number of proteins undergo a transition from a soluble to a coalescent state which can correspond to different physical states ranging from liquid and liquid crystal-like to solid (Fig. 2). This phenomenon – which has been designated under a variety of terms of which protein coalescence is perhaps the most satisfactory – encompasses a variety of biological processes that have been so far considered in isolation and now receive a more unified consideration because all involve new properties conferred by the collective behavior of proteins [Saarikangas and Barral 2016]. At the molecular level, these protein coalescence processes involve often but not always protein domains that tend to be intrinsically disordered as described here above. The structurally dynamic character of the IDPs or protein domains appears to be associated with their ability to form such coalescent states. Such phase transitions can correspond to the formation of so-called membrane-less organelles in which the coalescence of proteins (and often associated RNAs) leads to the formation of cellular compartments which are not bound by a lipid membrane. These protein coalescence structures (which receive different names like granules, droplets or protein bodies) often form in response to an environmental change (typically a stress, like nutrient depletion or heat). Some exist only transiently and dissolve when the environmental signal subsides, others become permanent and even, as we shall see, can be transmitted to the cell progeny as genetic material. Coalescent structures can be composed of up to several hundreds of different protein (and RNA) types or result from the collective behavior of a single protein type. Protein coalescence processes serve different cellular functions including sensing, filtering, protective storage, protein inactivation, amplification of signal or enzymatic activity and memory. While protein coalescence is often an adaptive phenomenon, collective protein behavior also leads to pathological situations, such as neurodegenerative protein deposition diseases like Parkinson's, Alzheimer's, and Huntington's diseases, which typically result from the collective behavior and coalescence of proteins. Physically, these protein flocks can correspond to liquid droplets and involve transient interactions between largely unstructured proteins, but they can also be highly ordered or disordered solids that assemble and disassemble at various rates. While the exact relation between the different condensed states is often unclear, liquid condensates can further maturate into liquid crystal-like and solid states. When an ordered solid state is involved, it most often corresponds to a protein state termed amyloid, an ordered arrangement of protein subunits into a fibrillar polymer with a cross-β structure formed by the stacking of β-strand structure elements perpendicular to the fibril axis.

An interesting classification of these protein coalescence phenomena based not so much on the physical state of the protein condensate but on its biological properties has been proposed [Saarikangas and Barral 2016]. This classification distinguishes *adaptopods* that form in response to a signal but dissolve when the inducing signal is removed from *mnemons* which are signal-induced but self-maintained (remain after the signal is removed). Mnemons thus fix a new state irreversibly. The Whi3 protein found in yeast provides such an example of proteins that – by collective behavior – act as a device capable of encoding the memory of past events. In yeast cells, exposure to the pheromone from the opposite mating-type induces an arrest of the cell cycle; the cell stops dividing and differentiates a mating structure termed *shmoo* (because it resembles the cartoon strip creature created by Al Capp). If, however, no mating partner is reached within a reasonable time, the cell cycle resumes and cells become refractory to a new pheromone exposure (they will not be fooled again into growth arrest and shmoo-formation). This acquired behavior results from the pheromone-induced coalescence of the Whi3 proteins which thus "encodes memory of deceptive encounters" [Caudron and Barral 2014, p. 100]. This mechanism of cellular memory based on collective protein behavior was reported to also occur in long-term memory in invertebrates and mammals. In yeast, mnemons are defined by the fact that they are not inherited in cell divisions. The Whi3 mnemon is physically retained in the mother cell that permanently becomes refractory to pheromone stimulation, but the daughter cells do not inherit memory of the event and are competent for mating.[21] This distinction in heritability of the protein coalescent structure defines a last category of protein assemblies termed *prions* that are transmitted to cell progeny and thus represent *genetic* elements that will be described in the following section.

These examples illustrate that in addition to their individual activities, collective behavior of proteins in the form of phase transition in a variety of modes entitles emergent properties that are developed in a multitude of cellular processes ranging from adaptation to memory and inheritance.

9 Inheritable Collective Folds, Proteins as Genes

Prions illustrate a specific case of phase transition, in which the coalescent state becomes transmissible. Prions are infectious proteins, they propagate their fold. The term "prion" was coined in the context of the study of fatal neurodegenerative diseases in different mammals, such as Creutzfeldt-Jacob disease and Kuru in humans and Mad Cow disease in cattle [Colby and Prusiner 2011]. But there are many other prion proteins notably in fungi where they were initially identified as non-Mendelian genetic elements [Harvey, Chen, and Jarosz 2018]. These fungal prion proteins are not pathological; they have adaptive, functional roles in cells. For the mammalian PrP prion, the prion state corresponds to a molecular mishap, a *misfolding*. Prion diseases, like other neurodegenerative diseases such as Parkinson's and Alzheimer's, are misfolding diseases. In contrast, in the case of fungal prions, the prion fold is the native fold. Importantly, prion folding can only be achieved cooperatively and not by an isolated molecule. The information contained in the DNA sequence can only be actualized when several protein monomers interact to collectively attain the prion fold as an assembly of molecules. They undergo a phase transition of the kind described in the above paragraph into a solid and highly ordered fibrillar amyloid aggregate. Once formed, this fold becomes self-propagating, in the sense that it directs the folding of other protein units to the same prion fold. The prion fold directly transmits this structural information by contact and imposes it on other molecules of the same kind. The prion template grows into a fibril by incorporation of monomers and ultimately breaks down to smaller fragments, the extremities of which then act as new templates for transmission of the fold, thus ensuring replication of the propagating entity.

For reasons of personal taste and because it is the first prion whose high-resolution structure has been solved, we will turn to the specific example of the *s* (or [Het-s]) prion of the filamentous fungus *Podospora anserina* to illustrate the

---

21 In the yeast-jargon, the daughter cell buds off by mitosis from the mother cell, so formally the relation of the two is in no way genetically analogous to the mother–daughter relation in humans.

prion principle [Wasmer et al. 2008]. Like a machine connected to other machines in various configurations, this protein stands at the crossroads of several biological phenomena and depending on the perspective that is adopted can be viewed as an inactive mutant, a protein defining biological individuality or an ultra-selfish genetic element. A description of the complex natural history of this protein can be found elsewhere [Daskalov and Saupe 2015]. It might be sufficient to state here that fungi expressing this protein can display two alternate phenotypes termed *s* and *s** that differ in their reactivity towards a genetically distinct strain termed *S*. *s** strains are able to fuse with *S* (*s** and *S* are compatible), while *s* strains cannot and produce a cell death reaction when in contact with *S* (*s* and *S* are incompatible). This difference in phenotype is not defined by a difference in DNA sequence but results from an epigenetic difference, residing only in the folding state of the *s* protein. It is important to understand that this fold contains transmissible information, hence genetic information. This information is transmitted horizontally from cell to cell and inherited vertically in cell divisions and in sexual reproduction. Experiments analogous to the classical work of Oswald Avery, Colin McLeod and Maclyn McCarty that established DNA as the 'transforming principle' (the genetic molecule) have been carried out with the *s* prion protein and amount to defining the protein as genetic material [Fink 2005]. Formally, the fold – derived from the collective behavior of these proteins – is a gene.

In most proteins, the folding is delayed typically by microseconds to seconds, but here, with this prion fold, the fold can be considerably delayed. The prion folding can in fact be deferred indefinitely (or at least never be actualized in the life span of the organism). The proteins in the unfolded and folded states are identical in terms of chemical composition, they are the same and yet they differ and this difference is the fold; in one state information is actualized, in the other not. This fold, this information now actualized is self-propagating. The prion protein affects itself, but the other itself, its *deferred* self (prion conversion is an auto-affection but also a hetero-affection). This biological system, with some of its specificities, bends and stretches the vocabulary and nomenclature of modern biology to its limits by making the genotype/phenotype divide undecidable.[22]

22 The textual analogy in molecular biology is a classical one and perhaps the *différance* proposed by Jacques Derrida in the linguistic context can be of help to describe the ambiguous signifier-to-signified relation in this system. Derrida gradually developed *différance* to describe the process by which a language (or any system of code) historically constitutes itself as a system of differences, producing spacing between elements that get defined by difference in relation to each other. The term carries a radical polysemy as it contains altogether the notions of temporal delay (*différé*), of alterity of dissemblance (*différence*), of allergic or polemic alterity (*différend*) and finally the notion of production of difference (*-ance* suffix). The information specifying the protein fold is present in the DNA of the corresponding gene (the sequence makes sign for the protein). But the relation of signifier-to-signified in the sign is deferred. This sign associates a single-but-multiple signified (unfolded non-prion/folded prion states of the *same* protein) to a given signifier (sequence). The

10 In the Beginning There Was ~~the~~ No Verb, Proteins, and the Origin of Life

Several aspects in the understanding of the collective behavior of proteins lead to some interesting ideas regarding the possible origin of life (however, it might be defined). Very basic requirements for cellular life are some form of templating underlying heritability and self-perpetuation of structure and some form of confinement with at least partial definition of an inside and an outside, the roles of template and boundary being attributed classically and respectively to nucleic acids (DNA and/or RNA) and the lipid bilayer membrane. Some researchers currently contemplate the hypothesis that proteins in the form of amyloid condensates might represent such initial self-templating molecules [Maury 2015]. It has been possible to obtain spontaneous formation and replication of amyloid peptides in Miller soup pre-biotic conditions [Greenwald, Friedman, and Riek 2016; Rout et al. 2018]. In these experiments, peptides are extended by amino acid addition selected by direct structural templating, not through decoding of a triplet sequence. The same is true for certain types of α-helical peptides, raising the possibility that direct templating in protein synthesis predates development of the genetic code. Also, in the frame of this view, amyloid (collective) folds would represent ancient, primitive folds and subsequent evolution of monomeric folded proteins a further step. Formation of complex folds would thus represent emancipation from this primitive rather monotonous state. Complex monomeric folding requires generally a certain length of the protein chain (a folded domain generally requires about 20–30 amino acids in length to fold). The increase in length of the protein chain might represent a later evolutionary stage compared to the small length peptide condensates that could form in prebiotic conditions. It is interesting to consider that this path might be taken in reverse during age-dependent

---

sequence signifies the prion fold, but the fold is deferred. In an anticipation of the last section of this text, we note that the temporal delay component of the *différance* is presented as a delay of the realization of desire ("*Différer* in this sense is to temporize, to take recourse, consciously or unconsciously, in the temporal and temporizing mediation in a detour that suspends the accomplishment or fulfillment of 'desire'" [Derrida 1982, p. 8] ("Différer en ce sens, c'est temporiser, c'est recourir, consciemment ou inconsciemment, à la médiation temporelle et temporisatrice d'un détour suspendant l'accomplissement ou le remplissement du 'désir'" [Derrida 1972, p. 8])). The fold is deferred, this difference is historically produced and by external reference (as a co-construction with the environment), there is *différance* in and of the fold. This prion system provides illustrations of how alterity can emerge, how to produce difference from the same [Daskalov and Saupe 2015]. In this system, an alterity of dissemblance (in sequence or structure) is differentiated into (produces) an alterity of allergy (a conflict, incompatibility). The dualistic oppositions in this system (*s/s**) and (*s/S*) are only decisive by external reference. There is *différance* in as much as the fold is delayed and in as much as it is productive; it is what actualizes the biological alterity between *s* and *S*. In passing, just like the 'a' in Derrida's *différance* is inaudible, the *s/S* difference is graphical and inaudible unless it is outspoken.

protein misfolding, in which proteins due to deficiency in protein quality control mechanisms or chemical alteration form amyloid aggregates at the expense of the normal folded states. It is Ernst Haeckel's ontogeny-recapitulates-phylogeny in reverse [Haeckel 1866].

The protein-based phase separation processes, mentioned in a previous section, offer a form of macromolecular organization that separates an interior from an exterior milieu and just as some cell organelles – which are considered as remnants of endosymbiotic cells – are bound by lipid bilayers, others are not and it is now envisioned that phase-separated protein bodies might have played a role as protocells at the origin of life [Stroberg and Schnell 2017]. Phase-separated droplets are capable of filtering out or concentrating certain chemicals and ions and are capable of mitosis-like division processes. In this form of individualization, there is formation of two compartments but no actual distinct material frontier separating them (as in the case of the lipid bilayer). It is the intrinsic interaction of the phase-separated entities that provides the compartmentation function.

While these views of the origin of life remain obviously speculative to the highest degree, they provide an alternative to the perspective strongly centered on the main informational molecule in extant organism, DNA, and to draw on Lily Kay's use of the biblical "in the beginning was the verb" metaphor framing the importance of information theory in the historical development of molecular biology [Kay 2000], maybe one could say that the verb was not made flesh but was the flesh right from the start. As an article on the origin of the word "protein" explains, "The name 'protein' [was] derived from the Greek adjective πρωτεἱος, meaning of the first rank or position" [Vickery 1950, p. 387]. The etymology relates to the fact that the described substance was supposed to be a primary stuff for animal nutrition produced by plants, so in fact does not imply temporal primacy, but Valère Novarina proposes that in some instance erroneous etymologies might be more effective than real ones [Novarina 2017, p. 22].

11 Proteins as Monads and Desiring Machines

The all-encompassing materialistic philosophy of Gilles Deleuze and Félix Guattari was defined by some as a biophilosophy and analyzed for instance as containing a critique of the neo-Darwinian modern synthesis [Marks 2006]. But aside and maybe complementary to this aspect, what is highly original in Deleuze's and Guattari's reception of molecular biology is their attention to protein activity and folding. In an anti-genealogic gesture, they find more interest in the actual living machine than in its genealogical origin. There are explicit references in both Deleuze's *Le Pli* (*The Fold*) and Deleuze's and Guattari's *Mille Plateaux* (*A Thousand Plateaus*) and *L'Anti-Œdipe* (*Anti-Oedipus*) to protein folding. In *Le Pli*, Deleuze proposes, following Leibniz, a description of organic development based on folding/unfolding processes and places protein folding at the very root of such organic developmental processes [Deleuze 1988, p. 15]. Deleuze subscribes to the Leibnizian machinistic view of organic life: "But the machines of nature, namely, living bodies, are still machines in their smallest parts *ad infinitum*" [Leibniz 1898, p. 254],[23] and in this way he distinguishes *machinism* from *mechanism*: "A mechanism is faulty not for being too artificial to account for living matter, but for not being mechanical enough, for not being adequately machined" [Deleuze 2006, p. 8].[24] Protein complexes with their many folds and motions conform to the baroque aesthetics as defined by Deleuze in his essay on Leibniz and the Baroque [Deleuze 1988, p. 6] and conform to Deleuze's definition of organic machines as baroque machines [Deleuze 1988, p. 12] (Fig. 1). Following Deleuze in his reading of Leibniz and putting aside the theological dimension of Leibniz' *La Monadologie*, it might be possible in some aspects to qualify proteins as *monads*.[25] While of course, in chemical and physical terms, proteins do not represent "simple substances" that cannot be further subdivided, it might be argued that any further subdivision below the level of the individual protein cannot be achieved without losing biological meaning, a fragment of the protein does not make sense any more, it is not made of machines but of (chemical) parts. As such, proteins might qualify as the Leibnizian "Atoms of Nature" [Leibniz 1991, p. 125]. From these elementary machines, by a series of higher order organization steps, macromolecular complexes, organelles, cells, organs and organisms can be assembled. Deleuze stresses that this organization, this formation of organisms, can only be understood if one assumes that the elementary components are already folded, organized.[26] Another relevant property of monads is a certain level of perfection and autonomy ("they have in them a certain perfection [...];

23 "Mais les machines de la nature, c'est à dire les corps vivants, sont encore des machines dans leurs moindres parties, jusqu'à l'infini" [Leibniz 1991, p. 162].

24 "Le tort du mécanisme ce n'est pas d'être trop artificiel pour rendre compte du vivant, mais de ne pas l'être assez, de ne pas être assez machine" [Deleuze 1988, p. 12].

25 "The Monad, of which we shall here speak, is nothing but a *simple* substance, which enters into compounds. By 'simple' is meant 'without parts'" [Leibniz 1898, p. 217]. ("La Monade, dont nous parlerons ici, n'est autre chose, qu'une substance *simple* qui entre dans les composés; simple, c'est-à-dire, sans parties" [Leibniz 1991, pp. 123–124].)

26 "On the other hand, the formation of the organism would remain an improbable mystery, or a miracle, even if matter were to divide infinitely into independent points. But it becomes increasingly probable and natural when an infinity of indeterminate states is given (already folded over each other), each of which includes a cohesion at its level, somewhat like the improbability of forming a word by chance with separate letters, but with far more likelihood with syllables or inflections" [Deleuze 2006, p. 7]. ("D'autre part, la formation de l'organisme resterait un mystère improbable ou un miracle si la matière se divisait même à l'infini en points indépendants, mais devient de plus en plus probable et naturelle quand on se donne une infinité d'états intermédiaires (déjà repliés) dont chacun comporte une cohésion, à son niveau, un peu comme il est improbable de former au hasard un mot avec des lettres séparées, mais beaucoup plus probable avec des syllabes ou des flexions" [Deleuze 1988, p. 10].)

they have a certain self-sufficiency [...] which makes them the sources of their internal activities" [Leibniz 1898, p. 229]),[27] which is in line with the ability to fold autonomously into a unique fold associated with function. They are perfect in the sense that they attain an energy minimum. Then importantly monads have *perceptions* and *appetites*.[28] The concept of allostery by which a protein perceives and changes in response to ligands (or physical conditions) illustrates that sentience, understood as "the capacity to exhibit a variety of potential internal states, which respond to the immediate state of the environment" [Kovac 2006, p. 564], extends *cognition* to the level of proteins (at the individual protein level, at the level of the higher order macromolecular complexes or the collective phase-transition behaviors). Jacques Monod – one of the main sources of reading on molecular biology for Deleuze and Guattari together with François Jacob's *The Logic of Life* (*La Logique du Vivant*, 1970) – was apparently the first to use the term "cognition" for substrate discrimination and enzyme function. Deleuze describes chemical affinities of enzymes as "perceptions and elections" and plays with the temptation to term these "souls" [Deleuze 1981]. Ladislav Kovac, in a remarkable text entitled "Life, chemistry and cognition," frames the contours of what he terms "cognitive biology," a concept centered on the notion that cognition is ultimately and fundamentally embodied [Kovac 2006, p. 563]. Kovac insists that cognition is a dual process of sensation and action and convincingly installs the proteins at the very basis of a hierarchical, multilayer, nested sentience. The central concept allowing this articulation between sensation and action is allostery, that is, the ability of a protein to undergo a conformational change in response to ligand binding (or in a broader sense to other physical cues such as light, heat, or mechanical forces) and to display changed functional properties in response to this signal.[29] Current biotechnological projects exploit the "cognitive ability" of proteins (and/or nucleic acids) to construct various Boolean logic gates in order to create biocomputing nanorobots that can be interfaced with living biological systems [Katz 2015].

Furthermore, a monad is defined as a reflection of the entire universe taken from a certain perspective ("as every Monad is, in its own way, a mirror of the universe" [Leibniz 1898, p. 253]).[30] Richard Lewontin states that organisms are molded

27 "elles ont en elles une certaine perfection [...], il y a une suffisance [...] qui les rend sources de leurs actions internes" [Leibniz 1991, pp. 133–134].

28 "*Perception*" and "*Appétition*" [Leibniz 1991, pp. 129, 131].

29 "Recognition is followed by an action. A ligand is a signal. In contrast to standard chemical interactions, binding energy is not fully dissipated as heat, but is used partly for molecular work – a specific pre-programmed change in the conformation of the protein. In this way, the signal is transmitted from one site on the protein to another. The transmission takes place in four-dimensional space, as it involves time as a coordinate, and this process gives biochemistry its vectoriality. [...] It is appropriate to regard most protein molecules as molecular engines. Hence, molecular cognition consists of molecular sensation – which has two inseparable aspects, recognition and signification – and molecular action" [Kovac 2006, p. 565].

30 "toute Monade étant un miroir de l'univers à sa mode" [Leibniz 1991, p. 161].

by and are thus molds of their environments [Lewontin 2001]. Molecular evolution shapes protein sequences/folds in response to specific biotic and abiotic cues in the environment, as such proteins seen as monads reflect a certain point of view on the universe. Perhaps the case of ancestral protein reconstruction experiments can serve to illustrate this principle. Sequences of very ancient enzymes dating back as far as LUCA (Last Universal Common Ancestor) can be expressed, produced, and structurally characterized. By analyzing their properties, thermal stability, thermal optimum, pH dependence, and substrate specificity, one can devise information about the physico-chemical conditions on earth a billion years ago [Akanuma 2017]. A single protein reflects its *Umwelt* [Uexküll 1909], mirrors its own perspective on the universe.

In the two volumes of *Capitalisme et Schizophrénie* (*Capitalism and Schizophrenia)*, Deleuze and Guattari place the concept of *desiring machine* as the basis of desiring-production. Desiring machines are connected with each other, machines branched on the flux of another machine, these loose and changing connections between elementary machines build up organisms and at higher levels social structures [Andoka 2012]. In their development, a critical step occurs in Section 2 of Chapter 4 of *Anti-Oedipus*, "beyond vitalism and mechanism" [Deleuze and Guattari 2000]. Both vitalism and mechanism deprive the machine of a direct relation to desire.[31] In mechanism, desire is an emergent effect of the abstracted structural unity explaining the organism. In vitalism, machines are subordinated to the desire of the individual unity of the organism. The biologist frames the understanding of life with the concept of function, a concept genuine to biology, says Peter Wolynes [Ferreiro, Komives, and Wolynes 2018], which might provisionally replace desire in the above sentences. Deleuze and Guattari aim at shattering both perspectives by putting into question both the individuality of the organism and the structural unity of the machine. What is achieved by dissolving the mechanistic and vitalistic perspectives is the direct

31 "From machines, mechanism abstracts a *structural unity* in terms of which it explains the functioning of the organism. Vitalism invokes an *individual and specific unity* of the living, which every machine presupposes insofar as it is subordinate to organic continuance, and insofar as it extends the latter's autonomous formations on the outside. But it should be noted that, in one way or another, the machine and desire thus remain in an extrinsic relationship, either because desire appears as an effect determined by a system of mechanical causes, or because the machine is itself a system of means in terms of the aims of desire. The link between the two remains secondary and indirect, both in the new means appropriated by desire and in the derived desires produced by the machines" [Deleuze and Guattari 2000, p. 284]. ("Le mécanisme abstrait des machines une *unité structurale* d'après laquelle il explique le fonctionnement de l'organisme. Le vitalisme invoque une *unité individuelle et spécifique* du vivant, que toute machine suppose en tant qu'elle se subordonne à la persistance organique et en prolonge à l'extérieur les formations autonomes. Mais on remarquera que, d'une manière ou d'une autre, la machine et le désir restent ainsi dans un rapport extrinsèque, soit que le désir apparaisse comme un effet déterminé par un système de causes mécaniques, soit que la machine soit elle-même un système de moyens en fonction des fins du désir. Le lien reste secondaire et indirect entre les deux, aussi bien dans les nouveaux moyens que le désir s'approprie que dans les désirs dérivés que suscitent les machines" [Deleuze and Guattari 1972, p. 337].)

connection between the machine and desire.[32] This may have been overlooked, but Deleuze and Guattari *directly* define proteins as the elementary desiring machines: "Proteins are both products and units of production; they are what constitutes the unconscious as a cycle or as the autoproduction of the unconscious – the ultimate molecular elements in the arrangement of the desiring-machines and the syntheses of desire" [Deleuze and Guattari 2000, p. 290].[33] Central to this point of view is the aforementioned concept of allostery that articulates recognition and action, converting the Leibnizian *perceptions* and *appetites* to desiring-production. Even the most trivial biotechnological application of protein function/desire (such as the use of lipases in washing powder) illustrates how the mechanistic or vitalistic unity can be undone and machine connections reshuffled and reorganized. An interesting mirror image of proteins seen as desiring machines can be found in the development of a chemical concept termed *frustration*. There is frustration in a protein sequence segment when fold stability conflicts with the plasticity required to perform various binding functions. Which regions of a protein sequence are frustrated (present a conflict between stability and plasticity) can be determined and Peter Wolynes proposes that analyzing protein frustration is a way to understand the goals of a protein, to "distinguish meaning from noise" and that "analyzing frustration is a tool to search for meaning," as if finding where frustration is will help in defining desire [Ferreiro, Komives, and Wolynes 2018, p. 71].

We aimed here at discussing in which sense proteins can be viewed as active matter, in both their individual and collective behavior, Deleuze and Guattari might say on their molecular and molar side. In the process, the materialism developed by these post-structuralist philosophers rooted in the Leibnizian theory of "les petites perceptions" ("the small perceptions") as Deleuze defines it [Deleuze 1981] represents a useful guide. Many problematic aspects of contemporary biology could be thought of in the terms developed in *L'Anti-Œdipe* and *Mille Plateaux*. Organismal unity in all phylogenetic branches is fundamentally reframed by metagenomics,

32 "Once the structural unity of the machine has been undone, once the personal and specific unity of the living has been laid to rest, a direct link is perceived between the machine and desire, the machine passes to the heart of desire, the machine is desiring and desire, machined. Desire is not in the subject, but the machine in desire – with the residual subject off to the side, alongside the machine, around the entire periphery, a parasite of machines, an accessory of vertebro-machinate desire" [Deleuze and Guattari 2000, p. 285]. ("Une fois défaite l'unité structurale de la machine, une fois déposée l'unité personnelle et spécifique du vivant, un lien direct apparaît entre la machine et le désir, la machine passe au coeur du désir, la machine est désirante et le désir, machiné. Ce n'est pas le désir qui est dans le sujet, mais la machine dans le désir – et le sujet résiduel est de l'autre côté, à côté de la machine, sur tout le pourtour, parasite des machines, accessoire du désir vertébro-machiné" [Deleuze and Guattari 1972, p. 339].)

33 "Les protéines sont à la fois produites et unités de production, c'est elles qui constituent l'inconscient comme cycle ou l'autoproduction de l'inconscient, ultimes éléments moléculaires dans l'agencement des machines désirantes et des synthèses du désir" [Deleuze and Guattari 1972, p. 345].

prevalence of horizontal transfer questions species genealogy, phylogenetic trees are replaced by reticulated (rhizome-like) trees [Lopez and Bapteste 2009], and exaptation has been documented down to the molecular level. The interesting aspect here is that in the late 1970s, at the time these works appeared, molecular biology offered a much simpler, cleaner, and ordered picture of the living world. Nevertheless, the reading of this picture by Deleuze and Guattari is much more in line with the multiple, destructured, reticulated image emerging from our twenty-first-century molecular biology.

Bibliography

Akanuma S. Characterization of Reconstructed Ancestral Proteins Suggests a Change in Temperature of the Ancient Biosphere. Life 2017, 7 (3), 33, DOI: 10.3390/life7030033.

Andoka F. Machine désirante et subjectivité dans l'Anti-Œdipe de Deleuze et Guattari. Philosophique 2012, 15, 85–94, DOI: 10.4000/philosophique.659.

Andrault R. The Machine Analogy in Medicine: A Comparative Approach of Leibniz and his Contemporaries. In: Smith M, Nachtomy O, eds. Machines of Nature and Corporeal Substances. Dordrecht, Springer, 2011, 95–114.

Anfinsen CB. Principles that Govern the Folding of Protein Chains. Science 1973, 181 (4096), 223–230, DOI: 10.1126/science.181.4096.223.

Baldwin RL. Clash between Energy Landscape Theory and Foldon-Dependent Protein Folding. PNAS 2017, 114 (32), 8442–8443, DOI: 10.1073/pnas.1709133114.

Bergson H. L'Évolution Créatrice. Paris, PUF, 2013 [1907].

Bokori-Brown M, Martin TG, Naylor CE, Basak AK, Titball RW, Savva CG. Cryo-EM Structure of Lysenin Pore Elucidates Membrane Insertion by an Aerolysin Family Protein. Nature Communications 2016, 7, 11293, DOI: 10.1038/ncomms11293.

Butterfield GL, Lajoie MJ, Gustafson HH, et al. Evolution of a Designed Protein Assembly Encapsulating its own RNA Genome. Nature 2017, 552 (7685), 415–420, DOI: 10.1038/nature25157.

Caudron F, Barral Y. Mnemons: Encoding Memory by Protein Super-Assembly. Microbial Cell 2014, 1 (3), 100–102, DOI: 10.15698/mic2014.01.134.

Colby DW, Prusiner SB. Prions. Cold Spring Harbor Perspectives in Biology 2011, 3 (1), a006833, DOI: 10.1101/cshperspect.a006833.

Cooper S, Khatib F, Treuille A, Barbero J, Lee J, Beenen M, Leaver-Fay A, Baker D, Popović Z, Players F. Predicting Protein Structures with a Multiplayer Online Game. Nature 2010, 466 (7307), 756–760, DOI: 10.1038/nature09304.

Daskalov A, Saupe SJ. As a Toxin Dies a Prion Comes to Life: A Tentative Natural History of the [Het-s] Prion. Prion 2015, 9 (3), 184–189.

Deleuze G. Le Pli. Leibniz et le Baroque. Paris, Les Editions de Minuit, 1988.

Deleuze G. The Fold. Leibniz and the Baroque. London, Continuum, 2006.

Deleuze G. Cours sur Spinoza du 6 Janvier 1981, 2de Partie, Transcrition: Véronique Delannay et Jean-Charles Jarrell, Université Paris 8. (Accessed June 11, 2020, at http://www2.univ-paris8.fr/deleuze/article.php3?id_article=9).

Deleuze G, Guattari F. L'Anti-Oedipe. Paris, Les Editions de Minuit, 1972.

Deleuze G, Guattari F. Mille Plateaux. Paris, Les Editions de Minuit, 1980.

Deleuze G, Guattari F. Anti-Oedipus. Minneapolis, University of Minnesota Press, 2000.

Deleuze G, Guattari F. A Thousand Plateaus. Minneapolis, University of Minnesota Press, 2005.
Derrida J. Marges de la Philosophie. Paris, Les Editions de Minuit, 1972.
Derrida J. Margins of Philosophy. Chicago, The University of Chicago Press, 1982.
Dill KA, MacCallum JL. The Protein-Folding Problem, 50 years on. Science 2012, 338 (6110), 1042–1046, DOI:10.1126/science.1219021.
Eaton WA, Wolynes PG. Theory, Simulations, and Experiments Show that Proteins Fold by Multiple Pathways. PNAS 2017, 114 (46), E9759–E9760, DOI: 10.1073/pnas.1716444114.
Englander SW, Mayne L. The Case for Defined Protein Folding Pathways. PNAS 2017, 114 (31), 8253–8258, DOI: 10.1073/pnas.1706196114.
Fenton AW. Allostery: An Illustrated Definition for the 'Second Secret of Life'. Trends in Biochemical Sciences. 2008, 33 (9), 420–425, DOI: 10.1016/j.tibs.2008.05.009.
Ferreiro DU, Komives EA, Wolynes PG. Frustration, Function and Folding. Current Opinion in Structural Biology 2018, 48, 68–73, DOI: 10.1016/j.sbi.2017.09.006.
Fink GR. A Transforming Principle. Cell 2005, 120 (2), 153–154, DOI: 10.1016/j.cell.2005.01.004.
Frauenfelder H. Physical Concepts in Biology-Past and Future. Physical Biology 2017, 14 (1), 010402, DOI: 10.1088/1478-3975/aa559f.
Frauenfelder H, McMahon BH, Fenimore PW. Myoglobin: The Hydrogen Atom of Biology and a Paradigm of Complexity. PNAS 2003, 100 (15), 8615–8617, DOI: 10.1073/pnas.1633688100.
Greenwald J, Friedmann MP, Riek R. Amyloid Aggregates Arise from Amino Acid Condensations under Prebiotic Conditions. Angewandte Chemie Internationale Edition 2016, 55 (38), 11609–11613, DOI: 10.1002/anie.201605321.
Guo P, Noji H, Yengo CM, Zhao Z, Grainge I. Biological Nanomotors with a Revolution, Linear, or Rotation Motion Mechanism. Microbiology and Molecular Biology Reviews 2016, 80 (1), 161–186, DOI: 10.1128/mmbr.00056-15.
Habchi J, Tompa P, Longhi S, Uversky VN. Introducing Protein Intrinsic Disorder. Chemical Reviews 2014, 114 (13), 6561–6588, DOI:10.1021/cr400514h.
Haeckel E. Generelle Morphologie der Organismen. Berlin, Georg Reimer, 1866.
Harvey ZH, Chen Y, Jarosz DF. Protein-Based Inheritance: Epigenetics beyond the Chromosome. Molecular Cell 2018, 69 (2), 195–202, DOI: 10.1016/j.molcel.2017.10.030.
Henzler-Wildman K, Kern D. Dynamic Personalities of Proteins. Nature 2007, 450 (7172), 964–972, DOI: 10.1038/nature06522.
Huang PS, Boyken SE, Baker D. The Coming of Age of de Novo Protein Design. Nature 2016, 537 (7620), 320–327, DOI: 10.1038/nature19946.
Ionov L. Polymeric Actuators. Langmuir 2014, 31 (18), 5015–5024, DOI: 10.1021/la503407z.
Jackson SE, Suma A, Micheletti C. How to Fold Intricately: Using Theory and Experiments to Unravel the Properties of Knotted Proteins. Current Opinion in Structural Biology 2017, 42, 6–14, DOI: 10.1016/j.sbi.2016.10.002.
Jee AY, Dutta S, Cho YK, Tlusty T, Granick S. Enzyme Leaps Fuel Antichemotaxis. PNAS 2018, 115 (1), 14–18, DOI: 10.1073/Pnas.1717844115.
Judson HF. Max Delbruck. The Eighth Day of Creation. New York, Cold Spring Harbor Press, 1996.
Kaganovich D. There Is an Inclusion for That: Material Properties of Protein Granules Provide a Platform for Building Diverse Cellular Functions. Trends in Biochemical Sciences 2017, 42 (10), 765–776, 10.1016/j.tibs.2017.08.002.
Katz E. Biocomputing – Tools, Aims, Perspectives. Curr Opin Biotechnol 2015, 34, 202–208.
Kay LE. Who Wrote the Book of Life? A History of the Genetic Code. Stanford, Stanford University Press, 2000.
Kendrew JC, Bodo G, Dintzis HM, Parrish RG, Wyckoff H, Phillips DC. A Three-Dimensional Model of the Myoglobin Molecule Obtained by X-Ray Analysis. Nature 1958, 181 (4610), 662–666, DOI: 10.1038/181662a0.

Kovác L. Life, Chemistry and Cognition: Conceiving Life as Knowledge Embodied in Sentient Chemical Systems Might Provide New Insights into the Nature of Cognition. EMBO Reports 2006, 7 (6), 562–566, DOI: 10.1038/sj.embor.7400717.
Kryshtafovych A, Monastyrskyy B, Fidelis K, Moult J, Schwede T, Tramontano A. Evaluation of the Template-Based Modeling in CASP12. Proteins 2018, 86 (Suppl. 1), 321–334, DOI: 10.1002/prot.25425.
Laskey RA, Honda BM, Mills AD, Finch JT. Nucleosomes are Assembled by an Acidic Protein which Binds Histones and Transfers them to DNA. Nature 1978, 275 (5679), 416–420, DOI: 10.1038/275416a0.
Lehn JM. Toward Complex Matter: Supramolecular Chemistry and Self-Organization. PNAS 2002, 99 (8), 4763–4768, DOI: 10.1073/pnas.072065599.
Leibniz GW. The Monadology and Other Philosophical Writings. Oxford, The Clarendon Press, 1898.
Leibniz GW. La Monadologie. Paris, Le Livre de Poche, 1991 [1714].
Lewontin RC. The Triple Helix: Gene, Organism, and Environment. Cambridge, Harvard University Press, 2001.
Liu J, Nussinov R. Allostery: An Overview of Its History, Concepts, Methods, and Applications. PLoS Computational Biology 2016, 12 (6), e1004966, DOI: 10.1371/journal.pcbi.1004966.
Lopez P, Bapteste E. Molecular Phylogeny: Reconstructing the Forest. Comptes Rendus Biologies 2009, 332 (2–3), 171–182, DOI: 10.1016/j.crvi.2008.07.003.
Lyotard JF. La Condition postmoderne. Rapport sur le savoir. Paris, Les Éditions de Minuit, 1979.
Marks J. Molecular Biology in the Work of Deleuze and Guattari. Paragraph 2006, 29 (2), 81–97, DOI:10.1353/prg.2006.0016.
Maury CP. Origin of Life. Primordial Genetics: Information Transfer in a Pre-RNA World Based on Self-Replicating Beta-Sheet Amyloid Conformers. Journal of Theoretical Biology 2015, 382, 292–297, DOI: 10.1016/j.jtbi.2015.07.008.
Melvin RL, Salsbury FR Jr. Visualizing Ensembles in Structural Biology. Journal of Molecular Graphics and Modelling 2016, 67, 44–53, DOI: 10.1016/j.jmgm.2016.05.001.
Monod J, Changeux JP, Jacob F. Allosteric Proteins and Cellular Control Systems. Journal of Molecular Biology 1963, 6, 306–329, DOI:10.1016/s0022-2836(63)80091-1.
Needleman D, Dogic Z. Active Matter at the Interface between Materials Science and Cell Biology. Nature Reviews Materials 2017, 2 (9), 17048, DOI: 10.1038/natrevmats.2017.48.
Novarina V. Voie Negative. Paris, POL, 2017.
Ovchinnikov S, Park H, Varghese N, et al. Protein Structure Determination Using Metagenome Sequence Data. Science 2017, 355 (6322), 294–298, DOI: 10.1126/science.aah4043.
Palamini M, Canciani A, Forneris F. Identifying and Visualizing Macromolecular Flexibility in Structural Biology. Frontiers in Molecular Biosciences 2016, 3, 47, DOI: 10.3389/fmolb.2016.00047.
Renn JP, Junker M, Besingi RN, Braselmann E, Clark PL. ATP-Independent Control of Autotransporter Virulence Protein Transport via the Folding Properties of the Secreted Protein. Chem Biol 2012, 19 (2), 287–296, DOI: 10.1016/j.chembiol.2011.11.009.
Rout SK, Friedmann MP, Riek R, Greenwald J. A Prebiotic Template-Directed Peptide Synthesis Based on Amyloids. Nature Communications 2018, 9 (1), 234, DOI: 10.1038/s41467-017-02742-3.
Saarikangas J, Barral Y. Protein Aggregation as a Mechanism of Adaptive Cellular Responses. Current Genetics 2016, 62 (4), 711–724, DOI: 10.1007/s00294-016-0596-0.
Sanchez T, Chen DT, DeCamp SJ, Heymann M, Dogic Z. Spontaneous Motion in Hierarchically Assembled Active Matter. Nature 2012, 491 (7424), 431–434, DOI: 10.1038/nature11591.
Scriven M. Explanation and Prediction in Evolutionary Theory. Science 1959, 130 (3374), 477–482, DOI: 10.1126/science.130.3374.477.
Shin Y, Brangwynne CP. Liquid Phase Condensation in Cell Physiology and Disease. Science 2017, 357 (6357), eaaf4382, DOI: 10.1126/science.aaf4382.

Shinde UP, Liu JJ, Inouye M. Protein Memory through Altered Folding Mediated by Intramolecular Chaperones. Nature 1997, 389 (6650), 520, DOI: 10.1038/39097.

Stroberg W, Schnell S. On the Origin of Non-Membrane-Bound Organelles, and their Physiological Function. Journal of Theoretical Biology 2017, 434, 42–49, DOI: 10.1016/j.jtbi.2017.04.006.

Tompa P, Rose GD. The Levinthal Paradox of the Interactome. Protein Science 2011, 20 (12), 2074–2079, DOI: 10.1002/pro.747.

Tompa P, Schad E, Tantos A, Kalmar L. Intrinsically Disordered Proteins: Emerging Interaction Specialists. Current Opinion in Structural Biology 2015, 35, 49–59, DOI: 10.1016/j.sbi.2015.08.009.

Uexküll J von. Umwelt und Innenwelt der Tiere. Berlin, Springer, 1909.

Ullmann A. In Memoriam: Jacques Monod (1910–1976). Genome Biology and Evolution 2011, 3, 1025–1033, DOI:10.1093/gbe/evr024.

Uversky VN. Dancing Protein Clouds: The Strange Biology and Chaotic Physics of Intrinsically Disordered Proteins. Journal of Biological Chemistry 2016, 291.13, 6681–6688, DOI: 10.1074/jbc.R115.685859.

Van Speybroeck L, De Waele D, Van De Vijver G. Theories in Early Embryology. Annals of the New York Academy of Sciences 2002, 981 (1), 7–49, DOI: 10.1111/j.1749-6632.2002.tb04910.x.

Vernadsky V. The Biosphere and the Noosphere. American Scientist 1945, 33 (1), 1–12, DOI: 10.1016/j.geoforum.2005.01.004.

Vickery HB. The Origin of the Word Protein. YJBM 1950, 22 (5), 387–393.

Virchow R. Archiv für pathologische Anatomie und Physiologie und für klinische Medizin. Vol. 8. Berlin, Springer, 1855.

Walsby AE, Hodge MJS. Schrödinger's Code-Script, not a Genetic Cipher but a Code of Development. Studies in History and Philosophy of Biological and Biomedical Sciences 2017, 63, 45–54, DOI: 10.1016/j.shpsc.2016.12.002.

Wasmer C, Lange A, Van Melckebeke H, Siemer AB, Riek R, Meier BH. Amyloid Fibrils of the HET-s (218–289) Prion Form a Beta Solenoid with a Triangular Hydrophobic Core. Science 2008, 319 (5869), 1523–1526, DOI: 10.1126/science.1151839.

Xue B, Romero PR, Noutsou M, et al. Stochastic Machines as a Colocalization Mechanism for Scaffold Protein Function. FEBS Letters 2013, 587 (11), 1587–1591, DOI: 10.1016/j.febslet.2013.04.006.

Zhou M, Guo J, Cha J, et al. Non-Optimal Codon Usage Affects Expression, Structure and Function of Clock Protein FRQ. Nature 2013, 495 (7439), 111–115, DOI: 10.1038/nature11833.

Zimmer C. Scientists are Designing Artisanal Proteins for your Body. New York Times, December 26, 2017. (Accessed November 26, 2019, at https://www.nytimes.com/2017/12/26/science/protein-design-david-baker.html.)

Sonia Dheur is a molecular biologist, permanent investigator at the CNRS. After conducting research on gene expression and genetic and epigenetic inheritance at IBGC-CNRS UMR 5095, she relocated at UMR5319 Passages (CNRS) where she undertakes critical and reflexive interdisciplinary studies on the recent development of biology as well as studies on common forms of reflexivity in instituted cultural and scientific practices. During the last three years, she has set up and animated a seminar exploring the poetic dimension of scientific production.

Sven Saupe is a geneticist, director of research at the Institut de Biologie et Génétique Cellulaires, CNRS (UMR 5095). In biology, he is interested in genetic and cellular aspects of non-self recognition and cell death processes in fungi. He also undertakes interdisciplinarity investigations on certain social, philosophical and symbolic implications of genetics and molecular biology.

Index

Open Access. © 2021 Peter Fratzl et al., published by De Gruyter. BY-NC-ND This work is licensed under the Creative Commons Attribution-NonCommercial-NoDerivatives 4.0 International License.
https://doi.org/10.1515/9783110562064-017

www.ingramcontent.com/pod-product-compliance
Lightning Source LLC
Chambersburg PA
CBHW061156240926
51846CB00006B/1027
* 9 7 8 3 1 1 0 5 6 1 8 1 4 *